The Molecular Biology of Picornaviruses

NATO ADVANCED STUDY INSTITUTES SERIES

A series of edited volumes comprising multifaceted studies of contemporary scientific issues by some of the best scientific minds in the world, assembled in cooperation with NATO Scientific Affairs Division.

Series A: Life Sciences

Recent Volumes in this Series

Volume 17 – DNA Synthesis: Present and Future
edited by Ian Molineux and Masamichi Kohiyama

Volume 18 – Sensory Ecology: Review and Perspectives
edited by M. A. Ali

Volume 19 – Animal Learning: Survey and Analysis
M. E. Bitterman, V. M. LoLordo, J. B. Overmier, and M. E. Rashotte

Volume 20 – Antiviral Mechanisms in the Control of Neoplasia
edited by P. Chandra

Volume 21a – Chromatin Structure and Function
Molecular and Cytological Biophysical Methods
edited by Claudio A. Nicolini

Volume 21b – Chromatin Structure and Function
Levels of Organization and Cell Function
edited by Claudio A. Nicolini

Volume 22 – Plant Regulation and World Agriculture
edited by Tom K. Scott

Volume 23 – The Molecular Biology of Picornaviruses
edited by R. Pérez-Bercoff

Volume 24 – Humoral Immunity in Neurological Diseases
edited by D. Karcher, A. Lowenthal, and A. D. Strosberg

Volume 25 – Synchrotron Radiation Applied to
Biophysical and Biochemical Research
edited by A. Castellani and I. F. Quercia

The series is published by an international board of publishers in conjunction with NATO Scientific Affairs Division

A Life Sciences **B Physics**	Plenum Publishing Corporation New York and London
C Mathematical and Physical Sciences	D. Reidel Publishing Company Dordrecht and Boston
D Behavioral and Social Sciences	Sijthoff International Publishing Company Leiden
E Applied Sciences	Noordhoff International Publishing Leiden

The Molecular Biology of Picornaviruses

Edited by

R. Pérez-Bercoff
Laboratory of Virology and Molecular Biology
Institute of Anatomy
University of Zurich
Zurich, Switzerland

PLENUM PRESS • NEW YORK AND LONDON
Published in cooperation with NATO Scientific Affairs Division

Library of Congress Cataloging in Publication Data

Nato International Advanced Study Institute on the Molecular Biology of Picornaviruses, Maratea, Italy, 1978.
The molecular biology of picornaviruses.

(NATO advanced study institutes series: Series A, Life sciences; v. 23)
"Based on the proceedings of the NATO International Advanced Study Institute on the Molecular Biology of Picornaviruses, held in Maratea (Gulf of Policastro), Italy, September 8–18, 1978."
Includes bibliographical references and index.
1. Picornaviruses–Congresses. 2. Molecular biology–Congresses. I. Perez-Bercoff, R. II. Title. III. Series.
QR410.N37 1978 576'.64 79-13845
ISBN 0-306-40192-4

Based on the proceedings of the NATO International Advanced Study Institute on The Molecular Biology of Picornaviruses, held in Maratea (Gulf of Policastro), Italy, September 8–18, 1978.

A Division of Plenum Publishing Corporation
227 West 17th Street, New York, N.Y. 10011

Printed in the United States of America

WHY THIS N.A.T.O. MEETING?

WHY THIS BOOK?

> In the year 1961, a paleontologist discovered the foot prints of a dinosaurian, an iguanodon, which had lived in the United Kingdom a few hundred millions years ago. There were 13 pairs of foot prints, a heretofore unknown large number of reptilian fossil foot prints. Moreover, the steps, clearly preserved in limestone, were lying in a perfectly straight line. From these remarkable facts the paleontologist drew two remarkable conclusions: (a) the iguanodon was going somewhere, and (b) it was going there with a definite purpose.
>
> Our work on the effect of temperature on viral development was started a long time ago. A paleontologist searching in a library would discover 21 fossil papers preserved in various fossil periodicals. He would see that they are lying in an extremely tortuous line. From this remarkable fact, the paleontologist would draw two remarkable conclusions: (a) we were going nowhere, and (b) we had no definite purpose. Yet our aim was to disclose the mechanism by which supra- and infra-optimal temperatures influence viral development. Despite the fact that our way was twisted, we have nevertheless landed somewhere and I would like to tell you where we are on Friday, June the 8th, 1962.

Andre Lwoff
Cold Spring Harbor Symp. Quant. Biol. (1962), *27*, 159

This book is a sort of natural "fall-out" of the homonymous N.A.T.O. Advanced Study Institute held in Maratea (Gulf of Policastro), Italy, in September 1978, and it faithfully reflects the format and purposes of this conference.

In organizing this meeting, I knew all too well that "the way had been twisted, (but) we had nevertheless landed somewhere..."

The intent, therefore, was to provide for a fresh and original review of all relevant topics and issues in the field, following a comprehensive and coherent programme.

Such an ambitious goal could only be reached thanks to the unlimited collaboration of the lecturers: They were requested to produce nothing less than "freer, broader, speculative and personal "considerations of the subjects" they had to cover... And so they did: their presentations unfolded a fantastic picture, a most fascinating and meaningful identification of the field, its present problems and trends.

But participants at this conference contributed many valuable observations while discussing specific points. Unfortunately, more often than not, it proved impossible to identify them in the records. Accordingly, we incorporated here and there some of these "anonymous" contributions as part of the edited version of the texts.

To all of the participants, I would like to express my gratitude for their actively taking part in all the scientific (and social...) activities (lectures, round tables, posters, encounters) of this N.A.T.O. ASI. The result was a stimulating atmosphere, conducive to authentic scientific exchanges. Hence, the success of this meeting should be credited only to them.

Finally, on behalf of all participants, I would like to thank Professor Sissini, the Mayor of Maratea, and Mr. B. Vitolo, the Chairman of its Tourist Office: their enthusiastic support greatly contributed to make of our time in their wonderful seaside resort an unforgettable one.

R. Perez-Bercoff
Zurich, October 1978

Contents

SECTION I:

THE STRUCTURE OF THE PICORNAVIRION

THE PICORNAVIRION: STRUCTURE AND ASSEMBLY

DOUGLAS G. SCRABA

Department of Biochemistry, The University of Alberta

Edmonton, Alberta, Canada

INTRODUCTION

The mammalian picornaviruses (from pico-small, and rna-containing a ribonucleate genome) comprise a large group of agents, which can presently be classified according to a scheme such as that shown in Table I.

Although there are some differences among subgroups which are manifest in the properties of pH stability and buoyant density in cesium salts, the evidence at this time points to a fundamental similarity of structure and assembly for all of these viruses. Consequently, in this brief review attention will be concentrated on the cardioviruses and poliovirus with the implication that the characteristics of these agents will be applicable - with only minor variations - to all mammalian picornaviruses.

I. STRUCTURE OF THE VIRION

A. Physical-Chemical Properties

The picornavirus particle is composed of a molecule of single-stranded RNA (30% by weight) enclosed in a capsid of protein (70%). There is no good evidence for the presence of carbohydrate or lipid in the virion (1). As viewed in the electron microscope by negative staining, the particle is isometric with a dry diameter of 27-28 nm. In solution it behaves as a spheroid (frictional ratio of 1.05-1.10) with a diameter of about 30 nm, and containing some 0.25 g water per gram of dry virus (2).

Table I

VERTEBRATE PICORNAVIRUSES

Genus Enterovirus	: Polio (3 serotypes) Coxsackie A (23) Coxsackie B (6) Echo (31) Enteroviruses of mice,swine,cattle Enterovirus 70 (conjunctivitis virus)	Sedimentation coefficient ~155 S Buoyant density (CsCl) ~1.34 g/ml Virions stable at pH 3-10 Empty capsids produced *in vivo*
Genus cardiovirus	: EMC ME Mengo Columbia-SK MM } Serologically very closely related	Sedimentation coefficient ~155 S Buoyant density ~1.34 g/ml Virions labile 5<pH<7 in the presence of 0.1 M Cl^- or Br^- No empty capsids *in vivo*
Genus Rhinovirus (Human)	: More than 120 serotypes	Sedimentation coefficient ~155 S Buoyant density ~1.40 g/ml Virions labile pH<5 Empty capsids produced *in vivo*
Genus Aphtovirus	: Foot-and-Mouth Disease Virus 7 serotypes	Sedimentation coefficient ~145 S Buoyant density ~1.43 g/ml Virions labile pH < 6.5 Empty capsids produced *in vivo*

Summarizing the best available values, the picornavirion has a sedimentation coefficient ($S^o_{20,w}$) of 150-160S, a diffusion coefficient ($D^o_{20,w}$) of 1.44-1.47 x 10^{-7} cm^2/sec, and a partial specific volume ($\bar{v}$) of 0.68-0.70 ml/g. Substituting these values in the Svedberg equation, one arrives at a "molecular weight" for the particle of about 8.3-8.5 x 10^6 (2, 3).

B. The RNA Component

The picornavirus genome is a single polyribonucleotide chain of molecular mass (Na^+ form) of 2.4-2.7 x 10^6 daltons (4, 5). No unusual nucleotides have been detected therein and, with the exception of some human rhinovirus genomes, approximately equimolar amounts of adenylate, cytidylate, guanylate and uridylate are present (6). The nucleotide residues are not, however, uniformly distributed along the genome. At the 3'-end is a tract of 20-50 adenylate residues which are required for the infectivity of isolated virion RNA (7, 8), and in the case of the cardioviruses and FMDV there is a cluster of 100-200 cytidylate residues located near the 5'-end (9, 10). Recently, a small protein (MW ~ 4000), called VPg, has been found to be covalently linked to the 5'end of the virion RNA of polio, EMC and FMD viruses (11, 12, 13).

This protein may play important roles in the replication of the genome and/or assembly of the virion (14).

C. The Protein Component

When total protein is extracted from the virus particle and subjected to analytical ultracentrifugation or density gradient sedimentation, it behaves as a relatively homogeneous entity with a molecular mass of approximately 30,000 daltons. However, as early as 1963 Maizel reported that the protein component of the poliovirion could be separated electrophoretically into several non-identical polypeptide species (15). With the discovery that the electrophoretic mobility of a polypeptide in polyacrylamide gels containing sodium dodecyl sulfate (SDS) is inversely proportional to the logarithm of its molecular weight (16), it became possible to determine simultaneously both the number of different polypeptides which comprise the capsid protein and their molecular weights. Also, if the virions have been produced in cells maintained in medium containing radioactive amino acids, measurement of the amount of radioactivity associated with each polypeptide enables one to calculate the number of copies of each in the capsid. The results of such analyses for selected picornaviruses are summarized in Table II. More complete data and references are given by Rueckert (17).

Table II

POLYPEPTIDE COMPOSITIONS OF PICORNAVIRAL CAPSIDS

Polypeptide Designation[b]		Molecular mass[a]/Estimated number of copies per virion						
		Poliovirus (Type 1)	Coxsackie Virus (Type B3)	Bovine Enterovirus	ME	Mengo	Rhinovirus (Type 1A)	FMDV (Type O)
VPO	ε	41/-	38/1-2	-	41/2	39/1-2	37/2	40/4
VP1	α	35/-	29/60	34/60	33/60	33/59	34/60	34/60
VP2	β	28/-	25/60	28/60	31/58	30/58	30/58	30/60
VP3	γ	24/-	21/60	26/60	25/60	24/59	26/60	26/60
VP4	δ	5.5/-	5/60	9/30	7.3/58	7.4/58	7/58	13/30

[a] Values are given in 10^3 daltons

[b] The VPO-4 system of nomenclature has been used for the enteroviruses and FMDV; the ε-δ system has been used for the cardioviruses and rhinovirus 1A

A reasonable interpretation of these data is that the picornavirus capsid is composed of four major polypeptide species: VP1 or α (MW ∿ 34,000), VP2 or β (MW ∿ 30,000), VP3 or γ (MW ∿ 25,000) and VP4 or δ (MW 5000-8000); and that there are approximately 60 copies of each in a complete capsid (17). The molecular weights of VP4 shown above for BEV and FMDV may be overestimates because of the unreliability of molecular weight determination for polypeptides of this size by SDS gel electrophoresis (e.g. the molecular weight for the δ polypeptide of Mengo virus is 7350 based upon amino acid composition analyses, whereas its apparent molecular weight by SDS gel electrophoresis is 10,600 (18). Thus it is probable that both BEV and FMDV contain a full complement of VP4 polypeptides. The polypeptide VP0 (ε) is a precursor of VP2 and VP4 (β and δ), and is not genetically distinct (19). It is likely that all picornavirions contain one or more of these uncleaved precursors.

Individual capsid polypeptides have been isolated from cardioviruses and FMDV by electrophoretic or chromatographic procedures and their amino acid compositions determined (18, 20, 21). The composition characteristics common to all of these polypeptides are as follows. There is an excess of potentially acidic (asx, glx) over basic (lys, arg) amino acid residues, a low sulfur content (cys + met = 2-4 mole %), and a substantial number of prolines (5-8 mole %) and non-α-helix forming residues (val + ile + ser + cys + thr + gly ≈ 35-40 mole %). Approximately 50 mole % of the amino acid residues are apolar. Considering the Mengo virion, the excess of acidic amino acid residues is reflected in the low isoelectric points of the individual capsid polypeptides (isoelectric pH's for α, β, γ and δ are 5.1, 5.3, 5.8 and 4.6 respectively (22). Optical rotatory dispersion and circular dichroism measurements indicate an α-helical content of only 5-10% for the Mengo capsid polypeptides _in situ_ (2, 23), in keeping with the large proportion of helix-disrupting and non-helix-forming amino residues. The apolar character of the capsid polypeptides is manifest in their insolubility in aqueous solvents at neutral pH.

Amino- and carboxyl-terminal amino acid sequences have been determined for the Mengo capsid polypeptides (24), and these are given in Table III. Dansylation of intact poliovirions has revealed three amino terminals, these being asp, ser and gly (25). The three large polypeptides from FMD virions (types A, O and C) have been found to have the amino terminal residues asp, thr and gly (21, 26) and their carboxyl-terminal residues (type A) are gln, glu and leu (27). Although the correlation of the N- and C-terminals for individual FMD polypeptides has not been done, it is tempting to speculate that all picornavirus capsid polypeptides have the following structure:

$$\mathrm{X}\overset{\mathrm{VP4}}{\underset{\delta}{\text{———}}}\mathrm{ala} \qquad \mathrm{asp}\overset{\mathrm{VP2}}{\underset{\beta}{\text{———}}}\mathrm{gln} \qquad {\mathrm{ser} \atop \mathrm{thr}}\overset{\mathrm{VP3}}{\underset{\gamma}{\text{———}}}{\mathrm{gln} \atop \mathrm{glu}} \qquad \mathrm{gly}\overset{\mathrm{VP1}}{\underset{\alpha}{\text{———}}}\mathrm{leu}$$

Table III

AMINO- AND CARBOXYL-TERMINAL SEQUENCES OF THE MAJOR CAPSID POLYPEPTIDES OF MENGO VIRUS

Polypeptide	Amino-terminal sequence	Carboxyl-terminal sequence
δ	Blocked	-Leu-Leu-Ala
β	Asp-Gln-Asn-Thr-Glu-Glu-Met-Glu-Asn-Leu	-(Val,Leu)-Arg-Gln
γ	Ser-Pro-Ile-Pro-Val-Thr-Ile-Arg-Glu-His	-(Gln)
α	Gly-Val-Glu-Asn-Ala-Glu-Lys-Gly-Val-Thr-	-(Val,Gly,Ala)-Val-Leu

D. Morphology and Architecture of the Virion

In 1959 Finch and Klug examined crystals of poliovirus by X-ray diffraction and concluded that the virion possessed icosahedral symmetry. Extrapolating from the experimental data, they also suggested that the capsid was made up on 60 identical asymmetric structure units with a diameter of 60-65 Å. Based upon the then accepted values of 6.7×10^6 for the particle weight of the virion and 2×10^6 for the molecular weight of the RNA, they calculated a molecular weight for the structure unit of 80,000 (28).

These conclusions were placed in some doubt when it was subsequently shown that: (a) turnip yellow mosaic virus, which produced an X-ray diffraction pattern virtually identical to that of poliovirus, had a capsid composed of 180 identical structural polypeptides arranged in quasiequivalent hexamer-pentamer clusters to give 32 morphological units or capsomeres (29); (b) the poliovirus capsid was composed of not one but several non-identical polypeptide species (15); and (c) the assumed particle weight and t the RNA molecular weight for poliovirus were not correct (2, 3, 4).

Electron microscopy of negatively stained preparations of virus particles - which has proved extremely valuable in elucidating the capsid architecture of a wide variety of animal, plant and bacterial viruses - has not been very informative in the case of the mammalian picornaviruses because their capsids are compact structures and essentially impermeable to the electron dense salts commonly used as negative stains (e.g. see Figure 2a). Thus the morphological structure units cannot be distinguished easily, and microscopists have variously suggested 32-, 42- or 60-capsomere models for the poliovirus capsid, based (usually) upon the appearance of one or two "favorably" stained and oriented particles in a field containing a large number of virions (30, 31, 32).

The problem of picornavirus capsid architecture was resolved by Rueckert and his collaborators (33,34) who, instead of examining the intact virion, studied the dissociation products which result when purified cardioviruses are incubated at pH 5 - 6.5 in the presence of 0.1 M chloride ions. By analytical ultracentrifugation and SDS-polacrylamide gel electrophoretic analyses they found that the primary product of such a dissociation is a protein moiety which has a sedimentation coefficient of 13-14S, a molecular weight of about 425, 000 and which contains equimolar proportions of the viral polypeptides α, β and γ . The small viral polypeptide species, δ, and the intact RNA are precipitated during the dissociation. The 14S material may be further dissociated by treatment with 2 M urea into components which sediment at ∼ 5S, have a molecular weight of about 86,000 and which also contain equimolar proportions of α, β and γ . These data are compatible with a model for the cardiovirus capsid in which the fundamental structure unit comprises one molecule of each of the non-identical polypeptides α, β and γ (whose combined molecular weights are about 89,000). The structure units, which have been called protomers, are joined together by hydrophobic interactions in clusters of five (the 14S pentamers), one of which is centered at each of the 12 vertices of a simple icosahedron. These results have been confirmed for the Mengo virion by electron microscopic examination of dissociating virus and size measurements of the isolated 14S and 5S products (35). (A similar model has been proposed for FMDV (36), but in this case dissociation by acidic pH occurs along the sides of icosahedral facets, creating trimers of (VP1, 2, 3) which have a sedimentation coefficient of 12S and a molecular weight of 265,000).

It may therefore be concluded that the asymmetric structure unit of the picornaviral capsid is composed of three polypeptides and conforms to the prediction of Finch and Klug. It has a molecular mass of approximately 86,000 daltons, a diameter of about 68 Å (35) and is repeated 60 times in the virus capsid. This corresponds to the simplest icosahedral lattice, with a triangulation number of 1 (37). Considering the 14S pentamers as capsomeres, there would be 12 capsomeres per virion. The location of the δ (VP4) polypeptides in the capsid has not yet been established, but it is possible that they are distributed over the internal surface and in direct contact with the virion RNA (38, 39; see below).

E. The External Surface of the Virion

When entero- or rhinovirions are incubated in the cold with susceptible cells and the temperature is then raised, a substantial portion of irreversibly altered virus particles are eluted. These "A-particles" lack the smallest capsid protein (VP4 or δ) and are unable to re-attach to cells. Their RNA component is, however intact and infectious (40, 41). *In vitro* treatment of virions with acid, alkali, heat or urea (42, 43, 44) also produced non-infectious particles which have lost VP4 (δ) and are antigenically similar to the A-particles and to the empty capsids produced during normal infection (C-antigenicity). These observations led to the suggestion that the VP4 (δ) proteins are located on the external surface of the intact virion, confer upon it native (D) antigenicity, and are responsible for its attachment to cellular receptors (45, 46, 47).

An alternative explanation is that inactivation of picornaviruses reflects a concerted conformational rearrangement of the capsid proteins, and the loss of VP4 (δ) is a consequence rather than a cause of inactivation (17, 48). Support for this idea has been provided by isoelectric focusing studies of poliovirus and a human rhinovirus. Poliovirions were resolved into two interconvertible forms with isoelectric pH's of 4.5 and 7.0, the latter form being infectious. Treatments which inactivated the virus (attachment and elution from cells or UV irradiation) produced only the non-infectious acidic form (49). Similarly, when a purified human rhinovirus preparation was subjected to isoelectric focusing in a sucrose gradient two forms of the virion were separated: (a) isoelectric pH = 6.5, 150 S, infectious, D antigenicity; and (b) isoelectric pH = 4.5, 140S, non-infectious, indirect evidence for C antigenicity. Both forms, however, contained intact RNA and the full complement of capsid polypeptides (50). Further evidence that a substantial conformational alteration of entero- and rhinovirus capsids accompanies the D to C antigenicity change has been obtained by examining the accessibilities of individual capsid proteins in intact virions, A-particles and naturally-occurring empty capsids to lacto-peroxidase-catalyzed iodination or alkylation with acetic anhydride (51, 52). The results of these experiments are summarized in Table IV. Since under the experimental conditions employed the radioactive reagents formed covalent bonds with tyrosine residues (^{125}I) or free amino groups (^{3}H-acetic anyhdride) which were exposed on the exterior of virions, and since VP4 was labeled with either reagent if virions were first disrupted, these data demonstrate that VP4 is likely to have an internal location in the virus capsid.

Assuming that VP4 (δ) can be eliminated from consideration as the virion capsid component responsible for attachment to cellular receptors, is it possible that one of the other structural polypeptide species performs this function, or does "the entire surface of the virion particle play a role in its ... ability or lack of ability to attach to host cells" (48) ? Considering FMDV, treatment of intact virions with trypsin or chymotrypsin caused a cleavage in the VP1 polypeptides (53) and resulted in a greatly decreased ability of the virus particles to attach to cells (∿1000-fold loss of infectivity) and to elicit the production of neutralizing antibodies in guinea pigs (54, 55). A correlation of these data with the observation that only VP1 was iodinated when FMD virions were incubated with ^{125}I and chloramine-T (56) suggests

Table IV

ACCESSIBILITY OF PICORNAVIRUS CAPSID POLYPEPTIDES TO CHEMICAL LABELING AGENTS[a]

Virus Type	Particle	Polypeptides Labeled by Reaction with	
		^{125}I(lactoperoxidase)	^{3}H-acetic anhydride
Polio	Virion	VP1 (VP2, VP3[b])	VP1 (VP2, VP3[b])
	A-particle	VP1, VP2 (VP3[b])	VP1, VP2 (VP3[b])
Human Rhinovirus-2	Virion	VP2 (VP1, VP3[b])	VP1 (VP2, VP3[b])
	A-particle	VP1, VP2, VP3	VP1, VP2, VP3
Bovine Enterovirus	Virion	VP1	-
	Empty Capsid	VP0, VP1, VP3	-

[a]Data taken from references 51 and 52.

[b]Small, but significant, amounts of radioactivity found in these polypeptides compared to the amounts associated with corresponding polypeptides when virions were disrupted before labeling.

that both the immunogenic and cell attachment sites are associated with this single polypeptide species. Iodination experiments with Mengo virions revealed that the α (VP1) polypeptides comprise much of the external surface, with the β (VP2) polypeptides being exposed to a lesser extent. In tests with specific antisera, antibody-binding sites for both α and β polypeptides could be demonstrated to be exposed on the external surface of the viral capsid; however, only anti-α antibodies could block attachment of virions to cells (57). These data are presented in Table V.

The general features of picornavirus capsid architecture which have been revealed by chemical labeling and immunological experiments are as follows.

(a) The VP4 (δ) polypeptides occupy internal locations in the capsid of the intact virion.

(b) The D to C antigenic conversion which accompanies the interaction of virions with host cell receptors involves a major conformational change in the capsid, and may be accompanied by the dissociation of the VP4 polypeptides.

(c) A single polypeptide species(VP1 (i.e. α) in the case of polio, BEV, FMDV and Mengo; VP1 or VP2 in HRV-2) occupies most of the external surface of the capsid and available evidence suggests that this species alone is responsible for the specific attachment of the virion to a cellular receptor.

Table V

DISTRIBUTION OF THE CAPSID POLYPEPTIDES WITH RESPECT TO THE SURFACE OF THE MENGO VIRUS PARTICLE

Polypeptide	Tyrosine Residues Accessible to Lactoperoxidase-catalyzed Iodination of Intact Virions[a]	Antigenic Determinants Exposed on the Surface of Intact Virions[b]	Antigenic Determinants Exposed when Virions are Disrupted[b]	Specific Antiserum Capable of Blocking Attachment of Virions to Cells[c]
α	+	+	+	+
β	+	+	+	-
γ	-	-	+	-
δ	-	-	+	-

[a]After 1 min exposure to ^{125}I and lactoperoxidase the ratios of radioactivity incorporated per tyrosine residue normalized to the value for the α polypeptides were: α = 1.00; β = 0.31; γ, δ < 0.06.

[b]From the results of immunodiffusion and complement fixation tests with antisera specific for each of the capsid polypeptides.

[c]From the results of plaque-neutralization (L-cells) and hemagglutination-inhibition (human erythrocytes) tests.

F. Chemical Crosslinking Studies

According to Rueckert's model (33, 34, 35), the asymmetric structure unit (protomer) of the picornavirus capsid is composed of one molecule each of the polypeptides α, β and γ (VP1, 2 and 3). In cardioviruses these are associated into pentamers by hydrophobic interactions (disrupted by urea), and the 12 pentamers which comprise the capsid are linked by electrostatic interactions (disrupted by chloride ions at pH 6). This model poses several intriguing questions. For example, are the three polypeptides in the protomer interwoven into a single trimeric unit or does each occupy a discrete spatial domain? Are specific polypeptide species involved in the hydrophobic and electrostatic interactions between protomers and pentamers? What is the location and distribution of the δ (VP4) polypeptides in the capsid? Are there specific polypeptide species which interact directly with the virion RNA?

The spatial relationships of proteins in such complicated assemblies as ribosomes and membranes have been at least partially elucidated by reaction with bifunctional crosslinking reagents and subsequent analysis of the polypeptide composition of complexes by gel electrophoresis procedures (58). We have been using this approach to further probe the molecular anatomy of the Mengo virion and to attempt to answer some of the questions listed above. While such studies are not complete, some useful information has been obtained (59).

Figure 1 illustrates the results of an experiment where Mengo polypeptides in intact virions were crosslinked via free amino groups by dimethylsuberimidate. Subsequent separation, cleavage

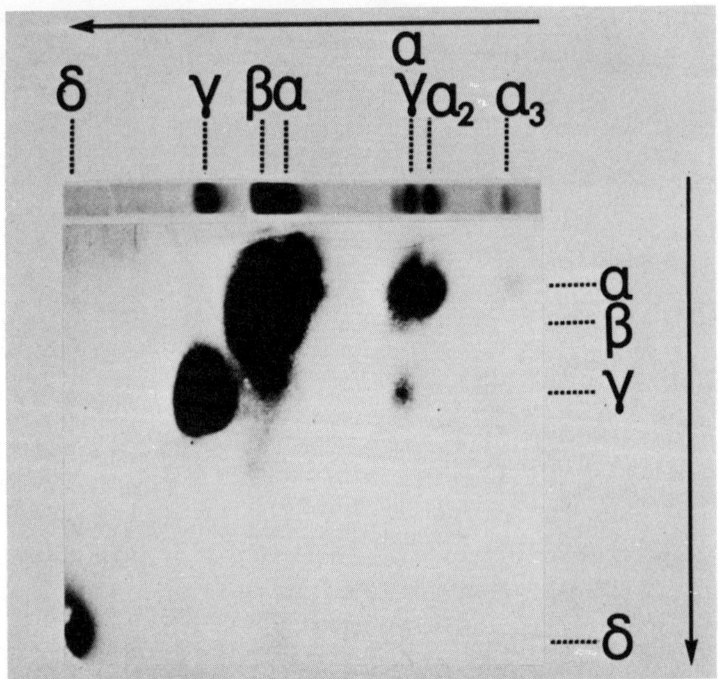

Figure 1. Analysis of Mengo capsid protein complexes formed by the crosslinking reagent dimethylsuberimidate (DMS). ^{3}H-amino acid-labeled virions were incubated with 20 mM DMS at room temperature for 24 hr, then disrupted by boiling in 2% sodium dodecylsulfate. Capsid proteins and higher-molecular weight complexes were separated by electrophoresis in SDS-containing polyacrylamide gels (60). A photograph of one such gel, with protein bands made visible by Coomassie Blue staining is shown at the top of the figure (δ stains very poorly with this dye). An identical gel was treated with NH_4OH and CH_3COOH to cleave the crosslinks, then place on top of an SDS-polyacrylamide slab gel, and electrophoresis in the second dimension was carried out. After the gel was dried, ^{3}H- labeled protein spots were visualized by autoradiography. The arrows indicate the direction of migration of proteins during electrophoresis.

and analysis of the complexes showed that their compositions were $\alpha\gamma$, α_2 and α_3.

A summary of the results of a number of crosslinking experiments is given in Table VI.

As might be inferred from the studies on the surface of the picornavirus capsid, the crosslinking data indicate that the α, β and γ polypeptides do indeed occupy discrete spatial locations in the protomer. Had these proteins been extensively intertwined, primarily αβγ and $(\alpha\beta\gamma)_n$ complexes would have been observed. The aggregation of protomers into $(\alpha\beta\gamma)_5$ pentamers involves interactions among the α polypeptides exclusively, since no β_n or γ_n complexes were formed by the crosslinking reagents. The location of the δ polypeptide remains mysterious. Even though each δ polypeptide has 2 lysine residues (18) available for crosslinking, no complexes involving δ were detected.

Figure 2 illustrates the results obtained when virions were incubated with various crosslinking reagents, then dissociated by incubation with NaCl at pH 6 (35), and examined in the electron microscope.

The micrographs show that virions treated with the crosslinkers DMA or DMS resembled native virions in being dissociable into $(\alpha\beta\gamma)_5$ pentamers, whereas DSP-treated virions were not dissociable. Referring to Table VI, our interpretation of this result is that the protomer structure is maintained by α-γ-β contacts, while the α-β interactions (detected only by crosslinking with DSP) are between polypeptides in neighboring pentamers.

G. A Model for the Mengo Virus Capsid

The experimental data pertaining to the structure of the Mengo virus particle can be summarized as follows.

(a) Virus Dissocation (35):

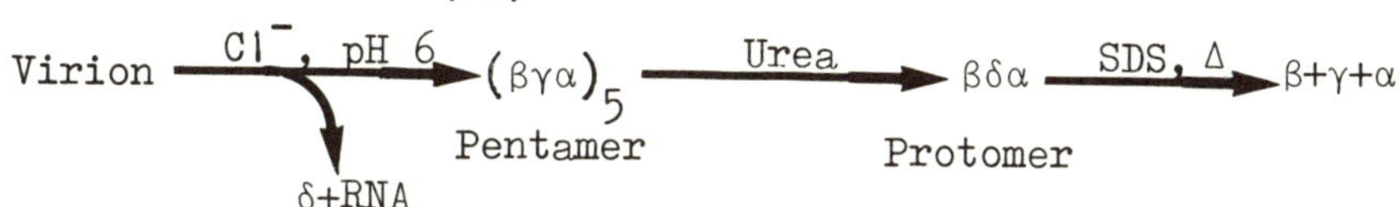

(b) Radioiodination (57):

Tyrosine residues exposed on external surface - α, β
Tyrosine residues buried within capsid - γ, δ

(c) Reaction with Specific Antibodies (57):

Accessible combining sites - α, β
Inaccessible combining sites - γ, δ
Attachment to cellular receptors - α

Table VI

CHEMICAL CROSSLINKING OF POLYPEPTIDES IN THE MENGO VIRION

Crosslinking Reagent	Maximum Linkage Distance (nm)	Molecular Weight of Complex[a]	Probable Composition	Composition Confirmed[b]
Dimethyladipimate (DMA)	0.8	58,000	$\alpha\gamma$	
		65,000	α_2	
Dimethylsuberimidate (DMS)	1.1	54,000	$\beta\gamma$	
		58,000	$\alpha\gamma$	yes
		65,000	α_2	yes
		86,000	$\alpha\beta\gamma$	
		95,000	α_3	yes
Dithiobis(succinimidyl propionate) (DSP)	1.5	54,000	$\beta\gamma$	yes
		62,000	$\alpha\beta$	yes
		86,000	$\alpha\beta\gamma$	yes
		97,000	α_3	yes

[a]Molecular weights estimated from migration during electrophoresis in SDS-polyacrylamide gels.

[b]Composition confirmed by chemical "reversal" of the crosslinks followed by electrophoresis in the second dimension.

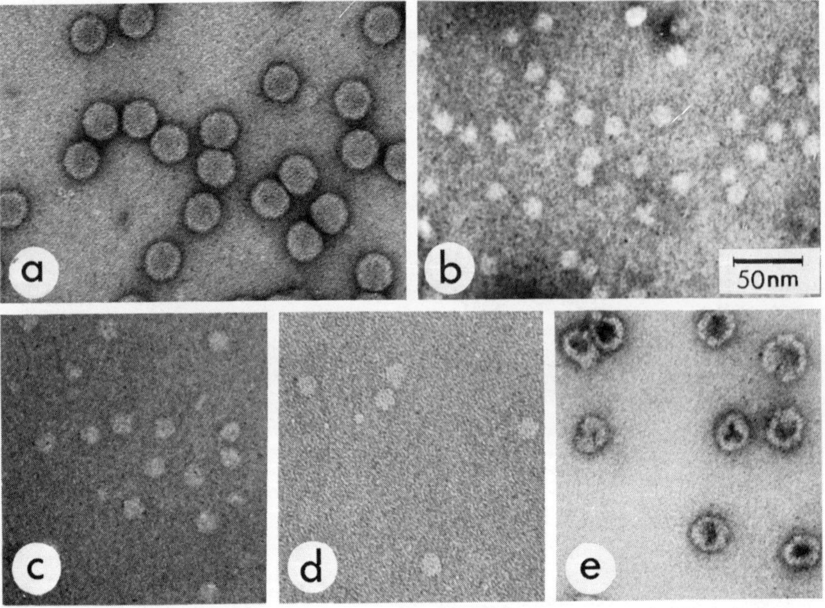

Figure 2. Electron micrographs of: (a) untreated Mengo virions; (b) virions dissociated into 14<u>S</u> pentamers by incubation with 0.14 <u>M</u> NaCl at pH 6; (c) virions treated with DMA prior to dissociation; (d) virions treated with DMS prior to dissociation; and (e) virions treated with DSP prior to dissociation. All preparations were negatively stained with 1% uranyl formate.

(d) Crosslinks formed by Bifunctional Reagents (59):

Within protomers - $\beta\gamma$, $\gamma\alpha$, $\beta\gamma\alpha$
Within pentamers - α_2, α_3
Between pentamers - $\beta\alpha$

(e) Gene order (61): $^{N}\delta\text{-}\beta\text{-}\gamma\text{-}\alpha^{C}$

These data have been incorporated into a schematic model of the Mengo virion which is presented in Figure 3. The model is incomplete in that the location of the δ polypeptides have not been established,nor have the interactions of the virion RNA with the capsid protein been determined.

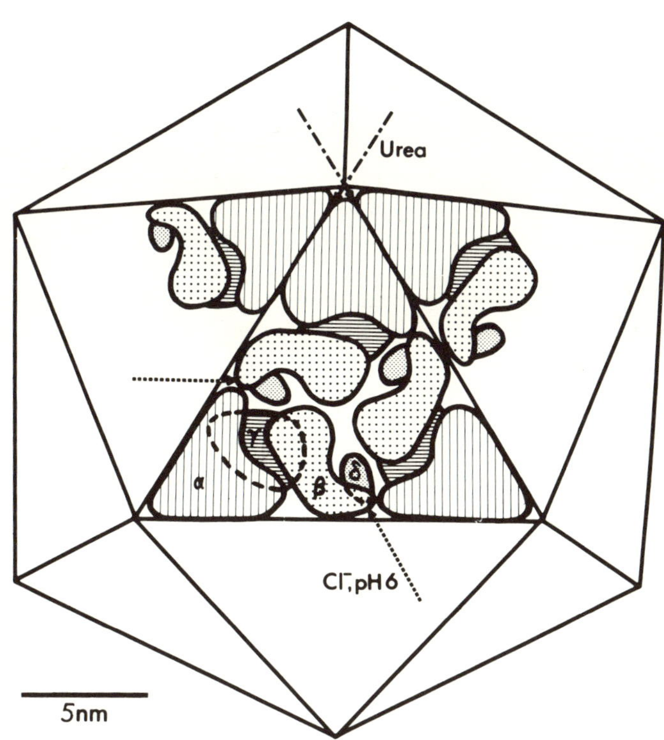

Figure 3. A model of the Mengo virion showing the relationships among the capsid polypeptide components. The shapes of the polypeptides are completely arbitrary. This is a surface view; the thickness of the capsid is 5-7 nm (35). Arrows indicate the non-covalent bonding contacts disrupted by chloride ions at pH 6 and by urea.

II. ASSEMBLY OF THE VIRION

A. The Generation of Capsid Polypeptides

Studies of the infectious process of a variety of picornaviruses (summarized in references 17 and 62) have established a number of general features. Soon after the initiation of infection there begins a rapid decrease in host cell protein synthesis, and after 3-4 hr it is virtually stopped. The virion RNA, after Vpg has been removed (14), acts as a messenger RNA. Host ribosomes begin translation at a single site near the 5'-end of this RNA (63) and proceed continuously to a termination site near its 3'-end. The initial translation product is a "polyprotein" of molecular weight ∿270,000 (19), and probably represents the entire information content of the viral genome. Primary cleavages of the nascent polyprotein (which are probably produced by a cellular protease (64)) generate a capsid polypeptide precursor molecule, one or more stable polypeptides, and a precursor of one (or more) of the components of a viral RNA replication complex. Secondary cleavages, which may be mediated by both cellular and viral proteases (64, 65), give rise to the capsid polypeptides ε, γ and α (VPO, 3 and 1) and the stable non-capsid polypeptide E (NCVP4). A final morphogenetic cleavage of $\varepsilon \rightarrow \delta + \beta$ (VPO → VP2 + 4) accompanies virion assembly. Since infection causes a shut-off of host protein synthesis it has been possible to efficiently label the virus-specific proteins *in vivo* with radioactive amino acids and to establish their precursor-product relationships by SDS-polyacrylamide electrophoretic analyses of infected cell lysates (pulse-chase experiments). Furthermore, the existence of a single initiation site has permitted "pactamycin mapping" of the virus-specific polypeptides on the viral genome. Pactamycin inhibits initiation of protein synthesis but does not interfere with chain elongation; thus the further a particular polypeptide-coding region is from the initiation site the more frequently it will be translated by ribosomes during the run-off period (∿10 min) after the addition of the drug (66). Comparisons of the amounts of radioactive label incorporated in the absence and in the presence of pactamycin have been used to establish the genome locations for the stable viral polypeptides and their precursors (and have confirmed the precursor-product relationships). The cleavage patterns and translational maps for the polypeptides of EMC, Mengo, polio, FMD and rhinoviruses are virtually identical (17, 62).

The generation of the Mengo capsid polypeptides (which map in the 5'- region of the viral genome) is depicted below) 24, 61).

The capsid precursor A (MW = 114,000) arises from proteolysis of the polyprotein. The conversion of A→B may be autocatalytic, with the γ portion of A acting as a protease (67). Subsequent

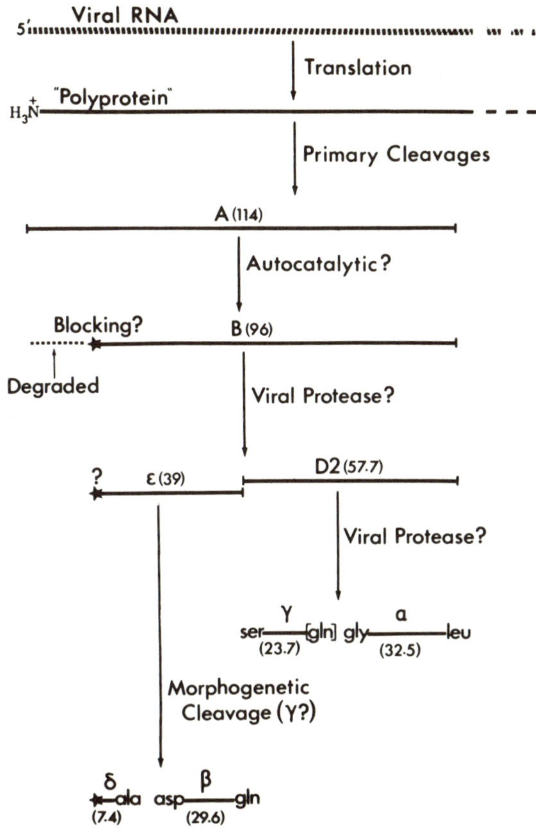

cleavages to give ε, γ and α may be mediated by a virus-coded protease with a specificity for glutamine residues (24, 65). The proteolytic conversion of ε → δ + β accompanies the RNA encapsidation process during virion assembly. The nature of the amino-blocking group on δ is not known (although it is apparently not acetyl or formyl), nor is the stage at which blocking occurs.

B. Virion Assembly

Information regarding the assembly of picornaviruses has been obtained largely from the characterization of subviral particles found in infected cells. In the case of poliovirus these include 6*S* and 13-14*S* particles, 80*S* "procapsids" and 125*S* "provirions" (68, 69, 70). The cardioviruses do not produce stable procapsids or provirions *in vivo*; however, 13*S* and 14*S* particles have been identified in EMC virus-infected cells (71), and recently a 50*S* particle has been isolated from Mengo virus-infected cells (73). Relationships among the various subviral particles and complete virions have been reviewed in detail by Rueckert (17). The diagram below illustrates hypothetical assembly sequences for Mengo and polioviruses.

The early stages in Mengo virus assembly are not clear since $(\varepsilon\gamma\alpha)_5$ is the smallest subviral particle identified thus far. Cleavage is likely to follow the association of the large capsid precursor protein into a pentamer since this species (i.e. B_5) has been identified in cells infected with EMC or rhinoviruses (72). The 50*S* particle incorporates radioactive amino acids before these are found in complete virions, and 50*S* particles accumulate when RNA synthesis (and virus production) are inhibited by cordycepin. Upon removal of the inhibitor there is a quantitative transfer of label to mature virions, suggesting that the 50*S* particle is an assembly intermediate (73). There is, however, an apparent equilibrium between the 50*S* and 14*S* species so the direct conversion of 14*S* pentamers into virions cannot be ruled out. The 50*S* particles may simply represent a storage depot for excess $(\varepsilon,\gamma,\alpha)_5$ subunits.

In poliovirus-infected cells, 80*S* procapsids accumulate when viral RNA synthesis is blocked by guanidine, and label from procapsids is transferred to virions when this block is released (69). The question of whether or not poliovirus RNA is incorporated directly into a preformed capsid remains open; there is one report that in poliovirus-infected monkey cells, 14*S* particles accumulate in the presence of guanidine and that this material is converted directly into virions when guanidine is removed (74).

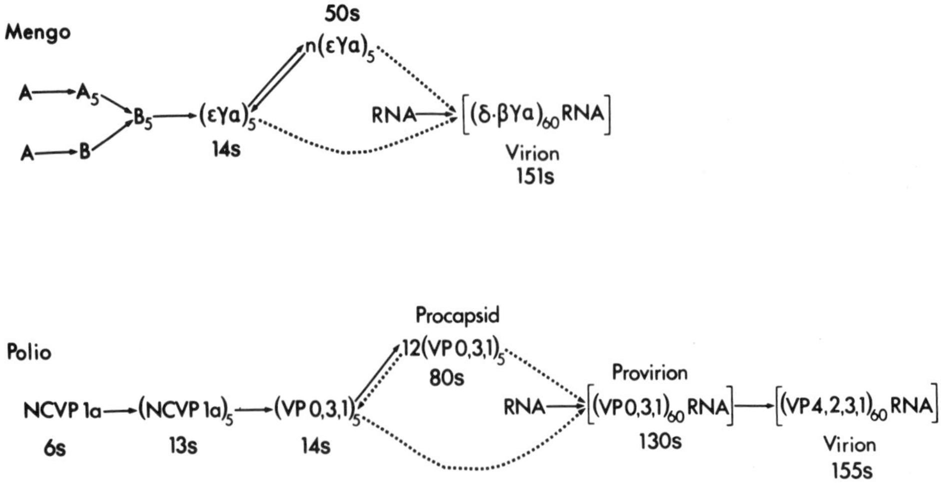

Mengo
50s
$n(\varepsilon\gamma\alpha)_5$
$A \rightarrow A_5$
$A \rightarrow B$
$B_5 \rightarrow (\varepsilon\gamma\alpha)_5$
14s
$RNA \rightarrow [(\delta\cdot\beta\gamma\alpha)_{60}RNA]$
Virion
151s

Polio
Procapsid
$12(VP\,0,3,1)_5$
80s
Provirion
$NCVP\,1a \rightarrow (NCVP\,1a)_5 \rightarrow (VP\,0,3,1)_5$
6s
13s
14s
$RNA \rightarrow [(VP\,0,3,1)_{60}RNA] \rightarrow [(VP\,4,2,3,1)_{60}RNA]$
130s
Virion
155s

ACKNOWLEDGEMENTS

I wish to express my thanks to a number of colleagues, past and present, who have contributed to the work done on Mengo virus in the Department of Biochemistry, University of Alberta; Tak Mak, Barry Ziola, Eva Paucha, Garry Lund, Joyce Hordern, Patrick Lee and Joan Leonard; and Professors John Colter, Cyril Kay and Larry Smillie. I would also like to thank the Medical Research Council of Canada for financial support.

REFERENCES

1. BURNES, A.T.H., FOX, S.M. and PARDOE, I.U. Evidence for the lack of glycoprotein in the encephalomyocarditis virus particle. J. Gen. Virol. (1973), 18,33-49.

2. SCRABA, D.G., KAY, C.M. and COLTER, J.S. Physicochemical studies of the three variants of Mengo virus and their constituent ribonucleates. J. Mol. Biol. (1967), 26, 67-79.

3. BURNESS, A.T.H. and CLOTHIER, F.W. Particle weight and other biophysical properties of encephalomyocarditis virus. J. Gen. Virol. (1970), 6, 381-393.

4. TANNOCK, G.A., GIBBS, A.J. and COOPER, P.D. A re-examination of the molecular weight of poliovirus RNA. Biochem. Biophys. Res. Commun. (1970), 38, 294-304.

5. BROWN, F., NEWMAN, J.F.E. and STOTT, E.J. Molecular weight of rhinovirus ribonucleic acid. J. Gen. Virol. (1970), 8, 145-148.

6. NEWMAN, J.F.E., ROWLANDS, D.J. and BROWN, F. A physicochemical subgrouping of the mammalian picornaviruses. J. Gen.Virol. (1973), 18, 171-180.

7. SPECTOR, D.H. and BALTIMORE, D. Requirement of 3'-terminal polyadenylic acid for the infectivity of poliovirus RNA. Proc. Nat. Acad. Sci. U.S.A. (1974), 71, 2983-2987.

8. BURNESS, A.T.H., PARDOE, I.U., DUFFY, E.M., BHALLA, R.B. and GOLDSTEIN, N.O. The size and location of the poly(A) tract in EMC virus RNA. J. Gen. Virol. (1977), 34, 331-345.

9. BROWN, F., NEWMAN, J., STOTT, J., PORTER, A., FRISBY, D., NEWTON, C., CAREY, N. and FELLNER, P. Poly(C) in animal viral RNAs. Nature (1974), 251, 342-344.

10. PEREZ-BERCOFF, R. and GANDER, M. The genomic RNA of Mengo virus. I. Location of the poly(C) tract. Virology (1977), 80, 426-429.

11. LEE, Y.F., NOMOTO, A., DETJEN, B.M. and WIMMER, E. A protein covalently linked to poliovirus genome RNA. Proc. Nat. Acad. Sci. U.S.A. (1977), 74, 59-63.

12. HRUBY, D.E. and ROBERTS, W.K. Encephalomyocarditis virus RNA, III. Presence of a genome-associated protein. J. Virol. (1978), 25, 413-415.

13. SANGAR, D.V., ROWLANDS, D.J., HARRIS, T.J.R. and BROWN, F. Protein covalently linked to foot-and-mouth disease virus RNA. Nature (1977), 268, 648-650.

14. NOMOTO, A., DETJEN, B., POZZATTI, R. and WIMMER, E. The location of the polio genome protein in viral RNAs and its implication for RNA synthesis. Nature (1977), 268, 208-213.

15. MAIZEL, J.V. Evidence for multiple components in the structural protein of type 1 poliovirus. Biochem. Biophys. Res. Commun. (1963), 13, 483-489.

16. SHAPIRO, A.L., VINUELA, E. and MAIZEL, J.V. Molecular weight estimation of polypeptide chains by electrophoresis in SDS-polyacrylamide gels. Biochem. Biophys. Res. Commun. (1967), 28, 815-820.

17. RUECKERT, R.R. On the structure and morphogenesis of picornaviruses. In Comprehensive Virology, Vol.6. Fraenkel-Conrat, H. and Wagner, R.R. eds. (1976), pp. 131-213. Plenum Press, New York.

18. ZIOLA, B.R. and SCRABA,D.G. Structure of the Mengo virion. III.Purification and amino acid compositions of the major capsid polypeptides. Virology, (1975), 64, 228-235.

19. JACOBSON, M.F., ASSO, J. and BALTIMORE, D. Further evidence on the formation of poliovirus proteins. J. Mol. Biol. (1970), 49, 657-669.

20. STOLTZFUS, C.M. and RUECKERT, R.R. Capsid polypeptides of Maus-Elberfeld virus. J. Virol. (1972), 10, 347-355.

21. ADAM, K.H. and STROHMAIER, K. Isolation of the coat proteins of foot-and-mouth disease virus and analysis of the composition of N-terminal end groups. Biochem. Biophys. Res. Commun. (1974), 61, 185-192.

22. CHLUMECKA, V., D'OBRENAN, P. and COLTER, J.S. Isoelectric focusing studies of Mengo variants, their protein structure units and constituent polypeptides. J. Gen.Virol. (1977), 35, 425-437.

23. KAY, C.M., COLTER, J.S. and OIKAWA, K. Circular dichroism studies on Mengo virus variants and their constituent ribonucleates. Can. J. Biochem. (1970), 48, 940-943.

24. ZIOLA, B.R. and SCRABA, D.G. Structure of the Mengo virion. IV. Amino and carboxyl-terminal analyses of the major capsid polypeptides. Virology (1976), 71, 111-121.

25. BURRELL, C.J. and COOPER, P.D. N-terminal aspartate, glycine and serine in poliovirus capsid proteins. J. Gen. Virol. (1973), 21, 443-451.

26. MATHEKA, H.D. and BACHRACH, H.L. N-terminal amino acid sequences in the major capsid proteins of foot-and-mouth disease virus types A, O and C. J. Virol. (1975), 16, 1248-1253.

27. BACHRACH, H.L., SWANEY, J.B. and VANDE WOUDE, G.F. Isolation of the structural polypeptides of foot-and-mouth disease virus and analysis of their C-terminal sequences. Virology (1973), 52, 520-528.

28. FINCH, J.T. and KLUG, A. Structure of poliomyelitis virus. Nature (1959), 183, 1709-1714.

29. FINCH, J.T. and KLUG, A. Arrangement of protein subunits and the distribution of nucleic acid in turnip yellow mosaic virus. II. Electron microscope studies. J. Mol. Biol. (1966), 15, 344-364.

30. MAYOR, H.D. Picornavirus symmetry. Virology (1964), 22, 156-160.

31. AGRAWAL, H.O. Studies on the structure of poliovirus. Arch. Gesamte Virusforsch. (1966), 19, 365-372.

32. HORNE, R.W. and NAGINGTON, J. Electron microscope studies of the development and structure of poliomyelitis virus. J. Mol. Biol. (1959), 1, 333-338.

33. RUECKERT, R.R., DUNKER, A.K. and STOLTZFUS, C.M. The structure of mouse-Elberfeld virus; a model. Proc. Nat. Acad. Sci. U.S.A. (1969), 62, 912-919.

34. DUNKER, A.K. and RUECKERT, R.R. Fragments generated by pH-dissociation of ME virus and their relation to the structure

of the virion. J. Mol. Biol. (1971), 58, 217-235.

35. MAK, T.W., COLTER, J.S. and SCRABA, D.G. Structure of the Mengo virion. II. Physicochemical and electron microscopic analysis of degraded virus. Virology (1974), 57, 543-553.

36. TALBOT, P. and BROWN, F. A model for foot-and-mouth disease virus. J. Gen. Virol. (1972), 15, 163-170.

37. CASPAR, D.L.D. and KLUG, A. Physical principles in the construction of regular viruses. Cold Spring Harbor Symp. Quant. Biol. (1962), 27, 1-24.

38. JOHNSTON, M.D. and MARTIN, S.J. Capsid and procapsid proteins of a bovine enterovirus. J. Gen. Virol. (1971), 11, 71-79.

39. MAIZEL, J.V., PHILLIPS, B.A. and SUMMERS, D.F. Composition of artificially produced and naturally occurring empty capsids of poliovirus type 1. Virology (1967), 32, 692-699.

40. CROWELL, R.L. and PHILLIPSON, L. Specific alterations of Coxsackie virus B3 eluted from HeLa cells. J.Virol. (1971), 8, 509-515.

41. LONBERG-HOLM, K. and KORANT, B.D. Early interaction of rhinoviruses with host cells. J. Virol. (1972), 9, 29-40.

42. KORANT, B.D., LONBERG-HOLM, K., NOBLE, J. and STASNY, J.T. Naturally occurring and artificially produced components of three rhinoviruses.Virology (1972), 48, 71-86.

43. KATAGIRI, S., AIKAWA, S. and HINUMA, Y. Stepwise degradation of poliovirus capsids by alkaline treatment. J. Gen. Virol. (1971), 13, 101-109.

44. LONBERG-HOLM, K. and YIN, F.H. Antigenic determinants of infective and inactivated human rhinovirus type 2. J. Virol. (1973), 12, 114-123.

45. BREINDL, M. VP4, the D-reactive part of poliovirus. Virology (1971), 46, 962-964.

46. PHILIPSON, L., BEATRICE, S.T. and CROWELL, R.L. A structural model for picornaviruses as suggested from an analysis of urea-degraded virions and procapsids of Coxsackie virus B3. Virology (1973), 54, 69-79.

47. STUDY GROUP ON PICORNAVIRIDAE. Picornaviridae. Intervirology (1975), 4, 303-316.

48. LONBERG-HOLM, K. and PHILIPSON, L. Early interaction between animal viruses and cells. In Monographs in Virology, Vol. 9, Melnick, J.L. ed. (1974), pp. 1-148. S. Karger, Basel.

49. MANDEL, B. Characterization of type 1 poliovirus by electrophoretic analysis. Virology (1971), 44, 554-568.

50. KORANT, B.D., LONBERG-HOLM, K., YIN, F.H. and NOBLE-HARVEY, J. Fractionation of biologically active and inactive populations of human rhinovirus type 2. Virology (1975), 63, 384-394.

51. CARTHEW, P. and MARTIN, S.J. The iodination of bovine enterovirus particles. J. Gen. Virol. (1974), 24, 525-534.

52. LONBERG-HOLM, K. and BUTTERWORTH, B. Investigation of the structure of polio-and human-rhinovirions through the use of selective chemical reactivity. Virology (1976), 71, 207-216.

53. BURROUGHS, J.N., ROWLANDS, D.J., SANGAR, D.V., TALBOT, P. and BROWN, F. Further evidence for multiple proteins in the foot-and-mouth disease virus particle. J. Gen. Virol. (1971), 13, 73-84.

54. ROWLANDS, D.J., SANGAR, D.V. and BROWN, F. Relationship of the antigenic structure of foot-and-mouth disease virus to the process of infection. J. Gen. Virol. (1971), 13, 85-93.

55. CAVANAGH, D., SANGAR, D.V., ROWLANDS, D.J. and BROWN, F. Immunogenic and cell attachment sites of FMDV; further evidence for their location in a single capsid polypeptide. J. Gen. Virol. (1977), 35, 149-158.

56. SANGAR, D.V., ROWLANDS, D.J., CAVANAGH, D. and BROWN, F. Characterization of the minor polypeptides in the foot-and-mouth disease virus particle. J. Gen. Virol. (1976), 31, 35-46.

57. LUND, G.A., ZIOLA, B.R., SALMI, A. and SCRABA, D.G. Structure of the Mengo virion. V. Distribution of the capsid polypeptides with respect to the surface of the virus particle. Virology (1977), 78, 35-44.

58. PETERS, K. and RICHARDS, F.M. Chemical cross-linking reagents and problems in studies of membrane structure. Ann. Rev. Biochem. (1977), 46, 523-551.

59. HORDERN, J.S., LEONARD, J.D. and SCRABA, D.G. Manuscript in preparation.

60. ZIOLA, B.R. and SCRABA, D.G. Structure of the Mengo virion. I. Polypeptide and ribonucleate components of the virus particle.

Virology (1974), 57, 531-542.

61. PAUCHA, E., SEEHAFER, J. and COLTER, J.S. Synthesis of viral-specific polypeptides in Mengo virus-infected L cells: evidence for asymmetric translation of the viral genome. Virology (1974), 61, 315-326.

62. REKOSH, D.M.K. The Molecular Biology of Picornaviruses. In The Molecular Biology of Animal Viruses, Vol.1. Nayak, D.P. ed. (1977), pp.63-110. Marcel Dekker Inc., New York.

63. EGGEN, K.L. and SHATKIN, A.J. In vitro translation of cardiovirus ribonucleic acid by mammalian cell-free extracts. J. Virol. (1972), 9, 636-645.

64. KORANT, B.D. Cleavage of viral precursor proteins in vivo and in vitro. J. Virol. (1972), 10, 751-759.

65. KORANT, B.D. Cleavage of poliovirus-specific polypeptide aggregates. J. Virol. (1973), 12, 556-563.

66. TABER, R., REKOSH, D. and BALTIMORE, D. Effect of pactamycin on synthesis of poliovirus proteins: a method for genetic mapping. J. Virol. (1971), 8, 395-410.

67. LAWRENCE, C. and THACH, R.E. Identification of a viral protein involved in post-translational maturation of the encephalomyocarditis virus capsid precursor. J. Virol. (1975), 15, 918-928.

68. PHILLIPS, B.A. The morphogeneis of poliovirus. Current Topics in Microbiology and Immunology (1972), 58, 156-174.

69. JACOBSON, M.F. and BALTIMORE, D. Morphogenesis of poliovirus. I. Association of viral RNA with coat protein. J. Mol. Biol. (1968), 33, 369-378.

70. FERNANDEZ-TOMAS, C.B. and BALTIMORE, D. Morphogenesis of poliovirus. II. Demonstration of a new intermediate, the provirion. J. Virol. (1973), 12, 1122-1130.

71. McGREGOR, S., HALL, L. and RUECKERT, R.R. Evidence for the existence of protomers in the assembly of encephalomyocarditis virus. J. Virol. (1975), 15, 1107-1120.

72. McGREGOR, S. and RUCKERT, R.R. Picornaviral capsid assembly: similarity of rhinovirus and enterovirus precursor subunits. J. Virol. (1977), 21, 548-553.

73. LEE, P.W.K., PAUCHA, E. and COLTER, J.S. Identification and

partial characterization of a new (50S) subviral particle in Mengo virus-infected cells. Virology (1978), 85, 286-295.

74. GHENDRON, Y., YAKOBSEN, E. and MIKHENJEVA, A. Study of some stages of poliovirus morphogenesis in MiO cells. J. Virol. (1972), 10, 261-266.

GENERAL ORGANIZATION AND STRUCTURE OF THE PICORNAVIRUS GENOME

PETER FELLNER

Searle Research Laboratories

Lane End Road, High Wycombe, Bucks, England

INTRODUCTION

The picornavirus group numbers over 200 distinct serotypes, isolated from a wide variety of human and animal sources, and it is likely that many others exist in nature (1), as yet unrecognised. Various sub-classifications of this large range of viruses have been proposed (2, 3). The most recent, and probably the most generally satisfactory classification (4, 5) arranges the picornaviruses into five genera (a) Cardioviruses, including encephalomyocarditis (EMC) and mengoviruses; (b) Human Rhinoviruses; (c) Equine Rhinoviruses; (d) Foot-and-Mouth Disease viruses; (e) Enteroviruses, including polio, Coxsackie viruses etc. The members of these different genera display differences in virus particle stability in the pH range 3 - 7, and in their particle density in caesium chloride, as well as in RNA base composition (see chapters 1 and 3).

However, all picornaviruses contain, characteristically, an RNA genome comprising about 32% of the total mass of each virus particle (6). This genomic RNA is a continuous single-stranded molecule (7) and is (+)-strand (8), i.e. it has the same 5'-3' polarity as the mRNA synthesised after virus infection, and it must itself be utilised as mRNA, at least initially upon infection. This is supported by the infectious nature of purified virion RNA, demonstrating that the RNA genome contains all the information necessary for de novo production of infectious virus, and that this can be directly translated following infection.

Early estimates of the size of picornavirus RNAs were somewhat disparate, ranging from as little as 1.7×10^6 daltons up to

3×10^6 daltons. More recent estimates have yielded, for all virus types which have been studied, fairly uniform values around 2.5×10^6 daltons. Thus Fenwick (9), by analytical ultracentrifugation of formaldehyde-treated RNA, obtained a value of 2.4×10^6 daltons for EMCV RNA. Poliovirus RNA molecular weight was estimated as 2.6×10^6 daltons by electronmicroscopy (10). Nair and Lonberg-Holm (11) estimated the molecular weight of rhinovirus RNA to be $2.4 - 2.5 \times 10^6$ daltons, and Tannock et al (12) determined the molecular weight of poliovirus RNA to be 2.56×10^6 daltons. Frisby (13) subjected EMCV RNA to polyacrylamide gel electrophoresis under denaturing conditions in the presence of formamide, and determined its molecular weight to be 2.50×10^6 daltons. He also demonstrated that ECHO (type 12), rhinovirus (type 2) and poliovirus (type 1) all sedimented at very similar rates to EMCV RNA, using isokinetic sucrose gradients, suggesting that all of these RNAs have closely similar molecular weights, and that each contains approximately 7700-7800 nucleotides.

The molecular weights of these RNAs, taken together with the known masses and compositions of picornavirus particles suggest that each particle contains only one RNA molecule.

The picornavirus genome has in the past proved attractive as a model for studying the translation of a eukaroytic mRNA (14), since virion RNAs may be conveniently obtained both highly purified and in significant amounts. However, it has become clear, since the advent of detailed structural investigations of these viral RNAs, that they possess special features not so far encountered in other eukaryotic mRNAs. The majority of this short review will be devoted to discussing the properties and possible functions of these unusual structural features.

I. THE 5'-TERMINUS OF PICORNAVIRUS RNA

Initial studies on the 5'-terminus of poliovirus RNA were reported by Wimmer (15). In this work, he attempted to specically label the 5'-terminal nucleotide with polynucleotide kinase, after treatment with phosphomonoesterase. The results of subsequent hydrolysis indicated that pAp was the predominant nucleotide present at the 5'-terminus of the RNA.

This finding was re-examined, following the discovery of the very widespread occurrence of the m_7G cap at the 5'-termini of nearly all eukaryotic cellular and viral mRNAs (reviewed by Shatkin, (16)). Nomoto et al.(17) and Hewlett et al. (18) demonstrated that both polio virion and mRNAs lack any m_7G cap, by examining the products arising upon complete nuclease hydrolysis of ^{32}P- labelled virion and post-infection polyribosomal RNAs. Both groups reported that the mRNA contains 5'-terminal pUp, recovered in molar amounts, but did not identify any 5'-terminal nucleotide arising from the virion RNA. Frisby et al. (19) also reported the absence of 5'-terminal m_7G in EMCV RNA, but were unable to positively identify the 5'-terminal nucleotide. The picornavirus virion and mRNAs thus appear to differ from all other eukaryotic cellular and viral mRNAs studied so far, in lacking a requirement of a 5'-terminal m_7G cap for translation to occur (with the exception of a satellite tobacco necrosis virus).

In view of the puzzling difficulty in determining the precise nature of the 5'-terminus of the virion RNA, RNase T_2 hydrolysates of labelled polio virion RNA were examined in greater detail (20, 21). These investigators detected an additional hydrolysis product, not present in the corresponding mRNA hydrolysates. This contained a ^{32}P-labelled nucleotide, linked to a protein or peptide which could be labelled with ^3H-amino acids 3 hours after infection. The linkage appeared to be covalent, being resistant to heating in sodium dodecyl sulphate and sucrose gradient centrifugation in very high concentrations of salt, but susceptible to hydrolysis with pronase, releasing pUp. Equally the protein was shown to remain associated with the intact virion RNA under conditions which might have been expected to dissociate non-covalent complexes, but could be released by nuclease treatment. Lee et al. (20) were able to determine that the molecular weight of this protein, termed VPg, is less than 7,000 daltons, but did not make more precise estimates. Sangar et al. (22) also detected a small protein covalently linked to FMDV virion RNA, which could be labelled with ^{35}S-methionine. Polyacrylamide gel electrophoresis of the labelled protein, released by nuclease treatment, enabled its molecular weight to be estimated as about 4,000 daltons. The amount of ^{35}S-methionine-labelled protein found attached to the virus RNA was consistent with the occurrence of one molecule of the protein per RNA molecule. More recently, a similar protein has also been detected attached to the 5'-terminus of EMCV RNA (23).

It was reported that VPg is linked to each nascent strand present within the poliovirus replicative intermediate form (20, 21). Flanegan et al. (21) find that the amount of VPg-pUp/unit weight arising from nuclease digestion of the replicative intermediate RNA is the same as that obtained from virion RNA. The intermediate has a mass about four times that of the virion RNA (consisting of a (-)-strand RNA and 6 nascent chains, on average half-completed (8). If one molecule of VPg is attached to the (-)-strand template (see below), then the remainder must be attached to the nascent (+)-strands.

This finding suggests that the presence of VPg might be necessary to prime initiation of transcription of the nascent chains by the viral replicase. Lee et al. (20) suggested alternatively that VPg may play a role in the assembly of virus particles, i.e. in linking the newly-synthesised RNA molecules to the procapsids. They, together with Fernandez-Muñoz and Lavi (24), also speculate that the absence of VPg from the viral mRNA renders it incompetent for encapsidation. This would explain why the mRNA is not a precursor of virion RNA (8). However, the subsequent finding (25) that the 5'-terminus of polio (-)-strand RNA, isolated from the replicative intermediate form, is also attached to VPg would indicate that this protein is more likely to have some other function, probably in initiation of replication, as first suggested.

Mapping of oligonucleotides arising upon T_1 ribonuclease digestion of the virion and polyribosomal mRNAs suggests that the nucleotide sequences are identical, apart from VPg attached to the 5'-terminal oligonucleotide from the virion RNA. The results of Flanegan *et al.* (21), Pettersson *et al.* (26) and Nomoto *et al.* (25, 27) show that the sequence of the 5'terminal oligomer of the virion RNA, attached to VPg, is identical with that of the free oligomer from the mRNA, in each case pUUAAAACAG. The sequence adjacent to the VPg attached to the (-)-strand is poly(U) (27, see below). These findings strongly suggest, but do not prove, that VPg is involved in priming in some way the initiation of all transcription, and that it is subsequently removed enzymatically from a portion of the newly synthesised RNA molecules. This fraction then functions as polyribosomal mRNA, and is not incorporated into virus particles.

II. POLY(C) TRACTS WITHIN PICORNAVIRUS RNAs

The presence of large poly(C) tracts has been detected within the genomes of some picornaviruses, in the first instance within encephalomyocarditis virus RNA. A large T_1 ribonuclease-resistant product was found to be released upon complete enzymatic hydrolysis of EMCV RNA (28). Further digestion of this product with RNase A yielded 87C, 1AAC, 1AC, 4U, 1G. The total chain-length of about 100 nucleotides was consistent with the mobility of this fragment upon polyacrylamide gel electrophoresis. Subsequently poly(C) tracts were detected by similar methods within the virion RNAs of two further cardioviruses, mengo and Maus-Elberfeld, and also within the three European types of FMD viruses, O, A and C (29). Quantitation of the amount of the poly(C)-containing T_1 ribonuclease product arising from EMCV RNA indicated that it occurred once in each molecule.

The sizes of the poly(C) tracts arising upon T_1 ribonuclease hydrolysis differed significantly between the different viruses. Encephalomyocarditis and mengoviruses, which appear to be closely related, both serologically (1) and by T_1 oligonucleotide mapping (30), both released a product of about 100 nucleotides, containing approximately 90 C residues. Maus-Elberfeld virus yielded a product of roughly 150 nucleotides, containing 110-130C residues. The FMDV types A-61 and C-997 also gave rise to products of about 150 nucleotides, containing 80-90% C whereas the types O-V1 and C-GC yielded products about 200 nucleotides in length, containing about 80% C (29).

More recently Harris and Brown (31) have studided the poly(C) tracts within virulent and avirulent SAT_1 serotypes of FMDV. The lengths of these were determined to be about 170 and 100 nucleotides respectively (the significance of this is further discussed below).

The approximate position of this poly(C) tract within the EMCV and FMDV genomes has been studied by two methods.

Harris and Brown (32) examined the location of the poly(C) tract within FMDV (SAT_1) RNA. The RNA was subjected to mild alkaline hydrolysis. The products, comprised of a continuous spectrum of fragments of different sizes, were fractionated by sucrose gradient centrifugation. The material thus fractionated was divided into six size classes, which were chromatographed on oligo(dT)-cellulose. Those fragments still containing the 3'-terminal poly(A) tract (discussed below) of FMDV RNA were retained. Upon subsequent digestion with T_1 ribonuclease, it was found that only the longest size class, fragments comprising >85% of the genome, contained the poly C tract as well as the 3'-terminal poly(A) tract, suggesting that the poly C tract is located within the 5'-terminal 15% of the genome. Perez-Bercoff and Gander (33) have studied the position of the poly(C) tract in mengovirus RNA using the same methods, and find that it is located in the 5'-terminal 10% of the RNA. Similar findings have been reported by Chumakov and Agol (34) concerning the location of the poly(C) tract within EMCV RNA, using consecutive chromatography of partial alkaline hydrolysates on poly(U)-and poly (I)-cellulose columns. Only fragments close in size to the complete genome were retained by both columns, indicating the 5'terminal location of the poly(C) tract.

Rowlands _et al_. (35) have established the location of the poly(C) tract within FMDV RNA by use of an alternative method. They have digested the RNA, annealed to oligo(dG), with RNase H. This specifically removes the heteroduplex poly(C).oligo(dG) region, releasing the remaining RNA fragments, which are about 400 and 7,000 nucleotides in length, respectively. This result taken together with the previous findings suggests that the poly(C) tract is located 400-500 nucleotides from the 5'-termini of picornavirus RNAs.

Consideration of the size of the largest of the virus-coded precursor polypeptides synthesised after EMCV infection (36), and the length of the EMCV RNA, indicates that about 1,000 nucleotides are not translated. It was suggested (28) that the poly(C) tract was likely to occur within this untranslated region, close to the 3'- or 5'-terminus, since no polyproline has been detected amongst the products of translation. However, although the poly(C) tract has been found to be located close to the 5'-terminus it is not yet entirely certain that it remains untranslated, since polyproline could be present within the transient largest precursor species, prior to post-translational cleavage.

Emtage et al. (37) have studied the nature of the products arising upon cDNA synthesis, using avian myeloblastosis virus reverse transcriptase, primed by oligo$(dG)_{10}$ and $(dT)_{12-18}$, in the presence of an EMCV RNA template. In each case, the products of synthesis were large and heterogeneous in size, with a mean length of about 2,000 nucleotides. Initiation of synthesis must occur at accessible primer-binding areas within the RNA sequence, most probably at the poly(C) and poly(A) tracts which are known to be single-stranded in native RNA (see below). Interestingly, the large products of both reactions substantially compete with each other in hybridising with EMCV RNA. The extent of competition leads to an estimate that the priming regions are only about 500 nucleotides apart in the primary sequence. Since it is known that in the linear genomic RNA, the poly(A) and poly(C) tracts are separated by about 7,000 nucleotides, this finding strongly suggests that the majority of the EMCV RNA template is circularised in the presence of the reverse transcriptase, and that the enzyme will continue synthesis of DNA chains complementary to the 3'-terminal area which in fact are initiated by oligo(dG) close to the 5'-terminus. The enzyme is known to function in this way during retrovirus replication.

The poly(C) tract within native EMCV RNA has been found to be predominantly single-stranded at pH 6.0 (38). This was determined by studying the susceptibility of C residues within the poly(C) tract to hydrolysis by sodium bisulphite, yielding U, using mild conditions under which only non-base-paired residues would be hydrolysed. The poly(C) tract was protected from conversion to poly(U) by prior annealing of poly(I) to native EMCV RNA.

It was suggested that poly(C) tracts might play a role in the replication of those viruses which contain it (28, 29). In particular, they speculated that such a poly(C) tract might function as a replicase-binding site, or a part of one, on the basis of two observations. First, Rosenberg et al. (39) reported that a partially purified EMC viral replicase preparation will synthesise poly(G) from a poly(C) template, but will not copy other synthetic primers. Second, the replicase of bacteriophage Qβ, which has subunits of similar size to those of the EMCV replicase (39) can also specifically use poly(C) as a template for poly(G) synthesis. However, the significance of this specificity is unclear, since Qβ

RNA does not contain any large poly(C) tracts (40); A. Porter, unpublished results).

Large poly(C) tracts are only encountered as T_1 ribonuclease-resistant products upon hydrolysis of cardiovirus and FMDV RNAs. However, if the poly(C) tract is involved in a basic virus function such as replication, it would be surprising if such a feature is not conserved, at least to some extent, throughout the Picornaviridae, since the replication processes are believed to be similar for all these viruses. If such poly(C) tracts were interrupted by several G residues, in addition to the small numbers of A and U residues present, then they would be forshortened and might not be detected amongst the very large numbers of smaller products arising upon complete T_1 ribonuclease digestion. However, recent results (D. Black, personal communication) suggest that this is unlikely to be the case, since the small numbers of A and U residues in the poly(C) tract of an FMDV RNA, arising upon T_1 RNase digestion, all occur at the extremities, with the poly(C) tract remaining completely uninterrupted by other nucleotides. This was determined by the use of rapid RNA sequencing methods directly on the isolated poly(C) tract, according to (41).

It is interesting that the presence or absence of poly(C) coincides with the classification of enteroviruses, cardioviruses, rhinoviruses and FMD viruses as separate genera, proposed in (4). Frisby (13) has suggested that this implies a non-involvement of the poly(C) tract in a universal function such as replication, and that it might have some specific role in assembly, as a nucleation site, interacting with the capsid proteins of those viruses which contain it. The capacity of poly(C) to form double-stranded helices at pH 5.7 or below might contribute to the known instability of cardioviruses and FMD viruses at acid pH, which has been employed as a criterion in the classification of the picornavirus genera (4).

Harris and Brown (31) have studied the RNAs of virulent and attenuated strains of FMDV (serotype SAT_1), in an attempt to define at the level of the genome the change(s) responsible for attenuation. The attenuated strain was derived from the virulent one by extensive serial passaging. Mapping and subsequent analysis of the T_1 ribonuclease oligonucleotides revealed only very small difference between the RNAs. However, the size of the poly(C) tract in the attenuated strain was determined to be about 100 nucleotides, whereas that from the virulent strain was about 170 nucleotides in length. This change did not arise from interposition of a single G residue in the poly(C) tract of the attenuated strain, but appeared very likely to result from a genuine large deletion. This interpretation was supported by slight size differences between the virus RNAs which could be detected. This result suggests that the poly(C) tract may be important in maintaining virulence in FMD viruses, but this cannot yet be fully established in view of the few single-site sequence differences which were also detected, and which might have affected other viral functions.

III. AMINOACYLATION OF A SITE WITHIN THE PICORNAVIRUS GENOME

It has been found that when mengovirus or EMCV RNAs are incubated with procaryotic or eukaryotic aminoacyl tRNA synthetase preparations, low levels of charging with specific amino-acids occur (42, 43).

Mengovirus RNA can be specifically acylated with histidine, using a mouse liver aminoacyl tRNA synthetase preparation (43). The virus RNA is somewhat hydrolysed during the incubation by endonucleases within the synthetase preparation. The principal amino-acylated product was shown to have a molecular weight in the range $0.20 \times 10^6 - 0.35 \times 10^6$ daltons, from its mobility upon electrophoresis through polyacrylamide-agarose gels. The authors find 0.18 moles of histidine are incorporated/mole of RNA. No other amino acids are incorporated, when employing a labelled mixture of them. They demonstrate that the charging is not due to the presence of a tRNA molecule specifically hydrogen-bonded to the viral RNA, as occurs with retroviruses (44) and certain plant viruses.

Similarly it was reported that EMCV RNA can be aminoacylated with serine, using synthetases from E.coli and bovine or rabbit liver, to the extent of 0.024-0.08 moles/mole of virus RNA (42). The possibility was excluded of a bound tRNA molecule acting as a substrate, and found that only serine was significantly incorporated out of 13 labelled amino acids which were employed. Five further amino acids were not tested. The aminoacylation does not occur at the 3'-terminus of the virus RNA, since the accepting ability is unaltered by prior periodate oxidation of the 3'-terminus, but takes place at an internal site, rendered accessible by the action of endonucleases contaminating the synthetases, as in the case of mengovirus RNA (42). During aminoacylation the virus RNA was hydrolysed, yielding a heterogeneous range of large fragments.

The significance of the presence of a site for aminoacylation within the genomes of these viruses is unclear. The aminoacylation occurs at a sequence within the interior which would not be available in the functional genome or mRNA. Also the observed levels of aminoacylation are low, however this may well be due to low yields of suitable substrate fragments arising from the uncharacterised endonuclease action of the synthetase preparations. In this connection, it is interesting that aminoacylation of FMDV RNA has not been detected when using a nuclease-free synthetase preparation (45). Lindley and Stebbing (42) also suggest that the presence of nucleotidyl transferase activity might be necessary, to generate suitable CCA ends for aminoacylation.

These findings lead to suggest that there exists within the genome a region which can specifically interact with tRNA binding enzymes, and this region might have a role as a replicase binding site, or in the regulation of translation, although the aminoacylation ability is likely to be an incidental artefact (42, 43). In the case of the small RNA bacteriophages, such as Qβ, the replicase contains host-cell coded polypeptides that normally function in the translational apparatus of the cell. If similar host-coded polypeptides are included in picornavirus replicases these might specifically interact with such a region within the genome. This may prove to be a widely distributed feature of RNA viruses, since it has also been found possible to specifically aminoacylate many plant virus RNAs directly at their 3'-termini. The nucleotide sequences of the 3'-terminal regions of a number of these plant virus RNAs have been determined (46). Some similarities of primary and secondary structure have been noted between these 3'-termini and certain tRNAs, although many differences are also apparent. These structures in plant viruses have also been suggested as having possible roles in replication or translational control.

IV. POLY(A) IN PICORNAVIRUS RNAs

Poly(A) sequences have been found to occur at the 3'-termini of most, although not all, eukaryotic cellular and viral mRNAs. The picornavirus RNAs, which act as mRNAs at least initially after infection, also contain, unsurprisingly, a 3'-terminal poly(A) tract in all cases which have been studied. These include:
(i) the cardioviruses; Columbia SK, EMC and mengo (47-54).
(ii) the Enteroviruses; polio, several Coxsackies and bovine enterovirus (30, 55-58). (iii) the FMD viruses (4, 30). (iv) the Rhinoviruses (59, 60).

In general, these poly(A) tracts have been isolated by complete digestion of the viral RNAs with both T_1 and pancreatic RNases. The 3'-terminal location is suggested by (a) end-labelling of the terminal adenosine residue with ^{3}H-borohydride (57) or ^{3}H-isoniazid (50); (b) the absence of any G or pyrimidine residues in highly-purified preparations of the poly(A) tract (50).

There is a considerable variation in the reported sizes of the poly(A) tracts within picornavirus RNAs, the published size estimates ranging from 15 to 150 nucleotides. However, recent work has suggested that the sizes of the poly(A) tracts are in fact relatively uniform throughout the picornavirus group (30, 50, 56).

Spector and Baltimore (55, 56) have studied the sizes of the poly(A) tracts of poliovirus and mengovirus RNAs. They find that the poliovirus poly(A) is heterogeneous, containing about 50-125 nucleotides from its mobility in non-denaturing SDS-containing polyacrylamide gels. This result agrees with the previous estimates (57, 58), which fall within this size-range. Spector and Baltimore also report that mengovirus poly(A) is only slightly shorter, with a mean length of 50-70 nucleotides. This disagrees with the earlier estimate of Miller and Plagemann (48) that the poly(A) of mengovirus RNA contained only 15-17 nucleotides. Other size estimates of poly(A) from EMC virus (52), Columbia SK (47) and rhinovirus (59) also fall within the range 50-100 nucleotides.

Frisby et al. (30) studied the heterogeneous poly(A) populations arising upon mapping T_1 RNase oligonucleotides from picornavirus RNAs by two-dimensional gel electrophoresis. Poly (A) tracts arising from several representatives of each of three picornavirus subclasses had very similar mobilities. The electrophoretic mobility corresponded to a mean length of 30-50 nucleotides in each case, although a small proportion of the poly(A) appeared to be as much as 100 nucleotides in length. They investigated the size of the poly(A) tract of EMCV RNA in greater detail (50), by electrophoresis through formamide-containing polyacrylamide gels in the presence of markers of defined chain-length. A mean size of about 40 nucleotides was observed for the poly(A) tract, somewhat lower than the size suggested by Spector and Baltimore (49) for poly(A) from the closely-related mengovirus, but greater than the recent estimates by Goldstein et al. (53) and Burness et al. (54). These authors find that the mean length of the poly(A) tract in EMC is 14-18 nucleotides, by use of three different methods, although some poly(A) containing up to 45 nucleotides is found upon polyacrylamide gel electrophoresis. The authors suggest that previous size estimates of poly(A) in picornaviruses have been too high, but the reasons for the discrepancy between these results and those reported by Spector and Baltimore (55, 56) and Frisby et al. (30, 50) remain unclear.

The function of the poly(A) tracts which occur in cellular mRNAs is still not clearly established. Possible roles have been suggested in (a) processing of precursors to mRNAs; (b) transport of mRNAs from the nucleus to the cytoplasm; (c) efficient translation of mRNAs; (d) protection of mRNAs against hydrolysis by exonucleases.

In considering the function of poly(A), its presence within picornavirus RNAs is of particular interest, firstly since its mode of synthesis differs from that of cellular mRNAs, and secondly since, unlike in cellular mRNAs, a functional poly(A) requirement has now been established, although the precise details are not yet known.

The poly(A) tracts in cellular mRNAs are added post-transcriptionally (61), but those in picornavirus RNAs are probably transcribed. Yogo and Wimmer (62, 63) have isolated a tract of poly(U) arising upon T_1 RNase hydrolysis of (-)-strand RNA, which was prepared from either the replicative form or the replicative intermediate of poliovirus. This contains about 60-120 nucleotides, i.e. it is approximately the same size as the poly(A) tract in the (+)-strand virus RNA (63). This strongly suggests that the poly(A) tract is copied off a poly(U) template, i.e. is genetically coded, rather than added post-transcriptionally. Yogo, Teng and Wimmer (64) also determined that the poly(U) is located at the 5'-terminus of the (-)-strand by labelling with polynucleotide kinase. This is consistent with the notion that the poly(A) and poly(U) tracts are transcribed from each other at successive steps in the replicative process.

Spector and Baltimore (55) demonstrated that the presence of at least a considerable proportion of the poly(A) tract is essential to maintain the infectivity of poliovirus RNA. They found that it was possible to selectively remove the poly(A) tract from poliovirus RNA by complexing the RNA with poly(dT) and then specifically hydrolysing the poly(A)-poly(dT) heteroduplex with highly-purified RNase H. This resulted in a reduction of approximately 15-fold in the infectivity of the RNA upon removal of 82% of the poly(A), even though no such loss accompanied incubation of the RNA with the enzyme alone. Removal of on average 60% of the poly(A) led to a six-fold reduction in infectivity. These results show that a poly(A) tract of significant length is necessary for the infectious process to occur, although they do not afford an estimate of the precise number. Elsewhere Spector and Baltimore (56) suggest that it might be as many as 40-50 nucleotides, corresponding to the smallest size fraction that they detect in poliovirus and mengo RNAs. Interestingly, the residual infectivity after poly(A) removal is 2-4 times greater than would be expected from the small numbers of molecules remaining still containing some poly(A), and plaques arising from this RNA yielded progeny with poly(A) tracts of normal length. This suggests that some additional mechanism exists in the infected cell which is capable of regenerating the poly(A) tract. The exact role of the poly(A) in infection is not known, but it cannot be excluded that it is identical to that of the poly(A) in cellular mRNAs, but in this case permitting the initial round of translation of the viral RNA immediately after infection.

V. 3'-TERMINAL NUCLEOTIDE SEQUENCES OF PICORNAVIRUS RNAs

The only extensive studies which have been carried out on the nucleotide sequences of picornavirus RNAs concern those sequences adjacent to the 3'-terminal poly(A) tract (65-67). In view of their location, it might be expected that these sequences could be involved in specific interactions with the viral replicases, or else act in some other way as control elements.

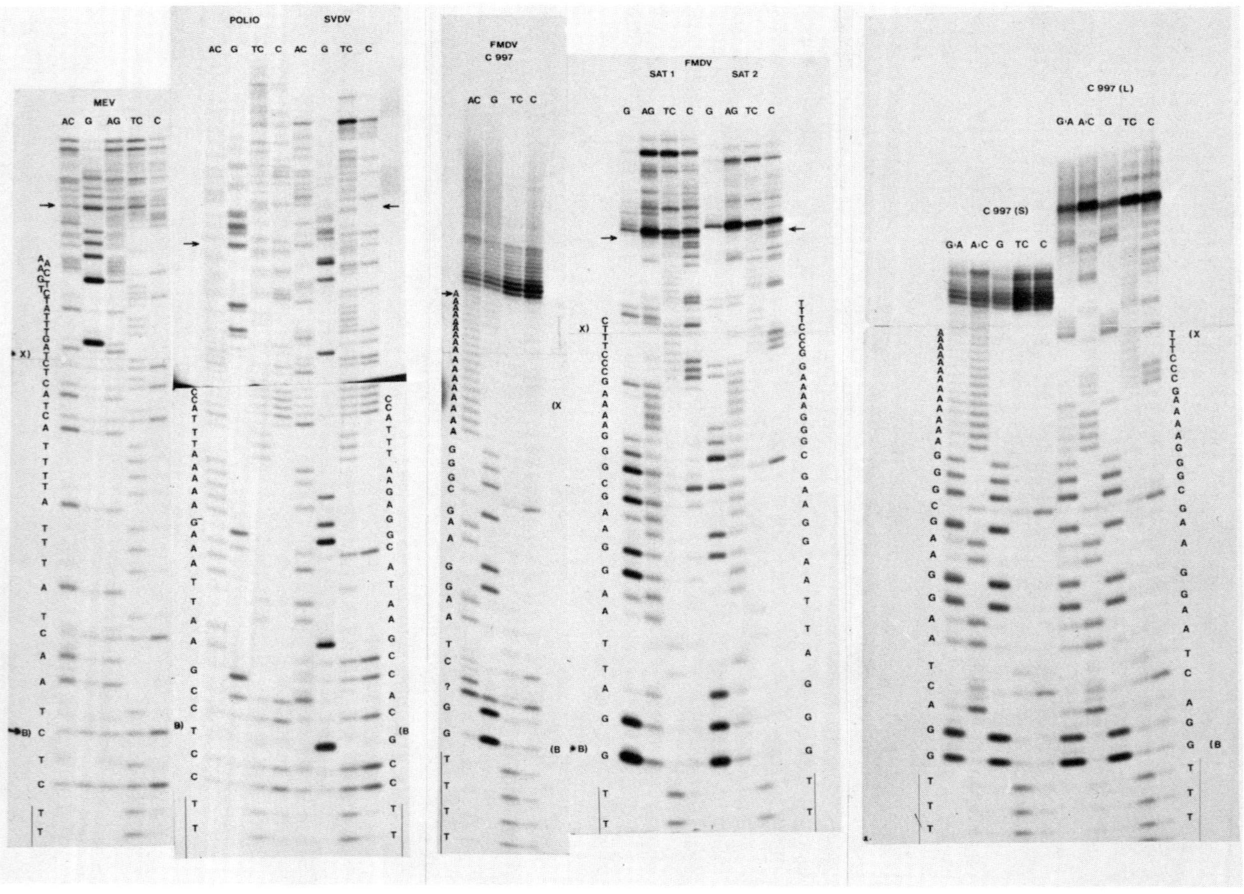

Figure 1. Sequence analysis of 3'-terminal cDNAs copied from picornavirus RNAs.

The following procedure has been used to determine these sequences:

The virus RNA is annealed to a ^{32}P-labelled primer oligo(dT)$_{10}$-X, where X is the nucleotide complementary to the first residue after the poly(A) tract in the viral RNA. The presence of this nucleotide has the effect of "anchoring" all the primer molecules at a defined position on the template. Then cDNA synthesis is carried out, from this primer, using the Klenow fragment of E.coli DNA polymerase in the presence of Mn^{2+}. The cDNA synthesised in this way is found to be 20-200 nucleotides in length depending on the viral RNA template used, and tends to fall into discrete size classes in each case, for unknown reasons. This DNA may then be subjected to the rapid gel sequencing procedure of Maxam and Gilbert (68) (Figure 1). In this way, Porter et al. (65, 66) have determined the nucleotide sequences adjacent to the poly(A) tract in nine virus RNAs, from members of three of the major subgroups of the picornaviruses. These sequences are presented in table 1.

The sequences of the RNAs from the cardioviruses EMCV and Maus-Elberfeld (129 and 54 nucleotides determined, respectively) are very closely related, displaying only 4 differences in the 3'-terminal 54 nucleotides.

Both RNAs contain 7 termination triplets within the 3'-terminal 30 nucleotides, encompassing all possible reading frames. Three further nonsense triplets occur up to position -125 in EMCV RNA, and at least 26 of the 3'-terminal nucleotides cannot therefore be translated, and therefore 5 nonsense triplets within the last 20 nucleotides are not normally used as termination codons.

The 3'-terminal 35 nucleotides of both RNAs include 26 purines. The sequence AAUAAA, which has been found close to the 3'-terminus of all other eukaryotic mRNAs which have been studied, also occurs at position -15 to -10 nucleotides. If this sequence fulfills a similar function in the cardiovirus and cellular mRNAs, it is unlikely to be involved in nuclear events or transport functions, since the infectious cycle of cardioviruses is entirely cytoplasmic.

Little possibility is apparent for formation of strong secondary structure within the 3'-terminal 129 nucleotides of EMCV RNA. Only one loop which involves nucleotides 71-103 is likely to have a reasonably high stability, although alternative structures involving base-pairing with other regions of the molecule cannot be excluded. However, the region close to the poly(A) tract is likely to be single-stranded and accessible, since the poly(A) tract can be selectively removed from the virus RNA by treatment with T_1 ribonuclease at an enzyme: RNA ratio of $1:10^4$. Porter et al.

Table 1. Comparison of the 3'-terminal nucleotide sequences of picornavirus RNAs.

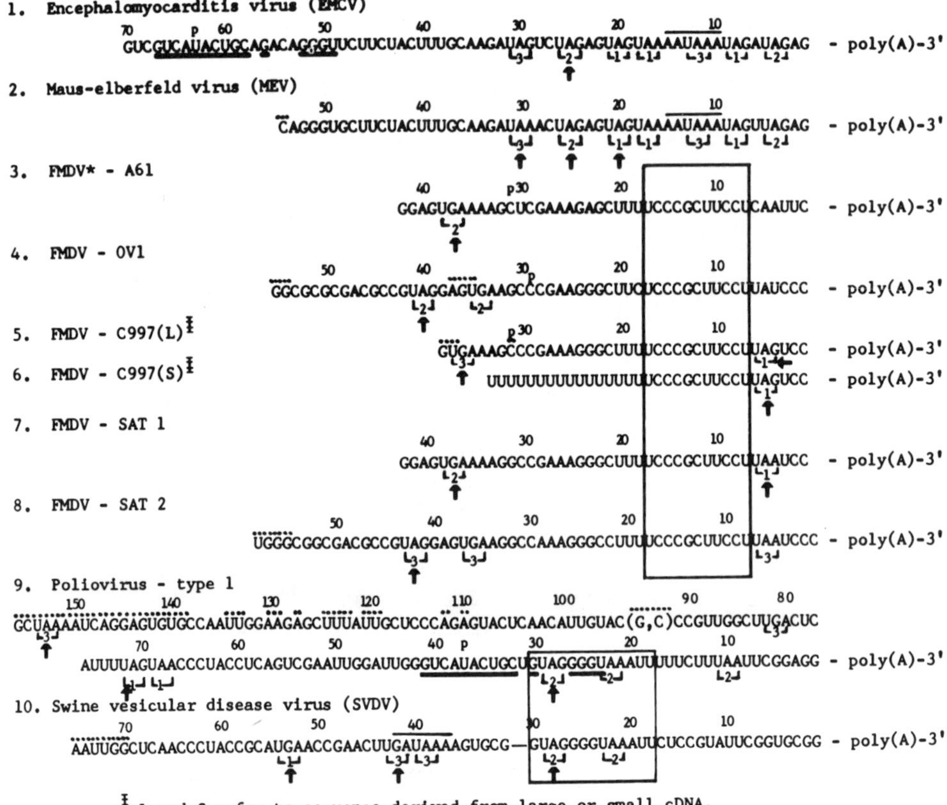

1. Encephalomyocarditis virus (EMCV)

GUCGUCAUACUGCAGACAGGGUUCUUCUACUUUGCAAGAUAGUCUAGAGUAGUAAAAUAAAUAGAUAGAG - poly(A)-3'

2. Maus-elberfeld virus (MEV)

CAGGGUGCUUCUACUUUGCAAGAUAAACUAGAGUAGUAAAAUAAAUAGUUAGAG - poly(A)-3'

3. FMDV* - A61

GGAGUGAAAAGCUCGAAAGAGCUUUUCCCGCUUCCUCAAUUC - poly(A)-3'

4. FMDV - OV1

GGCGCGCGACGCCGUAGGAGUGAAGCCCGAAGGGCUUCUCCCGCUUCCUUAUCCC - poly(A)-3'

5. FMDV - C997(L)[‡]

GUGAAAGCCCGAAAGGGCUUUUCCCGCUUCCUUAGUCC - poly(A)-3'

6. FMDV - C997(S)[‡]

UUUUUUUUUUUUUUUUUCCCGCUUCCUUAGUCC - poly(A)-3'

7. FMDV - SAT 1

GGAGUGAAAAGGCCGAAAGGGCUUUUCCCGCUUCCUUAAUCC - poly(A)-3'

8. FMDV - SAT 2

UGGGCGGCGACGCCGUAGGAGUGAAGGCCAAAGGGCCUUUUCCCGCUUCCUUAAUCCC - poly(A)-3'

9. Poliovirus - type 1

GCUAAAAUCAGGAGUGUGCCAAUUGGAAGAGCUUUAUUGCUCCCAGAGUACUCAACAUUGUAC(G,C)CCGUUGGCUUGACUC
AUUUUAGUAACCCUACCUCAGUCGAAUUGGAUUGGGUCAUACUGCUGUAGGGGUAAAUUUUUCUUUAAUUCGGAGG - poly(A)-3'

10. Swine vesicular disease virus (SVDV)

AAUUGGCUCAACCCUACCGCAUGAACCGAACUUGAUAAAAGUGCG—GUAGGGGUAAAUUCUCCGUAUUCGGUGCGG - poly(A)-3'

‡ L and S refer to sequence derived from large or small cDNA.

* Foot-and-mouth disease virus.

(66) suggest that a single-stranded structure at the 3'-terminus might be important in providing an easy access to the viral replicase during initiation of replication. On the other hand, the FMD virus RNAs do exhibit possibilities of strong secondary structure close to the 3'-terminal poly(A) (see below), suggesting that this is not a general feature of picornavirus RNAs.

The sequences have also been determined of 38-58 nucleotides present at the 3'-termini of five FMDV RNAs. These are set out in table 1. The following features are especially striking:

(a) Unlike the cardioviruses, and all other eukaryotic mRNAs, the sequence AAUAAA is not present, at least within the 3'-terminal 40 nucleotides. This argues against a role for this sequence as a universal signal for polyadenylation, as has been previously suggested (65).

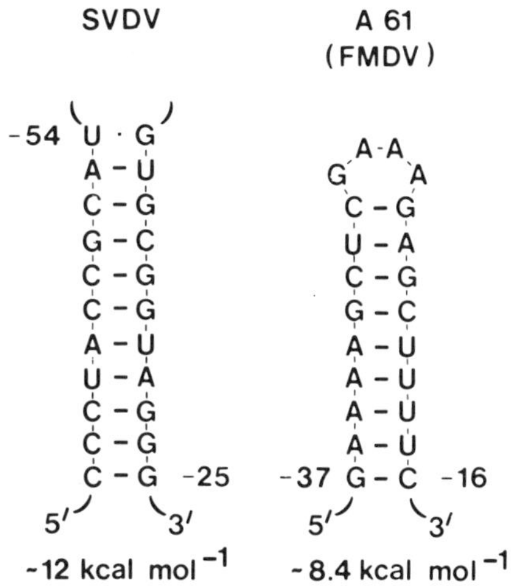

Figure 2. Possible secondary structures in two picornavirus RNAs.

(b) In sharp contrast to the cardioviruses, the FMDV RNAs have pyrimidine-rich segments (18 out of 20 nucleotides) adjacent to the poly(A), followed by a purine-rich segment, with few termination codons. The FMDV RNAs which have been studied display about 75% homology with each other within the 3'-terminal 20 nucleotides. The complete conservation of the undecamer UCCCGCUUCCU, in all strains, is especially striking. In four of the five serotypes (A-61, O-V1, SAT 1 and SAT 2) the 75% homology extends at least to the 3'-terminal 42 nucleotides, and at least to position -36 in strain C997.

(c) All of the FMDV RNAs are capable of forming stable hairpin structures close to the 3'-terminal poly(A), unlike the cardioviruses. An example is shown in Figure 2.

(d) In the case of C997 RNA only, an unusually short discrete subfraction of the transcribed DNA is found. This contains 33 nucleotides, and has been purified and sequenced. It contains a poly(dA) tract of 17 nucleotides, beginning 16 nucleotides from the oligo(dT) primer, suggesting that a subspecies of C997 RNA must exist containing a poly(U) tract at this position. Such a structure has not previously been found in the genomic RNA of an RNA virus, but is present at the 5'-terminus of (-) strands of picornaviruses, coding for the poly(A) tract. Its presence so close to the poly(A) tract of some of the C997 RNA molecules suggests a stable hairprin structure might form with the poly(A). Although its function in C997 RNA is unknown, its occurrence in other C serotypes of FMDV would indicate that it has some functional importance.

Finally extensive 3'-terminal sequences have been determined for RNAs of two enteroviruses, polio and swine vesicular disease virus (SVDV). The 3'-terminal 71 nucleotides (table 1) display about 65% homology. Like the FMDV strains, these RNAs do not contain the sequence AAUAAA (although the sequence GAUAAA is present in SVDV at position -38 to -43). The numbers of termination codons which are evident are intermediate between those of the cardioviruses and FMDV strains. The 3'-terminal sequence of SVDV RNA is capable of forming a strong secondary structure, like the FMDV RNAs, as shown in Figure 2. A region of poliovirus RNA (position -23 to -41) displays considerable homology with a section of the EMCV sequence (position -49 to -67), indicating possible common evolutionary ancestors even for viruses in different genera. Most interestingly, much of this conserved region is palindromic. An 11-nucleotide palindrome is also present in the RNA of FMDV strain A-61, centred on position 30. Palindromes have been previously detected in the 3'-terminal untranslated regions of cellular mRNAs (69, 71) and it has been suggested that they may provide binding sites for oligomeric proteins.

ACKNOWLEDGEMENT

I thank Dr. A. Porter for valuable discussions and for communicating results prior to publication.

REFERENCES

1. RUECKERT, R.R. In Comparative Virology, (1971), Maramorosch, K. and Kurstak, E., eds. Ch. 8. 255-306 Academic Press, New York.

2. WILDY, P. Classification and nomenclature of viruses. In Monographs in Virology, (1971), Vol. 5. Karger, Basel.

3. FENNER, F. In the Biology of Animal Viruses, (1971), 1, 20-21. Academic Press, New York.

4. NEWMAN, J.F.E., ROWLANDS, D.J. and BROWN, F. A physicochemical sub-grouping of the mammalian picornaviruses. J. Gen. Virol. (1973), 18, 171-180.

5. BURROUGHS, J.N. and BROWN, F. Physicochemical evidence for re-classification of the Caliciviruses. J. Gen. Virol. (1974), 22, 281-286.

6. BURNESS, A.T.H. Ribonucleic acid content of encephalomyocarditis virus. J. Gen. Virol. (1970), 6, 373-380.

7. MONTAGNIER, L. and SANDERS, F.K. Sedimentation properties of infective ribonucleic acid extracted from encephalomyocarditis virus. Nature (London), (1963), 197, 1178-1181.

8. BALTIMORE, D. The replication of picornaviruses. In the Biochemistry of Viruses, (1969), Levy, H.B., ed., p. 103-176, Marcel Dekker, New York.

9. FENWICK, M.L. The effect of reaction with formaldehyde on the sedimentation rates of ribonucleic acids. Biochem. J. (1968), 107, 851-859.

10. GRANBOULAN, N. and GIRARD, M. Molecular weight of poliovirus RNA. J. Virol. (1969), 4, 475-479.

11. NAIR, C.N. and LONBERG-HOLM, K.K. Infectivity and sedimentation of rhinovirus ribonucleic acid. J. Virol. (1971), 7, 278-280.

12. TANNOCK, G.A., GIBBS, A.J. and COOPER, P. A re-examination of the molecular weight of poliovirus RNA. Biochem. Biophys.

Res. Comm. (1970), 38, 298-304.

13. FRISBY, D.P. Structure and assembly of picornavirus. Ph.D. Thesis. (University of London), (1974).

14. SMITH, A.E. Control of Translation of animal virus messenger RNA. In Control Processes in Virus Multiplication, (1975), Burke, D.C. and Russell, W.C., eds., pp 183-224, Cambridge University Press.

15. WIMMER, E. Sequence studies of poliovirus RNA. I. Characterization of the 5'-terminus. J. Mol. Biol. (1972), 68, 537-540.

16. SHATKIN, A.J. Capping of eukaryotic mRNAs. Cell, (1976), 9, 645-653.

17. NOMOTO, A., LEE, Y.J. and WIMMER, E. The 5'-end of poliovirus mRNA is not capped with $m^7G(5')$ ppp (5') Np. Proc. Nat. Acad. Sci. U.S.A. (1976), 73, 375-380.

18. HEWLETT, M.J., ROSE, J.K. and BALTIMORE, D. 5'-terminal structure of poliovirus polyribosomal RNA is pUp. Proc. Nat. Acad. Sci. U.S.A. (1976), 73, 327-330.

19. FRISBY, D.P., EATON, M. and FELLNER, P. Absence of 5'-terminal capping group in encephalomyocarditis virus RNA. Nucleic Acids Research (1976), 3, 2771-2788.

20. LEE, Y.F., NOMOTO, A., DETJEN, B.M. and WIMMER, E. A protein covalently linked to poliovirus genome RNA. Proc. Nat. Acad. Sci. U.S.A. (1977), 74, 59-63.

21. FLANEGAN, J.B., PETTERSSON, R.F., AMBROS, V., HEWLETT, M.J. and BALTIMORE, D. Covalent linkage of a protein to a defined nucleotide sequence at the 5'-terminus of virion and replicative intermediate RNAs of poliovirus. Proc. Nat. Acad. Sci. U.S.A. (1977), 74, 961-965.

22. SANGAR, D.V., ROWLANDS, D.J., HARRIS, T.J.R. and BROWN, F. Protein covalently linked to foot-and-mouth disease virus RNA. Nature (London), (1977), 268, 648-650.

23. HRUBY, D.E. and ROBERTS, R.K. Encephalomyocarditis virus RNA. III. Presence of a genome-associated protein. J. Virol. (1978), 25, 413-415.

24. FERNANDEZ-MŪNOZ, R. and LAVI, U. 5'-termini of poliovirus RNA: difference between virion and non-encapsidated 35S RNA. J. Virol. (1977), 21, 820-824.

25. NOMOTO, A., DETJEN, B., POZZATTI, R. and WIMMER, E. The location of the polio genome protein in viral RNAs and its implication for RNA synthesis. Nature (London), (1977), 268, 208-213.

26. PETTERSSON, R.F., FLANEGAN, J.B., ROSE, J.K. and BALTIMORE, D. 5'-terminal nucleotide sequences of poliovirus polyribosomal RNA and virion RNA are identical. Nature (London), (1977), 268, 270-272.

27. NOMOTO, A., KITAMURA, N., GOLINI, F. and WIMMER, E. The 5'-terminal structures of poliovirion RNA and poliovirus mRNA differ only in the genome-linked protein VPg. Proc. Nat. Acad. Sci. U.S.A. (1977), 74, 5345-5349.

28. PORTER, A., CAREY, N.H. and FELLNER, P. Presence of a large poly(rC) tract within the RNA of encephalomyocarditis virus. Nature (London), (1974), 248, 675-678.

29. BROWN, F., NEWMAN, J., STOTT, J., PORTER, A., FRISBY, D., NEWTON, C., CAREY, N. and FELLNER, P. Poly(C) in animal viral RNAs. Nature (London), (1974), 251, 342-344.

30. FRISBY, D.P., NEWTON, C., CAREY, N.H., FELLNER, P., NEWMAN, J.F.E., HARRIS, T.J.R. and BROWN, F. Oligonucleotide mapping of picornavirus RNAs by two dimensional electrophoresis. Virology, (1976), 71, 379-388.

31. HARRIS, T.J.R. and BROWN, F. Biochemical analysis of a virulent and an avirulent strain of foot-and-mouth disease virus. J. Gen. Virol. (1977), 34, 87-105.

32. HARRIS, T.J.R. and BROWN, F. The location of the poly(C) tract in the RNA of foot-and-mouth disease virus. J. Gen. Virol. (1976), 33, 493-501.

33. PEREZ-BERCOFF, R. and GANDER, M. The genomic RNA of mengovirus I. Location of the poly(C) tract. Virology, (1977), 80, 426-429.

34. CHUMAKOV, K.M. and AGOL, V.I. Poly(C) sequence is located near the 5'-end of encephalomyocarditis virus RNA. Biochem. Biophys. Res. Comm. (1976), 71, 551-557.

35. ROWLANDS, D.J., HARRIS, T.J.R. and BROWN, F. A more precise location of the poly(C) tract in foot-and-mouth disease virus RNA. J. Virol. (1978), 26, 335-343.

36. BUTTERWORTH, B.E. and RUECKERT, R.R. Gene order of encephalomyocarditis virus as determined by studies with

pactamycin. J. Virol. (1972), 9, 823-828.

37. EMTAGE, J.S., CAREY, N.H. and STEBBING, N. Structural features of encephalomyocarditis virus RNA from analysis of reverse transcriptase products. Eur. J. Biochem. (1976), 69, 69-78.

38. GOODCHILD, J., FELLNER, P. and PORTER, A.G. The determination of secondary structure in the poly(C) tract of encephalomyocarditis virus RNA with sodium bisulphite. Nucleic Acids Res. (1975), 2, 887-895.

39. ROSENBERG, H., DISKIN, B., ORON, L. and TRAUB, A. Isolation and subunit structure of the polycytidylate-dependent RNA polymerase of encephalomyocarditis virus. Proc. Nat. Acad. Sci. U.S.A. (1972), 69, 3815-3819.

40. HINDLEY, J. Structure and strategy in phage RNA. Progr. Biophys. Mol. Biol. (1973), 26, 269-320.

41. DONIS-KELLER, H., MAXAM, A.M. and GILBERT, W. Mapping adenines, guanines and pyrimidines in RNA. Nucleic Acids Res. (1977), 4, 2527-2538.

42. LINDLEY, I.J.D. and STEBBING, N. Aminoacylation of encephalomyocarditis virus RNA. J. Gen. Virol. (1976), 34, 177-181.

43. SALOMON, R. and LITTAUER, U.Z. Enzymatic acylation of histidine to mengovirus RNA. Nature (London), (1974), 249, 32-34.

44. FARAS, A.J., DAHLBERG, J.E., SAWYER, R.C., HARADA, F., TAYLOR, J.M., LEVINSON, W.E., BISHOP, J.M. and GOODMAN, H.M. Transcription of DNA from the 70S RNA of Rous sarcoma virus. II. Structure of a 4S RNA primer. J. Virol. (1974), 13, 1134-1142.

45. CHATTERJEE, N.K., BACHRACH, H.L. and POLATNICK, J. Foot-and-mouth disease virus RNA - presence of a 3'-terminal polyribonucleic acid and absence of amino acid binding ability. Virology, (1976), 69, 369-377.

46. BRIAND, J.P., RICHARDS, K.E., BOULEY, J.P., WITZ, J. and HIRTH, L. Structure of the amino acid accepting 3'-end of high-molecular weight egg plant mosaic virus RNA. Proc. Nat. Acad. Sci. U.S.A. (1976), 73, 737-741.

47. JOHNSTON, R.E. and BOSE, H.R. Correlation of messenger RNA function with adenylate-rich segments in the genomes of

single-stranded RNA viruses. Proc. Nat. Acad. Sci. U.S.A. (1972), 69, 1514-1516.

48. MILLER, R.L. and PLAGEMANN, P.G.W. Purification of mengovirus and identification of an A-rich segment in its RNA. J. Gen. Virol. (1972), 17, 349-353.

49. SPECTOR, D.H. and BALTIMORE, D. Poly(A) on mengovirus RNA. J. Virol. (1975), 16, 1081-1084.

50. FRISBY, D.P., SMITH, J., JEFFERS, V. and PORTER, A. Size and location of poly(A) in encephalomyocarditis virus RNA. Nucleic Acids Res. (1976), 3, 2789-2810.

51. GILLESPIE, D., MARSHALL, S. and GALLO, R.C. RNA of RNA tumour viruses contains poly(A). Nature New Biol. (1972), 236, 227-231.

52. GILLESPIE, D., TAKEMOTO, K., ROBERT, M. and GALLO, R.C. Polyadenylic acid in visna virus RNA. Science (1973), 179, 1328-1330.

53. GOLDSTEIN, N.O., PARDOE, I.U. and BURNESS, A.T.H. Requirement of an adenylic acid-rich segment for the infectivity of encephalomyocarditis virus RNA. J. Gen. Virol. (1973), 31, 271-276.

54. BURNESS, A.T.H., PARDOE, I.U., DUFFY, E.M., BHALLA, R.B. and GOLDSTEIN, N.O. The size and location of the poly(A) tract in EMC virus RNA. J. Gen. Virol. (1977), 34, 331-345.

55. SPECTOR, D.H. and BALTIMORE, D. Requirement of 3'-terminal poly (adenylic acid) for the infectivity of poliovirus RNA. Proc. Nat. Acad. Sci. U.S.A. (1974), 71, 2983-2987.

56. SPECTOR, D.H. and BALTIMORE, D. Polyadenylic acid on poliovirus RNA. III. *In vitro* addition of polyadenylic acid to poliovirus RNAs. J. Virol. (1975), 15, 1432-1439.

57. YOGO, Y. and WIMMER, E. Polyadenylic acid at the 3'-terminus of poliovirus RNA. Proc. Nat. Acad. Sci. U.S.A. (1972), 69, 1877-1882.

58. ARMSTRONG, J.A., EDMONDS, M., NAGAZATO, H., PHILIPS, B.A. and VAUGHAN, M.H. Polyadenylic acid sequences in the virion RNA of poliovirus and eastern equine encephalitis virus. Science, (1972), 176, 526-528.

59. NAIR, C.N. and OWENS, M.J. Preliminary observations pertaining to polyadenylation of rhinovirus RNA. J. Virol.

(1974), 13, 535-537.

60. MacNAUGHTON, M.R. and DIMMOCK, N.J. Polyadenylic acid sequences in rhinovirus RNA species from infected human diploid cells. J. Virol. (1975), 16, 745-748.

61. WEINBERG, R.A. Nuclear RNA metabolism. Annu. Rev. Biochem. (1973), 42, 329-354.

62. YOGO, Y. and WIMMER, E. Poly(A) and poly(U) in poliovirus double-stranded RNA. Nature New Biol. (1973), 242, 171-174.

63. YOGO, Y. and WIMMER, E. Sequence studies of poliovirus RNA. III. Polyuridylic acid and polyadenylic acid as components of the purified poliovirus replicative intermediate. J. Mol. Biol. (1975), 92, 467-477.

64. YOGO, Y., TENG, M.H. and WIMMER, E. Poly(U) in poliovirus minus RNA is 5'-terminal. Biochem. Biophys. Res. Commun. (1974), 61, 1101-1109.

65. PORTER, A.G., MERREGAERT, J., van EMMELO, J. and FIERS, W. Sequence of 129 nucleotides at the 3'-terminus of encephalomyocarditis virus RNA. Eur. J. Biochem. (1978). In Press.

66. PORTER, A.G., FELLNER, P., BLACK, D.N., ROWLANDS, D.J., HARRIS, T.J.R. and BROWN, F. 3'-terminal nucleotide sequences in the genomic RNA of picornaviruses. Nature (London), (1978). In Press.

67. MERREGAERT, J., van EMMELO, J., DEVOS, R., PORTER, A., FELLNER, P. and FIERS, W. The 3'-terminal nucleotide sequence of encephalomyocarditis virus RNA. Eur. J. Biochem. (1978), 82, 55-63.

68. MAXAM, A. and GILBERT, W. A new method for sequencing DNA. Proc. Nat. Acad. Sci. U.S.A. (1977), 74, 560-564.

69. SEEBURG, P.H., SHINE, J., MARTIAL, J.A., BAXTER, J.D. and GOODMAN, H.M. Nucleotide sequence and amplification in bacteria of structural gene for rat growth hormone. Nature (London), (1977), 270, 486-494.

70. SHINE, J., SEEBURG, P.H., MARTIAL, J.A., BAXTER, J.D. and GOODMAN, H.M. Construction and analysis of recombinant DNA for human chorionic somatomammotropin. Nature (London), (1977), 270, 494-499.

71. PROUDFOOT, N. Sequence analysis of the 3'-non coding regions of rabbit α- and β-globin messenger RNAs. J. Mol. Biol. (1976), 107, 491-525.

STRUCTURE-FUNCTION RELATIONSHIPS IN THE PICORNAVIRUSES

F. BROWN

Animal Virus Research Institute

Pirbright, Surrey, England

INTRODUCTION

The recent report of the Picornavirus Study Group of the International Committee for Taxonomy of Viruses divided the picornaviruses into 4 genera: (1) enterovirus; (2) cardiovirus; (3) rhinovirus and (4) foot-and-mouth or aphtho virus (1 and see chapter 1 of this book). Several other viruses were left ungrouped because of lack of information. For the purposes of this paper only those viruses which have been allotted to the 4 genera referred to above will be discussed.

As Scraba has pointed out (see chapter 1), the evidence available at present points to a fundamental similarity in structure and assembly for the virus particles of the different genera. However there are some differences between the genera which are expressed in their differing stability below pH7, their buoyant density in caesium chloride and their sensitivity to photodynamically active dyes. There are also fundamental differences in the structure of the genomes of the members of the four genera which are described fully by Fellner (see chapter 2). The possible implications of these differences on the structure of the virus particles and on the RNAs are discussed.

Despite the overall similarity of all the members of the picornavirus group, each of the genera possesses unique characteristics. One of the most significant features of foot-and-mouth disease virus (FMDV) is its considerable antigenic variation, which presents important problems in vaccination programmes. A detailed analysis of the RNA and proteins of the virus has been undertaken in an attempt to gain an understanding of

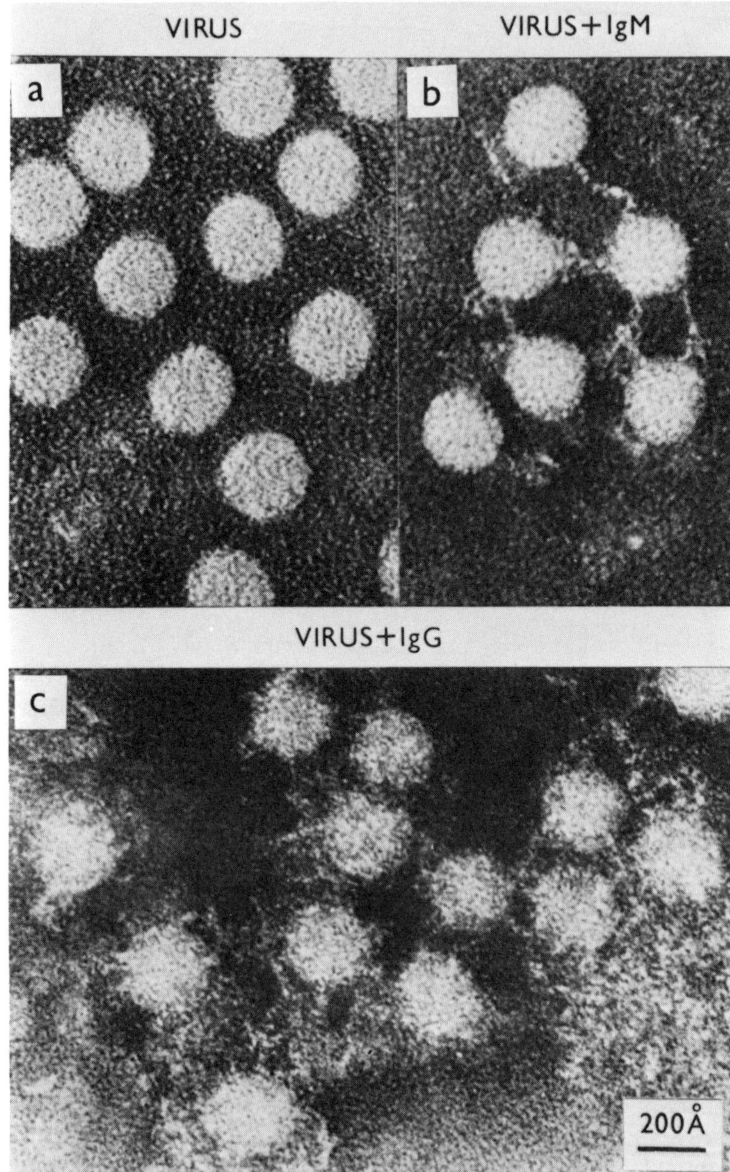

Figure 1. Electron micrographs of FMDV (a) alone; (b) complexed with IgM; (c) complexed with IgG.

the biochemical basis for this variation. The preliminary results of this study are presented and discussed here.

I. PHYSICOCHEMICAL PROPERTIES OF THE VIRUS PARTICLES

A. Morphology

The morphology of all the picornaviruses is very similar and the small differences in size which have been reported are probably as much an indication of the differences between laboratories as between viruses. Their appearance is singularly uninteresting and apart from an attempt to locate the immunizing antigen on the surface of FMDV by the use of antibody complexing (Figure 1, from ref. 2), the morphological examination has cast little light on the detailed structure of the virus particle. Nevertheless the differences in the stability of the enteroviruses and FMDV below pH7 can be demonstrated quite dramatically by comparing their appearance in the electron miscroscope after treatment with CO_2. Whereas swine vesicular disease virus particles remain intact, all the FMDV particles are disrupted (Figure 2).

B. Relationship Between Buoyant Density in Caesium Chloride and Stability Between pH3 and pH7

The remarkable difference in buoyant density between the entero- and cardioviruses (1.34g/ml) and FMDV (1.43g/ml) can best be explained by the difference in penetration of the Cs^+ ions and subsequent reaction with the RNA. If we accept values for the density of empty capsids and RNA of 1.30g/ml and 1.91g/ml respectively we can calculate that the density of a particle containing 30% RNA and 70% protein should be 1.48g/ml. This calculation will only be valid, however, if the RNA and protein react completely with the Cs^+ ions. Clearly the RNA in the entero- and cardioviruses must be protected from reaction with the Cs^+ ions by its interaction with the virus protein. Even the FMDV particle has a buoyant density less than the theoretical maximum. It is interesting to note, however, that the apparent density of FMDV particles increases as the time of centrifuging increase (3). When this value exceeds 1.47g/ml the virus particle disintegrates.

The stability of the virus particles below pH7 can be correlated with their density in caesium chloride. Thus the enteroviruses and cardioviruses are stable at pH3 whereas FMDV is disrupted below pH7. The rhinoviruses, which have a buoyant density of 1.40g/ml, occupy an intermediate position, being stable between pH7 and pH5 but unstable below pH5.

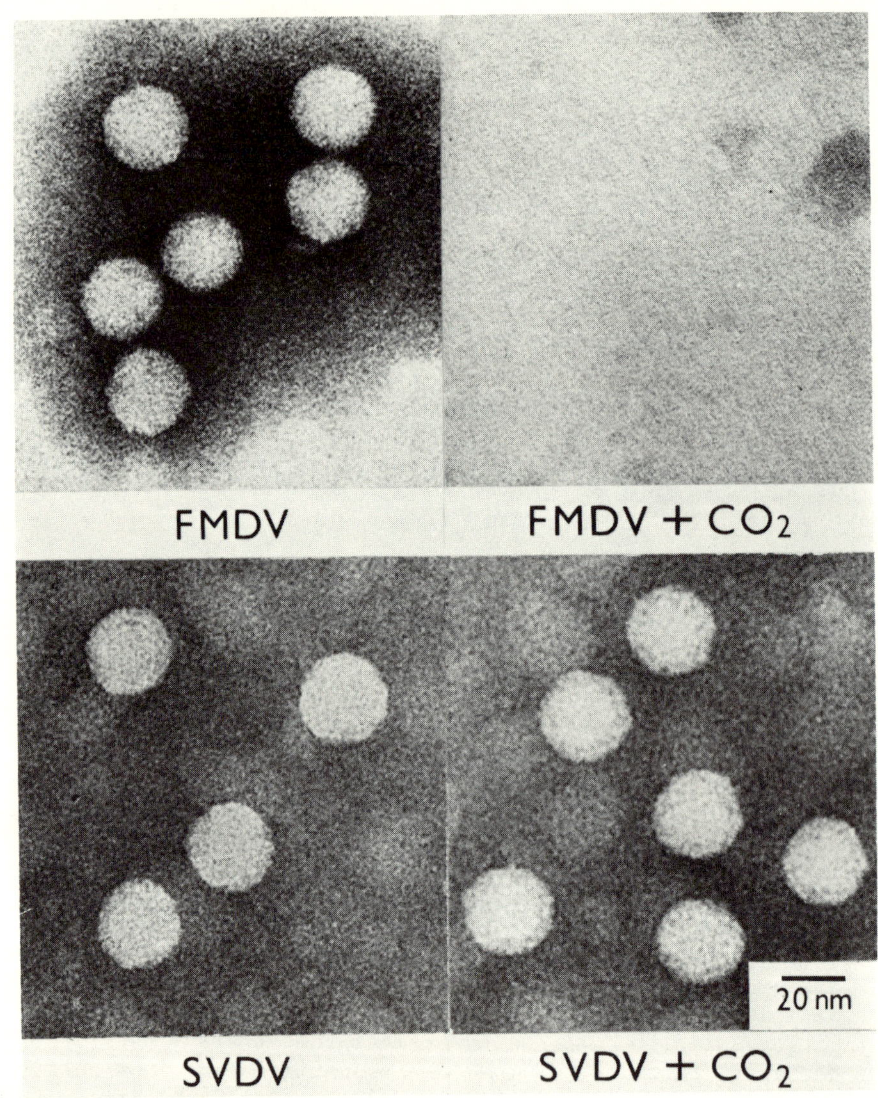

Figure 2. Electron micrographs of FMDV and swine vesicular disease virus before and after exposure to carbon dioxide.

C. Effect of Photodynamically Active Dyes on the Picornaviruses

FMDV is rapidly inactivated by low concentrations (2 μg/ml) of proflavine in the presence of visible light. This is due to inactivation of the RNA and eventually the particles disrupt into free RNA and 12S protein subunits (4). In contrast poliovirus is relatively resistant to photodynamically active dyes and any inactivation which occurs is due to oxidation of the protein coat. This observation is yet another manifestion of the porous nature of the FMDV particles compared with those of the other genera.

II. VIRUS RNA

The general structure and organisation of picornavirus RNA molecules are described by Fellner in chapter 2. I will therefore concentrate only on those features which appear to be different in the RNAs of the four genera.

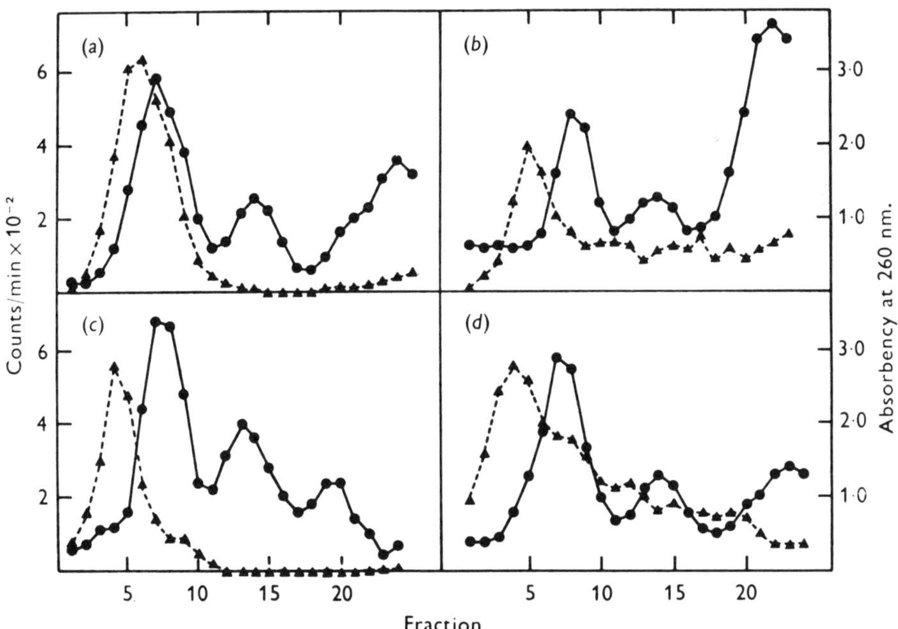

Figure 3. Sucrose density gradient centrifugation of the RNAs of (a) human rhinovirus HGP; (b) equine rhinovirus NM 11; (c) swine vesicular disease virus; (d) FMDV, type O-V1. ▲-----▲ , virus RNA; ●——● , ribosomal RNA.

A. Sedimentation Characteristics

The sedimentation profile of FMDV-RNA is markedly heterogeneous in contrast to those of the entero-, cardio- and rhinovirus RNAs (5 and Figure 3). The heterogeneous profile is obtained consistently, irrespective of whether the RNA is obtained from the virus particles by extraction with phenol-SDS or simply by lowering the pH below 7 in the presence of SDS. All the strains of virus which have been examined at Pirbright gave similar heterogeneous RNA profiles although those of an entero- and a rhinovirus RNA extracted with phenol under identical conditions in the same laboratory were homogeneous (5). Several suggestions have been made to account for the apparent heterogeneity, such as weak links within the molecule (6), cation attack (7), the effect of heat (8), the method used for isolation (9) and the presence of nucleases (10). There is general agreement that the RNA extracted from virus particles which have been kept at 37°C for several hours shows extensive degradation (Figure 4). The loss of RNA sedimenting at 35S and the corresponding increase in more slowly sedimenting molecules was equivalent to the loss of virus infectivity. Other properties of the virus particles such as their sedimentation rate and antigenic activity are unaffected.

For the purpose of RNA analysis in recent years we have used RNA extracted from fresh virus particles i.e. those produced early in the growth cycle before degradation by heat has occurred and in a recent paper Denoya _et al_. (11) have drawn attention to the importance of using fresh particles. With these particles homogeneous profiles of RNA can be obtained and treatment with pronase or heating at 60°C does not produce any alteration in the profile. Denoya _et al_. (11) also emphasised the value of purifying the virus and extracting the RNA as quickly as possible after harvesting.

These observations suggested that a nuclease might be associated with the virus particle. Brown and Wild (7) found that degradation of virus RNA occurred when virus particles were incubated at 37°C even in the presence of SDS, which is a potent inhibitor of ribonuclease activity. However Denoya _et al_. (33) have now found a nuclease activity located _inside_ the virus particles.

Gauntt (12) postulated the presence of a nuclease in rhinovirus particles on the basis of similar observation. However the paper by Denoya _et al_. (11) of a nuclease in FMDV records the first demonstration of an enzyme in a picornavirus.

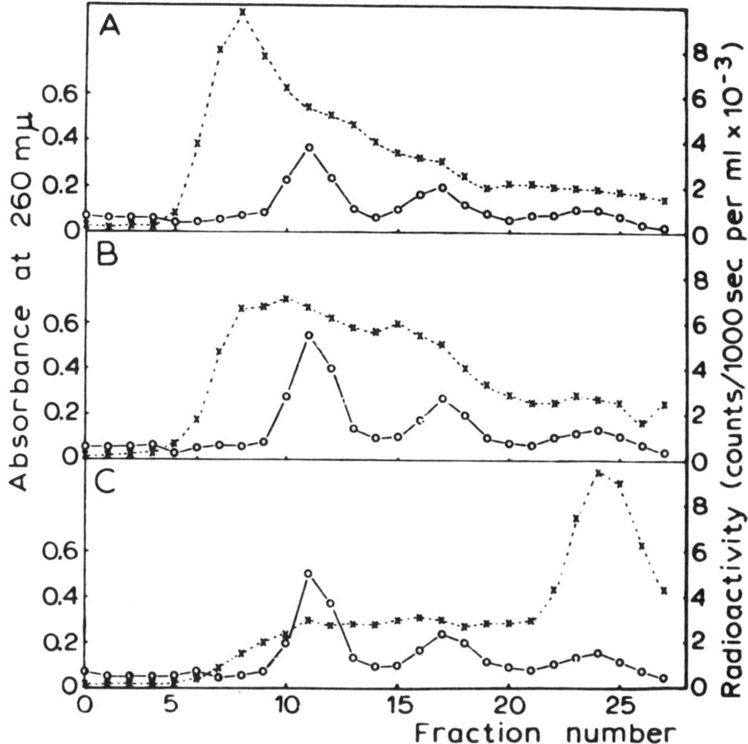

Figure 4. Sucrose density gradient centrifugation of ^{32}P-RNA extracted from FMDV which had been incubated at 37°C for (a) 0h; (b) 4h; (c) 8h. x - - x, ^{32}P; o — o, ribosomal RNA.

B. Polycytidylic Acid Tract

The poly(C) tract first described by Porter et al. (13) appears to be present only in the cardio- and FMD viruses (14). The length of the poly(C) tract varies between different isolates of the same virus. This was first noted for different FMDV serotypes (14) and has since been observed with a virulent and an attenuated strain of each of two different serotypes of the virus (15; Robson, K.J.H., unpublished observations). It may be significant that with each of the serotypes, the length of the poly(C) tract of the attenuated strain was only about one half of the length of that in the virulent strain.

Table 1. Length of the poly (C) tract in six encephalomyocarditis virus isolates.

Virus isolate	Animal of origin	poly (C) tract Number of nucleotides
DW	Pig	600
V 251	Pig	540
GS 8	Grey squirrel	225
RS 3	Red squirrel	175
V 297d	Pig	100
RRR	?	55
FMDV-A61	Cattle	150

The variability in the length of the poly(C) tract is not confined to FMDV-RNA. In some recent experiments at Pirbright, P. Stephenson and D.J. Rowlands have found that the length of the poly(C) tract varies between 50 and 500 nucleotides in six different isolates of encephalomyocarditis virus. The results of one dimensional polyacrylamide gel electrophoresis of the ribonuclease T_1 hydrolysates of the six RNAs are shown in Table 1.

Unlike the poly(A) tract of the picornavirus RNAs, which gives a heterogeneous band in one dimensional polyacrylamide gel electrophoresis, the poly(C) tract is usually a sharp, well defined band, suggesting that the length is homogeneous for an individual virus RNA. Similarly, in two dimensional electrophoretic separations, the poly(C) tract is found as a small spot whereas the poly(A) tract is present as an elongated spot.

As reported by Porter _et al_. (13), Brown _et al_. (14) and Frisby _et al_. (16), the poly(C) tract of the FMDV and cardiovirus strains contain a small number of adenylic acid and uridylic acid residues. D.N. Black (unpublished data) has now shown that in both FMD and EMC virus RNAs, the A and U residues are situated near the 5' end and the major part of the poly(C) tract is a continuous sequence of cytidylic acid residues.

As mentioned by Fellner (chapter 2), the poly(C) tract in both the FMDV and cardiovirus RNA is situated near the 5' end of the molecule. Rowlands _et al._ (17) showed that the poly(C) tract in one isolate of FMDV is situated about 400-500 nucleotides from the 5' end of the RNA. It is clearly important to establish whether this location is a feature common to both the cardiovirus RNAs and other members of the FMDV genus.

The presence of a poly(C) tract in the FMDV and cardiovirus genomes and its absence from the human rhino- and enterovirus RNAs raises questions regarding the translation of the different genomes. D.V. Sangar (unpublished data) has shown that the poly(C) residue is present on the polysomal RNA of FMDV infected cells and is presumably involved in translation. As a first step towards gaining an understanding of the role of the poly(C) tract in translation, the _in vitro_ translation products of the intact RNA and the L fragment to the 3' side of the poly(C) tract (see Figure 5) have been compared. In a preliminary experiment, D.J. Rowlands and D.V. Sangar (unpublished data) found that the L fragment codes for the same products as the untreated RNA with the exception of a protein with a molecular weight of _c_.5×10^3. This is about the size of the protein (VPg) covalently attached to the 5' end of the RNA.

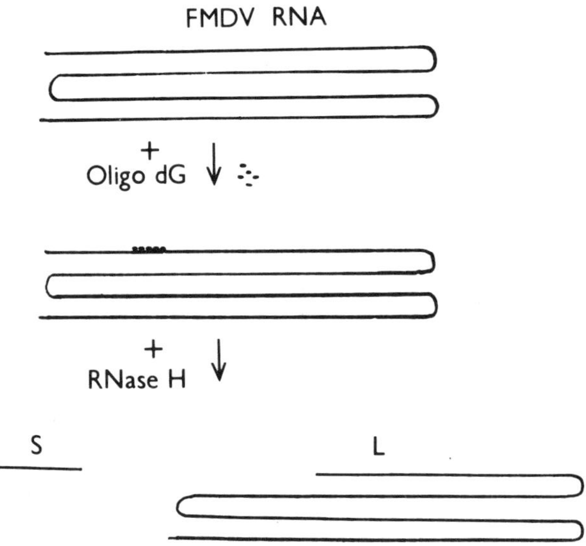

Figure 5. Diagrammatic representation of FMDV-RNA showing the position of the poly(C) tract.

A comparison of the tryptic peptide maps of VPg and this small molecular weight *in vitro* translation product should enable us to decide whether VPg is coded for by the region of the genome to the 5' side of the poly(C) tract.

III. VIRUS PROTEINS

All the picornaviruses contain 4 major polypeptides, three with molecular weights of *c*.30 x 10^3 (VP_1 - VP_3) and one which varies in molecular weight from 5.5 to 13.5 x 10^3 (see chapter 1). Calculations made from the amount and molecular weight of each polypeptide indicated that there were equal numbers of copies of VP_1, VP_2 and VP_3 for all the viruses and either the same or half the number of copies of VP_4. This difference clearly presents problems if we accept that there is a fundamental similarity in the architecture and morphogenesis of the virus particles in the different genera. However, most of the values for the molecular weight of the various VP_4s have been obtained by polyacrylamide gel electrophoresis and, as Stoltzfus & Rueckert (18) have pointed out, this method for determining the molecular weight of small polypeptides can produce erroneous results. These authors found, for example, that the molecular weight of the VP_4 of Maus-Elberfeld virus was 7.3 x 10^3 by filtration through agarose in the presence of 6M guanidine compared with 10 x 10^3 by polyacrylamide gel electrophoresis.

In a re-appraisal of the problem with FMDV, D.J. Rowlands (personal communication) has found a molecular weight of 8 x 10^3 by filtration through Sepharose in the presence of 6M guanidine compared with 13.5 x 10^3 by polyacrylamide gel electrophoresis. In a preliminary experiment using a different approach, A. King and T.R. Doel (personal communication) have calculated from the number of lysine residues in VP^4 (estimated by measuring the alteration in migration of the polypeptide in isoelectric focusing polyacrylamide gels following carbamylation) and the percentage of the amino acid in the polypeptide (determined by conventional amino acid analysis) that the molecular weight is not greater than 8 x 10^3. These estimates suggest that FMDV also contains 60 copies of VP_4 instead of the 30 previously reported. It would be worthwhile to make a similar re-assessment of the molecular weight of several VP_4 species by methods other than polyacrylamide gel electrophoresis.

A. *In Situ* Sensitivity to Trypsin of One of the Polypeptides of Foot-and-Mouth Disease Virus

FMDV particles differ from those of other picornaviruses in being susceptible to the action of proteolytic enzymes. A summary of the effect of trypsin on the virus is given in Table 2. The

Table 2. Effect of trypsin on the properties of FMDV

Property	Effect
Infectivity	Decreased up to 1000 fold
Attachment to cells	Decreased
Antigenicity	Decreased - up to 1000 fold
Reaction with IgG	Unaltered
Reaction with IgM	Abolished
Sedimentation rate	Unaltered
Morphology	Apparently unaltered
Structure of RNA	Unaltered
Structure of proteins	VP_1 cleaved

infectivity of the particles is reduced by about 1000-fold (although this varies between strains) but there is no alteration in their morphology. The reduced infectivity can be accounted for by the lowered ability of the particles to attach to susceptible cells (19-21). The RNA retains its infectivity i.e. the infectivity of the RNA extracted from the treated particles is as high as that of the untreated particles.

The antigenicity of the particles is also reduced, treated particles producing only about 0.1 - 1.0% of the neutralizing activity of untreated particles. The treated particles no longer react with IgM antibody, either in immunodiffusion experiments or immune complexing in the electron miscroscope (Figure 6a). However the treated particles still react with IgG antibody (Figure 6b) and a spur line is obtained when the untreated and treated virus particles are allowed to diffuse towards IgG antibody from adjacent wells (Figure 7).

Analysis of the virus proteins by polyacrylamide gel electrophoresis shows that only one of the capsid polypeptides is

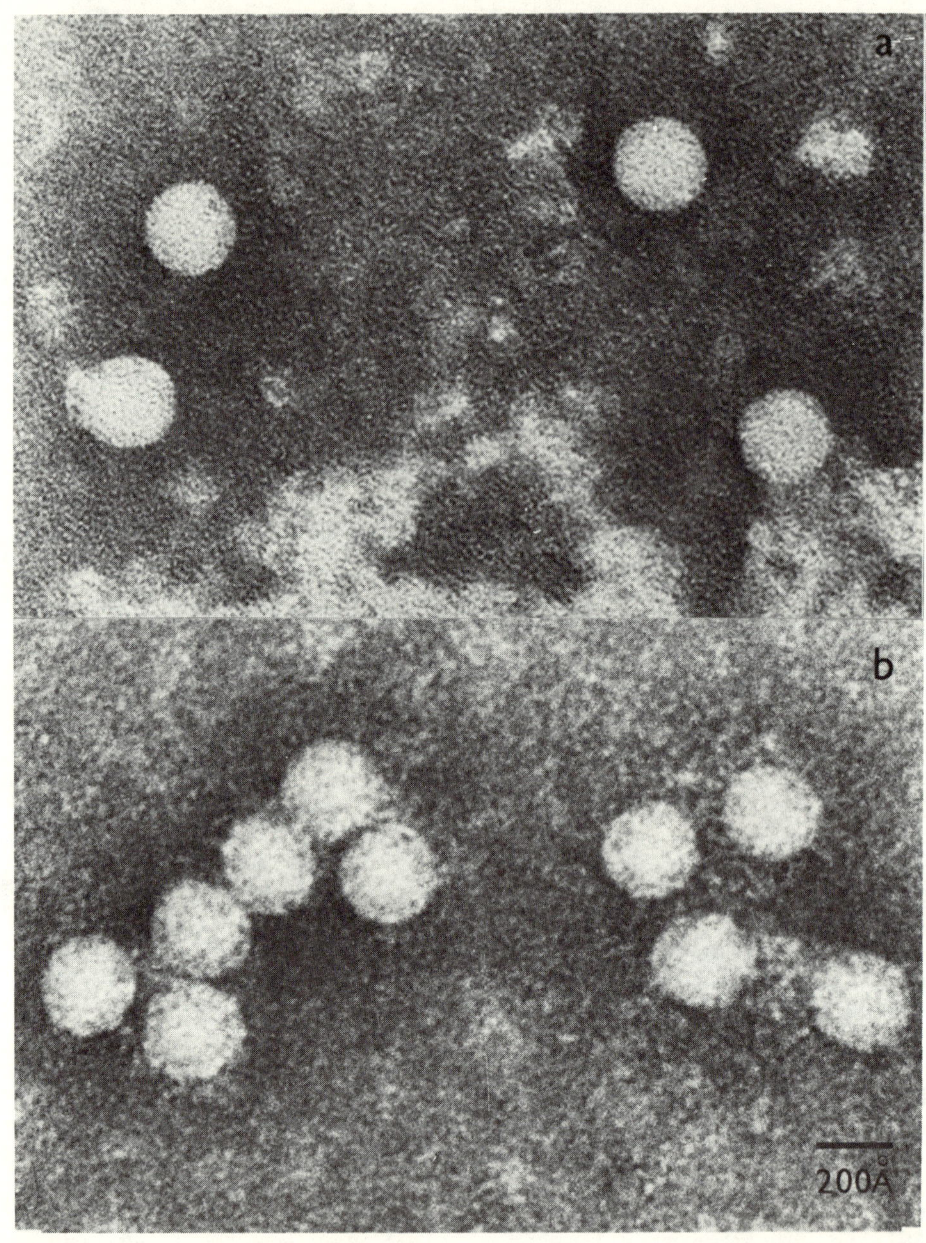

Figure 6. Electron micrograph of FMDV mixed with (a) IgM; (b) IgG.

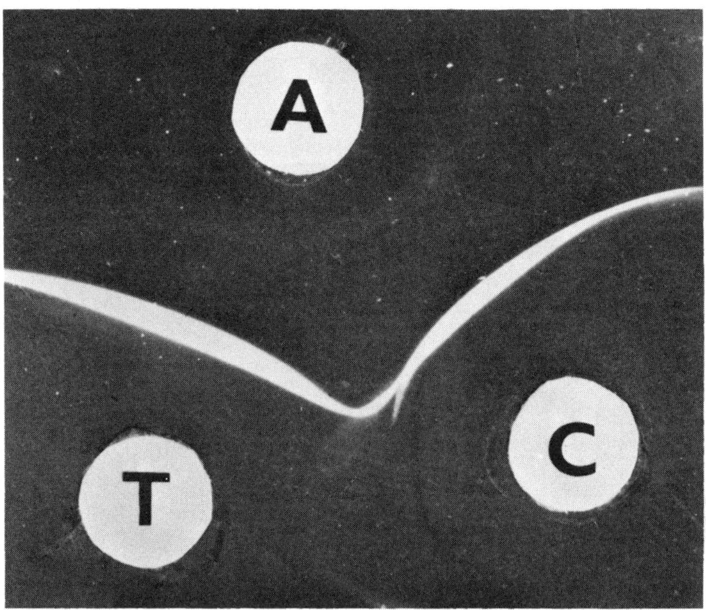

Figure 7. Effect of trypsin on the immunodiffusion line obtained with FMDV and IgG.

affected by the enzyme. In the example shown, polypeptide VP_1 is cleaved into two pieces with molecular weights of c.18 and 10 x 10^3 (Figure 8). These observations clearly implicate VP_1 in the production of neutralizing antibody. It is perhaps significant that VP_1 is the only polypeptide which becomes labelled when the virus particles are iodinated with potassium iodide in the presence of lactoperoxidase or chloramine T (22-23).

B. Antigenicity of the Individual Polypeptides of Foot-and-Mouth Disease Virus

The importance of polypeptide VP_1 in the production of neutralizing antibody was confirmed by observations first made by Laporte et al. (24) and confirmed later in a more comprehensive series of experiments (25). The individual polypeptides were obtained by disrupting the virus at 100°C in SDS-urea and then separating them by polyacrylamide gel electrophoresis. They were then inoculated into pigs or guinea-pigs. Only the trypsin sensitive polypeptide stimulated the production of neutralizing antibody and conferred protection.

It should be pointed out that the level of neutralizing

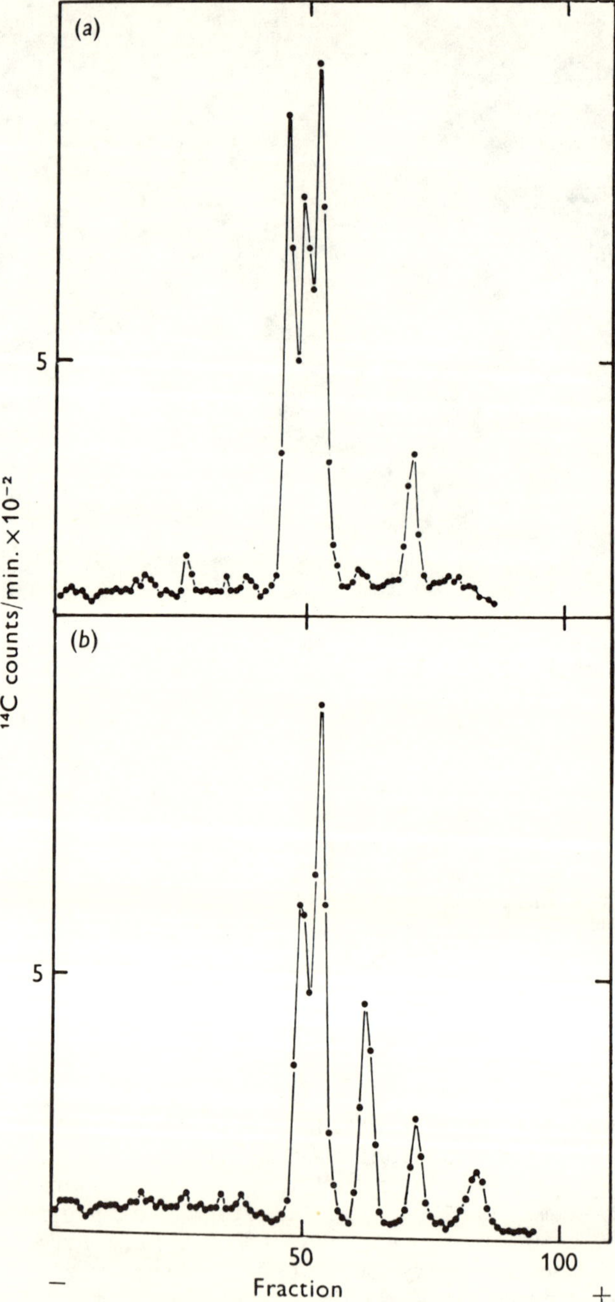

Figure 8. Effect of trypsin on the protein composition of FMDV. Polyacrylamide gel electrophoresis of (a) untreated virus proteins; (b) proteins from virus treated with trypsin.

antibody produced by inoculation of the separated polypeptide was very low compared with that produced by an equivalent weight of intact virus particle. This is hardly surprising because it has been known for many years that disruption of the virus particles into the 12S protein subunit by reducing the pH to about 6 also resulted in a considerable decrease in immunizing activity. Quite clearly the configuration of the polypeptide is important in determining the antibody response.

IV. VIRUS-INDUCED PROTEINS

The general pattern of *in vivo* translation in the picornaviruses is described by Lucas-Lenard in chapter 7 of this book. The same pattern seems to apply to all the picornaviruses that have been examined. However foot-and-mouth disease possesses some unique features which will be described in some detail.

In harvests produced by infecting BHK 21 cells with the virus, four virus structural units are found: (1) the infective particle sedimenting at 140S; (2) empty particles, of about the same diameter as the complete virus particles but possessing no RNA and sedimenting at 80S; (3) the 12S protein subunit consisting of a trimer of the three polypeptides VP_1, VP_2 and VP_3 and (4) the virus infection associated (VIA) antigen (26) which sediments at *c*. 3S, has a molecular weight of 56×10^3 and has been proposed as the inactive form of the virus specific RNA dependent RNA polymerase (27).

The first two products are also found in the entero- and rhinovirus genera but there are no particles corresponding to the 12S subunit or the VIA antigen in the other virus genera (although an artificial 14S particle can be produced from cardiovirus particles by treatment with 0.1M NaCl or NaBr at pH6). The VIA antigen is unusual in that it reacts in serological tests equally well with sera against all 7 serotypes of the virus. It assumed greater interest when it was found to co-run in polyacrylamide gel electrophoresis with a virus induced protein coded for by a section of the genome near the 3' end.

In their initial experiments, Sangar *et al*. (28) found four primary products in FMDV-infected BHK 21 cells but a re-assessment was made after tryptic peptide analysis of these products had shown that two of them, P56 and P100, gave similar maps. The second series of experiments (29) showed that, as with other picornaviruses, only three primary products 5' - P188, P52, P100 - 3' are present. The polypeptide P56 is apparently part of P100.

The interesting feature of P56 is that, besides co-electrophoresing with the VIA antigen, it also gives an identical tryptic peptide map (Figure 9). If it is correct that the VIA

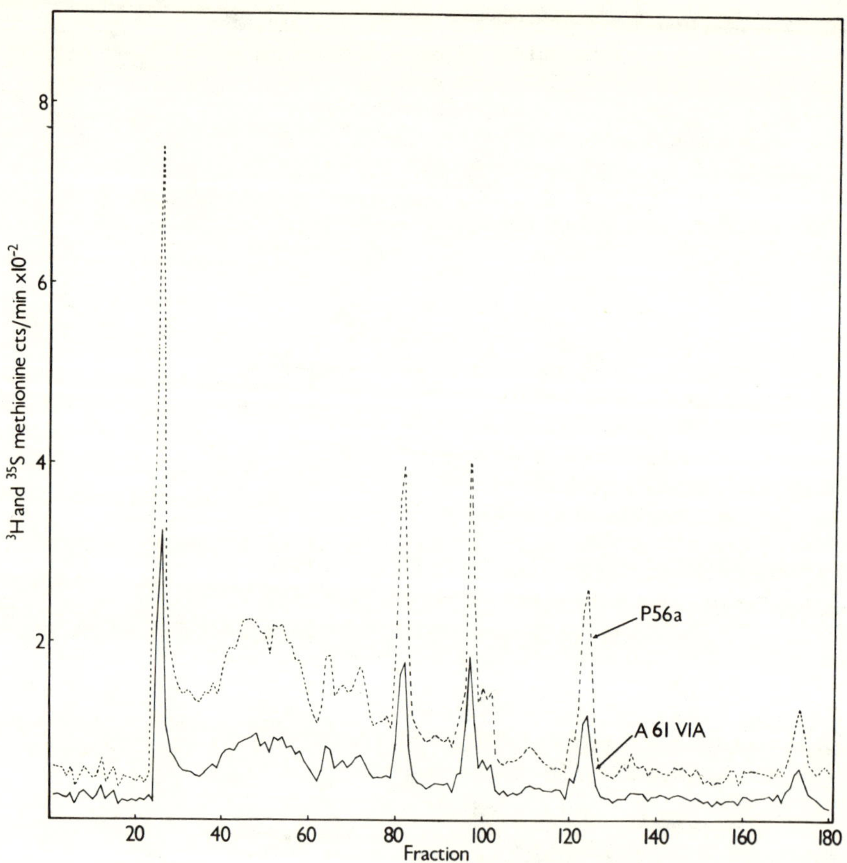

Figure 9. Tryptic peptide analysis of the VIA antigen and induced P56 polypeptide of FMDV, strain A61.

antigen is the RNA polymerase, this would mean that the part of the genome coding for the enzyme is located towards the 3' end of the RNA.

V. ANTIGENIC VARIATION

Antigenic variation is of considerable importance in foot-and-mouth disease. There are seven distinct serotypes of the virus. These are the European or classical types O, A and C, the Southern African Territories types SAT1, SAT2 and SAT3, and Asia 1. An animal recovering from infection with one type of the virus is not protected against infection with any of the remaining six serotypes. In addition there is variation within serotypes so that an animal immune to one sub-type may not be adequately protected against infection with another sub-type from the same serotype.

The variation measured by these tests is expressed on the virus capsid proteins which are coded for by the 30 to 40% of the virus RNA at the 5' end (28). It is not known, however, whether variation is limited to this region of the genome. We have been studying the biochemical basis for variation by analysing the RNA and primary products of several isolates of the virus.

A. The RNA Component

Dietzschold et al. (30) demonstrated by saturation hybridisation that the sequence homology between the RNAs of isolates of each of the European types of the virus was between 44 and 65%. The homology between sub-types of type O was only very slightly greater (31). Using a competition hybridisation test, however, Robson et al. (32) have found sequence homologies of 60 to 70% between types O, A, C and Asia 1. These viruses form a group distinct from the SAT serotypes because the homology between the two groups is 25 to 40%. Homology between the individual SAT serotypes is also 60 - 70% (Table 3).

There is more sequence homology between sub-types of an individual serotype of the virus. Robson et al. (32) found that two sub-types of serotype O isolated at an interval of 40 years had 80% sequence homology. A more extensive survey has now been made of sub-types of serotype A (Table 4). Homologies of 75% were obtained with isolates from different sub-types and values greater than 80% were found with isolates from the same sub-type. To relate the homology of the RNAs to the virus induced proteins, strain O^{V1} of type O and strains A61 and A Bage of type A (Table 5) were selected for tryptic peptide analysis of the virus induced proteins. The results of this analysis are described in the next section.

B. The Proteins

As indicated above, three primary products are found in BHK 21 cells infected with FMDV. The tryptic peptide maps of P88 and P52 of the three strains O^{V1}, A61 and A Bage show clearly that there are considerable differences in P88 but very few in P52 (Figures 10 - 13). Unfortunately in our preliminary experiments there were insufficient radioactive counts in the preparations of P100 for an accurate comparison to be made. However the tryptic peptide maps of the VIA antigens associated with these viruses were similar, indicating that more than half of the P100 polypeptides have a similar structure. These results indicate that the variation is

Table 3. Sequence homology of the RNAs of the seven serotypes of FMDV

Unlabelled ss RNA competitor	^{3}H ss RNA + unlabelled ds RNA						
	A_{10}	O_6	C_2	Asia 1/1	SAT 1/2	SAT 2/1	SAT 3/1
A_{10}	100	66	67	65	30	-	33
O_6	67	100	68	62	26	-	33
C_2	67	70	100	57	37	-	37
Asia 1/1	64	64	60	100	26	33	30
SAT 1/2	25	25	-	-	100	64	68
SAT 2/1	37	40	28	-	65	100	63
SAT 3/1	37	27	-	-	68	62	100

located in that region of the genome that codes for the structural polypeptides, suggesting that variation is probably occurring as a result of antibody pressure.

The extent of the variation in the individual polypeptides of the virus particle is clearly the next area for study. In preliminary experiments, K.J.H. Robson has shown that the VP4 polypeptides of O^{V1}, A61 and A Bage give similar products on hydrolysis with <u>Staphylococcus aureus</u> protease. She found the greatest amount of variability was on VP1. The results of an examination of the individual polypeptides of several different strains of the virus should enable us to locate more precisely the regions of the genome responsible for antigenic variation.

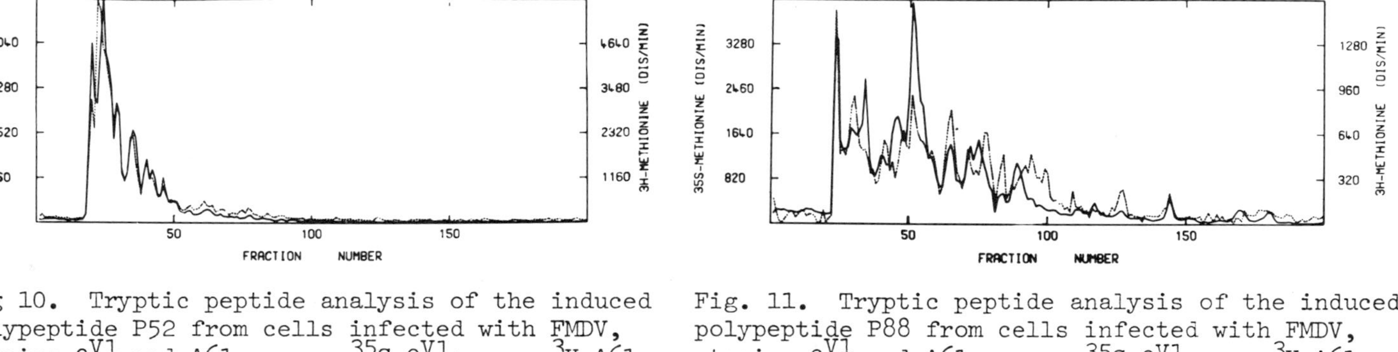

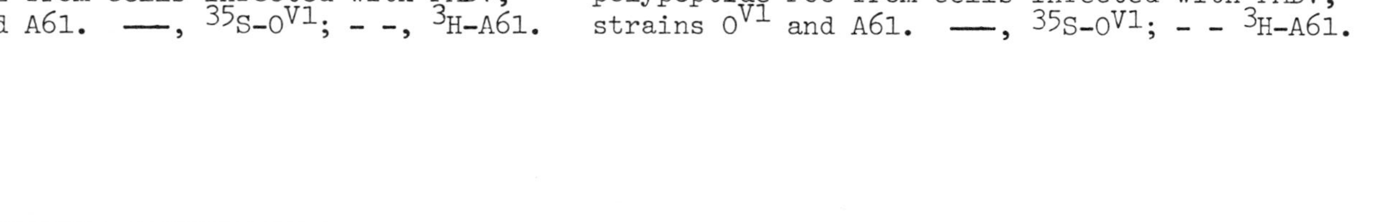

Fig 10. Tryptic peptide analysis of the induced polypeptide P52 from cells infected with FMDV, strains 0^{V1} and A61. ——, ^{35}S-0^{V1}; - -, ^{3}H-A61.

Fig. 11. Tryptic peptide analysis of the induced polypeptide P88 from cells infected with FMDV, strains 0^{V1} and A61. ——, ^{35}S-0^{V1}; - - ^{3}H-A61.

Fig. 12. Tryptic peptide analysis of the induced polypeptide P52 from cells infected with FMDV, strains A Bagé and A61. ——, ^{35}S-A Bagé; - -, ^{3}H-A61.

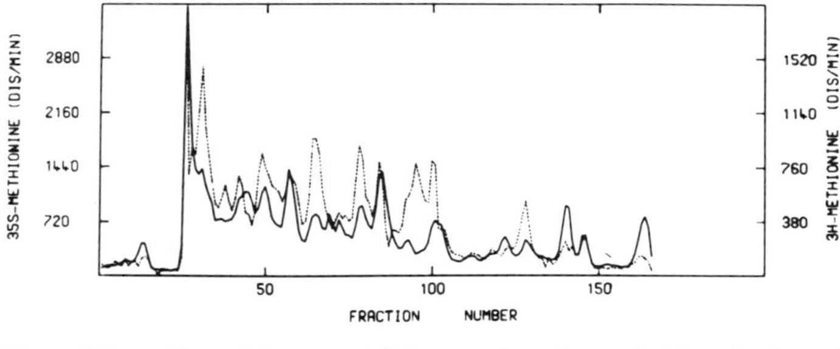

Fig. 13. Tryptic peptide analysis of the induced polypeptide P88 from cells infected with FMDV, strains A Bagé and A61. ——, ^{35}S-A Bagé; - -, ^{3}H-A61.

Table 4. Sequence homology of the RNAs of 5 isolates of FMDV, serotype A

Unlabelled ss RNA competitor	^{3}H ss RNA + unlabelled ds RNA				
	A_5	A_{22} Iraq 24/64	A_{22} Turkey 1/70	A_{22} Greece 1/72	A_{24}
A_5	100	74	76	77	87
A_{22} Iraq 24/64	74	100	87	87	76
A_{22} Turkey 1/70	74	86	100	87	76
A_{22} Greece 1/72	74	88	87	100	76
A_{24}	87	75	76	77	100

Table 5. Sequence homology of the RNAs of two A sub types and an O serotype of FMDV

Unlabelled ss competitor	^{3}H ss RNS and unlabelled ds RNA		
	A_{10}(61)	A_{24}(Bage)	O_6(V1)
A_{10}(61)	100	-	66
A_{24}(Bage)	84	-	63
O_6(V1)	67	-	100

ACKNOWLEDGEMENTS

I am greatly indebted to D.N. Black, B. Cartwright, T.R. Doel, B.M. Gorman, J.F.E. Newman, K.J.H. Robson, D.J. Rowlands and D.V. Sangar for their willingness to allow me to present some of their results before publication and for numerous discussions. I am also grateful to C.J. Smale, J. Ryder and P.A. Wallbridge for the electron micrographs and photographs.

REFERENCES

1. COOPER, P.D., AGOL, V.I., BACHRACH, H.L., BROWN, F., GHENDON, Y., GIBBS, A.J., GILLESPIE, J.H., LONBERG-HOLM, K., MANDEL, B., MELNICK, J.L., MOHANTY, S.B., RUECKERT, R.R., SCHAFFER, F.L. and TYRELL, D.A.J. Picornaviridae: second report. Intervirology (1978), 10, 165-180.

2. BROWN, F. and SMALE, C.J. Demonstration of three specific sites on the surface of foot-and-mouth disease virus by antibody complexing. J. Gen. Virol. (1970), 7, 115-127.

3. ROWLANDS, D.J., SANGAR, D.V. and BROWN F. Buoyant density of picornaviruses in caesium salts. J. Gen. Virol. (1971), 13, 141-152.

4. BROWN, F. and STEWART, D.L. The influence of proflavine on the synthesis of foot-and-mouth disease virus. J. Gen. Microbiol. (1960), 23, 369-379.

5. BROWN, F., NEWMAN, J.F.E. and STOTT, E.J. Molecular weight of rhinovirus RNA. J. Gen. Virol. (1970), 8, 145-148.

6. BREESE, S.S. Jr., A comparison of molecular weights of foot-and-mouth disease virus RNA fragments determined from lengths and s-rates. J. Gen. Virol. (1976), 31, 1-8.

7. BROWN, F. and WILD, T.F. The effect of heat on the structure of foot-and-mouth disease virus and the viral ribonucleic acid. Biochem. Biophys. Acta.(1966), 119, 301-308.

8. BROWN, F., CARTWRIGHT, B. and STEWART, D.L. The effect of various inactivating agents on the viral and ribonucleic acid infectivities of foot-and-mouth disease virus and on its attachment to susceptible cells. J. Gen. Microbiol. (1963), 31, 179-186.

9. ARLINGHAUS, R.B., POLATNICK, J. and VANDE WOUDE, G.F. Studies on foot-and-mouth disease virus ribonucleic acid synthesis.

Virology, (1966), 30, 541-550.

10. MATHEKA, H.D., TRAUTMAN, R. and BACHRACH, H.L. Carrier-free zone electrophoresis of infectious ribonucleic acid derived by phenol and sodium dodecyl-sulfate methods from purified foot-and-mouth disease virus. Arch. Biochem. Biophys. (1967), 121, 325-330.

11. DENOYA, C.D., SCODELLER, E.A., GIMENEZ, B.H., VASQUEZ, C. and LA TORRE, J.L. Foot-and-mouth disease virus. I. Stability of its ribonucleic acid. Virology, (1978), 84, 230-235.

12. GAUNTT, C.J. Fragmentation of RNA in virus particles of rhinovirus type 14. J. Virol. (1974), 13, 762-764.

13. PORTER, A., CAREY, N.H. and FELLNER, P. Presence of a large poly (rC) tract within the RNA of encephalomyocarditis virus. Nature (London), (1974), 248, 675-678.

14. BROWN, F., NEWMAN, J., STOTT, J., PORTER, A., FRISBY, D., NEWTON, C., CAREY, N. and FELLNER, P. Poly(C) in animal viral RNAs. Nature (London), (1974), 251, 342-344.

15. HARRIS, T.J.R. and BROWN, F. Biochemical analysis of a virulent and an avirulent strain of foot-and-mouth disease virus. J. Gen. Virol. (1977), 34, 87-105.

16. FRISBY, D.P., NEWTON, C., CAREY, N.H., FELLNER, P., NEWMAN, J.F.E., HARRIS, T.J.R. and BROWN, F. Oligonucleotide mapping of picornavirus RNAs by two-dimensional electrophoresis. Virology, (1976), 71, 379-388.

17. ROWLANDS, D.J., HARRIS, T.J.R. and BROWN, F. More precise location of the polycytidylic acid tract in foot-and-mouth disease virus RNA. J. Virol. (1978), 26, 335-343.

18. STOLTZFUS, C.M. and RUECKERT, R.R. Capsid polypeptides of Maus-Elberfeld virus. J. Virol. (1972), 10, 347-355.

19. WILD, T.F. and BROWN, F. Nature of the inactivating action of trypsin on foot-and-mouth disease virus. J. Gen. Virol. (1967), 1, 247-250.

20. WILD, T.F., BURROUGHS, J.N. and BROWN F. Surface structure of foot-and-mouth disease virus. J. Gen. Virol. (1969), 4, 313-320.

21. CAVANAGH, D., SANGAR, D.V., ROWLANDS, D.J. and BROWN, F. Immunogenic and cell attachment sites of FMDV: further evidence for their location in a single capsid polypeptide. J. Gen. Virol. (1977), 35, 149-158.

22. LAPORTE, J. and LENOIR, G. Structural proteins of foot-and-mouth disease virus. J. Gen. Virol. (1973), 20, 161-168.

23. TALBOT, P., ROWLANDS, D.J. BURROUGHS, J.N., SANGAR, D.V. and BROWN, F. Evidence for a group protein in foot-and-mouth disease virus particles. J. Gen. Virol. (1973), 19, 369-380.

24. LAPORTE, J., GROSCLAUDE, J., WANTYGHEM, J., BERNARD, S. and ROUZE, P. Neutralisation en culture cellulaire du pouvoir infectieux du virus de la fièvre aphteuse par des serums provenant de porcs immunisés à l' aide d' une proteine viral purifiée.C.R. Acad. Sci. Paris (1973), Serie D, 276, 3399.

25. BACHRACH, H.L., MOORE, D.M., McKERCHER, P.D. and POLATNICK, J. Immune and antibody responses to an isolated capsid protein of foot-and-mouth disease virus. J. Immunol. (1975), 115, 1636-1641.

26. COWAN, K.M. and GRAVES, J.H. A third antigenic component associated with foot-and-mouth disease infection. Virology (1966), 30, 528-540.

27. POLATNICK, J., ARLINGHAUS, R.B., GRAVES, J.H. and COWAN, K.M. Inhibition of cell-free foot-and-mouth disease virus ribonucleic acid synthesis by antibody. Virology (1967), 31, 609-615.

28. SANGAR, D.V., BLACK, D.N., ROWLANDS, D.J. and BROWN, F. Biochemical mapping of the foot-and-mouth disease virus genome. J. Gen. Virol. (1977), 35, 281-297.

29. DOEL, T.R., SANGAR, D.V., ROWLANDS, D.J. and BROWN, F. A re-appraisal of the biochemical map of foot-and-mouth disease virus RNA. J. Gen. Virol. (1978), (in press).

30. DIETZSCHOLD, B., KAADEN, O.R., TOKUI, T. and BOHM, B.O. Polynucleotide sequence homologies among the RNAs of foot-and-mouth disease virus types A, C and O. J. Gen. Virol. (1971), 13, 1-7.

31. DIETZSCHOLD, B., KAADEN, O.R. and AHL, R. Hybridisation studies with subtypes and mutants of foot-and-mouth disease virus type O. J. Gen. Virol. (1972), 15, 171-174.

32. ROBSON, K.J.H., HARRIS, T.J.R. and BROWN, F. An assessment by competition hybridization of the sequence homology between the RNAs of the seven serotypes of FMDV. J. Gen. Virol. (1977), 37, 271-276.

33. DENOYA, C.D., SCODELLER, E.A., VASQUEZ, C. and LA TORRE, J.L. Ribonuclease activities associated with purified

foot-and-mouth disease virus. Archives of Virology, (1978), 57, 153-159.

SECTION II:

EARLY EVENTS DURING THE INFECTIOUS PROCESS: SHUT-OFF OF CELLULAR SYNTHESIS

INHIBITION OF CELLULAR PROTEIN SYNTHESIS AFTER VIRUS INFECTION

JEAN M. LUCAS-LENARD

Biological Sciences Group, Biochemistry and Biophysics Section, University of Connecticut
Storrs, Connecticut 06268, U.S.A.

INTRODUCTION

This review is concerned with the effects of picornavirus infection on the biosynthesis of macromolecules in animal cells. Soon after infection of animal cells with a member of the picornavirus group, there is an inhibition of cellular RNA and protein synthesis. Later in infection, the synthesis of cellular DNA is also decreased. During this time, viral constituents are synthesized and assembly of viral progeny occurs. Late in infection, there is an irreversible loss of the capacity of the cell to synthesize any macromolecules, host or viral, and finally, the cell dies.

The same sequence of events occurs after infection of animal cells with other cytotoxic viruses. The inhibition of the synthesis of macromolecules, often referred to as "shut-off", appears to result from the specific effect of one or more viral gene products on the host cell. Although the shut-off phenomenon has been under investigation for nearly two decades, little is known about the mechanism by which the inhibition takes place.

Although the effect of virus infection on cellular RNA and DNA synthesis will be considered, this review will be concerned primarily with the inhibition of cellular protein synthesis. Several reviews on virus-induced changes in the synthesis of macromolecules have been written in the past few years (1,2,3,4), and thus the reference list is selective rather than exhaustive. This review will consist of three parts: the first will be concerned with the properties of the shut-off phenomenon. The second part will deal with theories that have been proposed to explain shut-off.

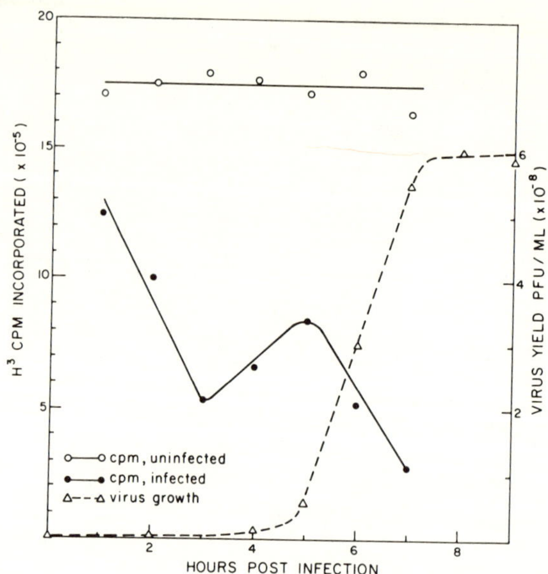

Figure 1. Rate of inhibition of cellular protein synthesis. Mouse L-929 cells (1 x 10^8) were infected with mengovirus (moi = 20). At hourly intervals yields of infectious virus were determined by plaque assay. Also, 5 x 10^6 cells were removed and pulsed with 1.0 μCi ^{3}H-amino acids for 35 min. The hot TCA precipitable counts incorporated were multiplied by a factor of 20 to convert to the incorporation of the total culture. From Manak (5).

And finally, in the third part the current status of the problem will be discussed. Where possible mengovirus will be used as the example of a typical picornavirus, since this virus is being studied in my laboratory.

I. THE INHIBITION OF SYNTHESIS OF CELLULAR MACROMOLECULES

A. Cellular Protein Synthesis Inhibition

General description. Within the first hour after infection of L-cells with mengovirus, there is an inhibition in the rate of amino acid incorporation into protein. This inhibition reaches a low point at around three hours after infection, increases slightly until about five hours after infection, and then falls (Figure 1). The rate of viral protein synthesis never reaches that of the uninfected, control culture in this system.

Polyacrylamide gel electrophoresis of the proteins labeled at various times after infection has revealed that early after infection there is a decreased synthesis of most cellular proteins. By four hours after infection there are some viral proteins making their appearance in the gels and by five hours, the predominant proteins synthesized are viral (Figure 2). The gel pattern clearly demonstrates a gradual inhibition of cellular protein synthesis and a near complete changeover from cellular to viral protein synthesis by 5 hours after infection.

The inhibition of protein synthesis is accompanied by a

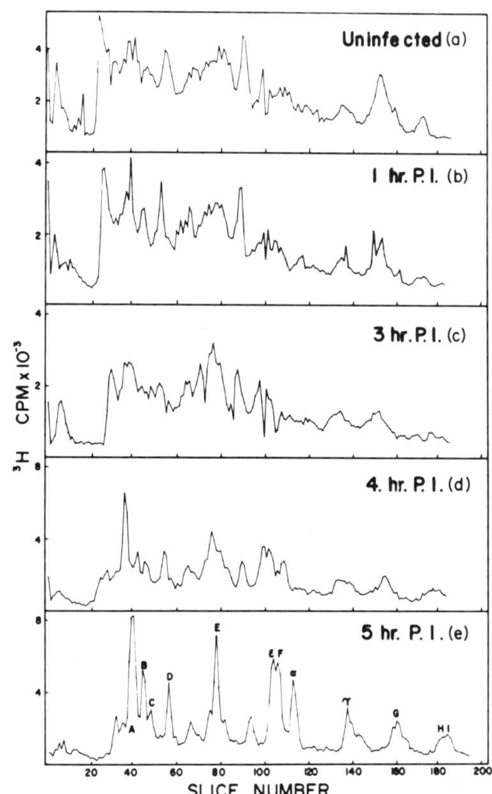

Figure 2. Polyacrylamide gel electrophoresis pattern of changeover from L-cell protein synthesis to mengovirus protein synthesis. L-cells mock infected or infected with mengovirus for the time periods indicated in the figure were labeled with a (^{3}H)-amino acid mixture for 25 min. After the labeling period the samples were analyzed as described by Nakai and Lucas-Lenard (6).

disaggregation of polysomes (7, 8, 9) (See Figure 3, panels a and b). Later in infection the assembly of a class of polysomes, generally longer than those of uninfected cells, is observed (7, 10) (See Figure 3, panels b, c and d). The appearance of these polysomes is attributed to the synthesis of viral RNA capable of functioning as message. The presence of new polysomes suggests that most of the pre-existing ribosomes are not inactivated as a consequence of virus infection, since they participate in viral protein synthesis.

Requirement for viral RNA translation for shut-off. Translation of the viral genome is required to bring about the inhibition of cellular protein synthesis. If the poliovirus genome is inactivated by ultraviolet irradiation prior to the infection of cells, no inhibition of cellular protein synthesis can be observed (11, 12). The UV treatment results in, among other alterations, the formation of uridine dimers in the RNA, thus blocking its translation, but presumably has little effect on the input viral capsid proteins. If cells are infected in the presence of cycloheximide (11) or puromycin, upon removal of the drugs, protein synthesis resumes at the control levels of uninfected cells for a

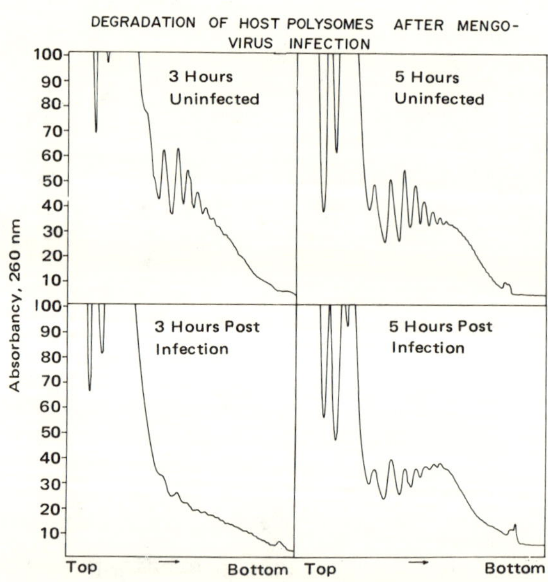

Figure 3. Sucrose gradient analysis of cytoplasmic extracts of infected and uninfected cells. Cells were harvested 3 h or 5 h after mock adsorption or virus adsorption. The details of this experiment are reported in Colby *et al* (9).

short time before the onset of the cellular protein synthesis inhibition. Thus, if protein synthesis is prevented, no viral inhibitor of cellular protein synthesis is formed.

Experiments by Helentjaris and Ehrenfeld (12) indicate that the inactivation of the shut-off ability by UV irradiated virus follows one-hit kinetics. A multiplicity of one plaque forming unit per cell is sufficient to shut-off both cellular protein and RNA synthesis. Also, a hit anywhere in the genome is sufficient to inactivate the genomes' capacity to cause shut-off. This is not an unexpected finding, considering that the synthesis of poliovirus proteins occurs by the cleavage of a single precursor polypeptide (13). It is possible that damage anywhere in the RNA would prevent the proper cleavage of any translation product as well as prevent the translation of sequences distal to the UV hit.

Involvement of a viral structural protein in shut-off. A structural protein may be involved in the shut-off of cellular protein synthesis. This conclusion was derived from experiments by Steiner-Pryor and Cooper (14), who observed that certain temperature sensitive mutants of poliovirus defective in the structural proteins of the virus particle were unable to inhibit cellular protein synthesis at the non-permissive temperature. Mutants defective in non-structural proteins, including some involved in RNA replication, displayed the normal shut-off phenotype, suggesting that RNA replication is not necessary for shut-off. Similar conclusions were derived from studies with inhibitors of viral RNA synthesis (15).

While it would seem that a strong case exists for the involvement of structural proteins in the shut-off phenomenon, there is one piece of evidence that does not completely agree with this finding. Cole and Baltimore (16) have isolated, purified and characterized a particular defective, interfering (DI) particle of poliovirus which, in essence, is a deletion mutant lacking the segment of RNA that codes for about one third of the capsid protein precursor. The component of the structural protein that is made is unstable and rapidly degraded. This DI particle, however, is as effective as wild type virus in the inhibition of cellular protein and RNA synthesis. It is possible that the unstable two-thirds of the capsid protein persists long enough to inhibit the synthesis of cellular macromolecules. Further studies are necessary to help resolve this problem.

B. Cellular RNA Synthesis Inhibition

The rate of cellular RNA synthesis also declines soon after infection of cells with a picornavirus. In mengovirus infected L-cells, RNA synthesis is inhibited rapidly to less than 10% of control (17). Between 2 and 3 hours, the rate of RNA synthesis increases and by about 6 hours nearly reaches the level of RNA synthesis in uninfected cells (17). This newly synthesized RNA is viral RNA.

Studies on the species of RNA that are inhibited after poliovirus infection have led to the conclusion that the synthesis of 45S nucleolar ribosomal precursor RNA is primarily depressed (18). The synthesis of non-ribosomal RNA was not affected at early times after infection (18). The appearance of mRNA in polysomes is also not depressed in infected cells for at least 1-3 hours after infection (9, 19, 20). The conversion of the 45S ribosomal RNA precursor to the 32S and 18S RNA is inhibited or slowed (18). The methylation of the 45S RNA was not inhibited, suggesting that the virus is not exerting its action at this level (18).

Although most experiments clearly indicate that early after infection ribosomal RNA is the first species of RNA to be inhibited, experiments using isolated nuclei (21, 22) from mengovirus infected L-cells and EMC virus infected mouse plasmacytoma cells indicate that the polymerase II activity (responsible for heterogeneous nuclear RNA and mRNA synthesis) is inhibited 1-2 hours before RNA polymerase I and III activities (responsible for rRNA and 4S and 5S RNA synthesis, respectively). No difference in activity and relative proportions of the three RNA polymerases, however was found after infection using solubilized enzymes assayed in the presence of exogeneous DNA as template (21). It is assumed that the inhibition in whole cells results from an initiation defect, since these measurements using nuclei and solubilized enzymes do not measure true initiation. The explanation as to why polymerase II should be inhibited before polymerase I when rRNA synthesis is clearly the first to be inhibited in whole cells remains to be worked out.

The inhibition of cellular RNA synthesis, like protein synthesis, is dependent on the expression of the viral genome, since the inhibition is prevented by irradiation of the infecting virus (17).

C. Inhibition of DNA Synthesis

The inhibition of cellular DNA synthesis by picornaviruses is expressed later in infection than the depression of cellular protein or RNA synthesis. In mengovirus infected L-cells (17, 23) the inhibition starts within 2 hours and is virtually completed by 6 hours. Between 2 and 5 hours after infection, the primary DNA synthetic event blocked by mengovirus is the multifocal initiation of new DNA chain synthesis (23). However, by 6 hours, replication fork movement is also retarded and there is a more generalized derangement in DNA synthesis (24).

II. POSSIBLE MECHANISMS TO EXPLAIN CELLULAR PROTEIN SYNTHESIS INHIBITION

A. Stability of Host Cell mRNA

The decrease in cellular protein synthesis could result from degradation or inactivation of cellular mRNA. Early studies (19, 20), however, showed that cellular mRNA present in polysomes from poliovirus infected HeLa cells had the same sedimentation characteristics as that from uninfected cells. Because no appreciable degradation of cellular mRNA was observed, it was concluded that cellular protein synthesis inhibition does not result from an appreciable increase in endonuclease activity which would degrade cellular mRNA.

The finding that most eukaryotic mRNAs have a poly(A) sequence at their 3'-terminus (See ref. 25 for a review) simplified the isolation procedure of mRNA considerably, since it could be bound to poly (U) or oligo dT embedded into filters or columns and recovered free of non-poly (A)-containing RNA such as rRNA. Using this technique the fate of pre-labeled cellular poly(A)-containing polysomal mRNA could easily be followed in uninfected and mengovirus infected L-cells (9). The pre-labeled cellular mRNA was found to associate with polysomes for up to at least 3 hours after infection, even though by this time there was considerable inhibition of cellular protein synthesis (Table 1).

By five hours after infection, a substantial percentage of the cellular mRNA in polysomes was replaced by viral RNA. Even at this time, however, the amount of mRNA in uninfected and infected cells remained the same (Table 2), suggesting that no major degradation of cellular mRNA had taken place. Examination by polyacrylamide gel electrophoresis of the pre-labeled cellular mRNA in polysomes from infected cells revealed no major degradation (Figure 4), confirming earlier observations.

Table 1. Distribution of poly(A)-containing RNA in sucrose gradients[a]

Expt	Time after adsorption (h)	Poly(A)counts/min associated with polysomes (%)	
		Uninfected cells	infected cells
1	3	60	58
2	3	52	56
1	5	50	38
2	5	55	38

a. The details of this experiment are described in Colby et al (9).

Table 2. Relative stability of poly(A)-containing RNA of infected (I) and uninfected (U) cells[a]

Time after adsorption	I/U (counts/min bound)
15 min	1.19
1.5 h	1.17
3 h	1.16
5 h	1.36

a. The host poly(A)-containing RNA was labeled with ^{3}(H)-uridine. At the time periods indicated, portions were removed and the labeled RNA was isolated and measured on poly(U) glass fiber filters. See Colby et al. (9).

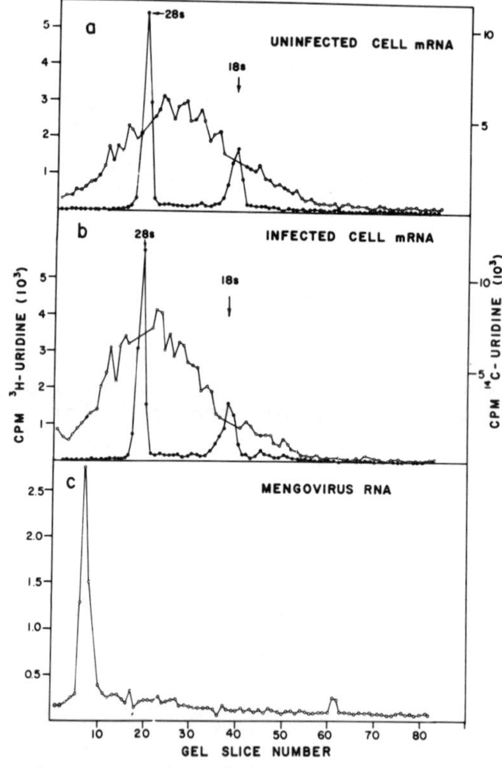

Figure 4. Polyacrylamide gel electrophoresis of poly(A)-containing RNA isolated at 3 h after infection or mock infection. O - O, Poly(A)-RNA; ● - ●, 28S and 18S ribosomal RNA markers. From Colby _et al_. (9).

Further evidence that cellular mRNA is not degraded after infection came from experiments using cellular mRNA extracted from infected cells to prime protein synthesis in extracts. From mouse plasmacytoma tumor cells Lawrence and Thach (26) isolated a poly(A)-containing 10S RNA fraction which in extracts was translated into a protein found in plasmacytoma cells, as determined by co-electrophoresis and tryptic mapping. This messenger RNA, when isolated from infected cells, was less active in translation than when isolated from uninfected cells by about 50%. However, mRNA isolated from pactamycin inhibited cells, was equally inhibited and thus the inhibition was not considered to result specifically from virus infection, but rather, from inhibition of initiation which may lead to mRNA degradation.

Recently, this question was re-examined, but with cellular mRNAs lacking the 3'-poly(A) sequence which includes the mRNAs for histones. In these experiments Gallwitz _et al_. (27) quantified the amount of translatable cytoplasmic histone mRNA at different times after infection of Ehrlich ascites tumor cells with mengovirus using a reticulocyte cell-free protein synthesizing system. Their results indicated that histone mRNA content disappears slowly after infection, reaching 50% of the control value at 4 hours after infection and 20% of the control value at the end of infection when cells started to lyse. In contrast histone synthesis in infected cells was inhibited much more rapidly: the degree of inhibition was 60% by 2 hours and about 90% at 3-4 hours after infection. These results showed that the relative histone synthesizing potential of virus-infected cells is substantially higher throughout infection than actually expressed in whole cells. Thus, it appears that the virus-induced shut-off of histone synthesis is not directly a consequence of inactivation or degradation of histone mRNA. These experiments, like those carried out with poly(A)-containing mRNA (9), showed that most of the histone message remains in polysomes for up to 8 hours post infection.

It has been postulated that the 3'-poly(A) sequence on eukaryotic mRNAs might be involved in mRNA stability since it has been found that these sequences on mRNAs in HeLa cells (28) are successively degraded with increasing age of mRNA molecules. On this basis Koschel (29) examined whether infection of HeLA cells with poliovirus leads to a more rapid degradation of poly(A) sequences which in turn might explain the shut-off of cellular protein synthesis. No detectable differences were observed in the poly(A) sequences in cellular cytoplasmic RNA from uninfected cells and cells infected with poliovirus for 4 hours. Both amount and size of these sequences were identical.

The 5' end of most eukaryotic mRNA molecules consists of a blocked oligonucleotide structure $m^7G(5')ppp(5')NmpNp...$ or $m^7G(5')ppp(5')NmpNmpNp...$ termined a "cap" (See ref.30 for a review). For efficient initiation of protein synthesis, many mRNA molecules require this structure. Poliovirus mRNA and probably the RNA of other picornaviruses lack this "cap" structure (30) and translation occurs in its absence. This difference in mRNA structure was thought to provide a molecular basis for discrimination of the two mRNA classes, so that poliovirus RNA could still be translated while host protein synthesis declined. Fernandez-Muñoz and Darnell (31) examined the 5' terminal structure of pre-labeled poly(A) containing HeLa cell mRNA in mock infected cells, cells infected for 2 hours with poliovirus and in puromycin treated cells. In all three cases, most, if not all, of the mRNA appeared capped both before infection or mock infection and after virus-induced inhibition of protein synthesis was complete. Possible minor modifications of bases in the host mRNA was also examined as a basis for translation inhibition. The results indicated little or no change in methylation of caps due to infection and no charge difference or differences in electrophoretic migration. The poliovirus mRNA was also examined with the idea that perhaps it might be modified after infection. It also remained unmodified--no caps were found on poliovirus RNA during the first 3 1/2 hours of infection (31).

Apparently related to the problem of host cell protein synthesis inhibition by viruses is the phenomenon of viral interference. In some cases, superinfection of a cell already infected with one virus does not affect the yield of the original virus. In other cases, however, the superinfecting virus completely inhibits translation of the original viral mRNA. Such is the case for superinfection of vesicular stomatitis virus (VSV)-infected cells by poliovirus (32). The kinetics and general properties of the shut-off of VSV protein synthesis appear to be the same as the shut-off of cellular protein synthesis after poliovirus infection (32).

The VSV genome codes for only five proteins (33), and soon after infection the five VSV proteins represent the only translation products of the infected cell, since this virus also inhibits cellular protein synthesis (34). The five VSV monocistronic mRNAs serving as templates for these proteins have been relatively well characterized (35) and thus this system lends itself to studies on the mechanism of superinfection. Ehrenfeld and Lund (36) examined the consequences of superinfecting VSV infected HeLA cells at two hours after infection with poliovirus. By 1.5 to 2 hours after poliovirus superinfection, virtually all VSV polysomes disaggregated. The pattern of protein synthesis as shown on polyacrylamide gels changed from VSV specific to poliovirus specific within 2.5 hours after infection. All five of the VSV proteins were inhibited to the same extent.

VSV mRNA synthesis continued at normal rates, and the mRNAs appeared to be undegraded as determined by their sedimentation rates in sucrose gradients. Their 3' poly(A) tract was intact, and their 5'-cap structures were normal. Also, at least 4 of these mRNAs were biologically active, since they were capable of stimulating the synthesis of VSV specific polypeptides in a wheat germ cell-free extract.

In summary, it appears that cellular mRNAs are not damaged after infection by picornaviruses. The mRNAs are intact, both 3' and 5' ends are normal, and the cellular mRNA is not detectably modified. The same observations have been made with regard to viral mRNAs in cells superinfected with a different virus, thus suggesting the possibility that the turn-off in each case occurs by the same mechanism.

B. Increased Permeability of Cell Membranes to Salts

Brief exposure of cultured animal cells to hypertonic medium results in a reversible inhibition of protein synthesis (37, 38). The inhibition is independent of the solute (NaCl, KCl, NH_4Cl, sucrose) used to increase the osmolarity of the medium and is accompanied by a complete disaggregation of polyribosomes (37). Studies by Saborio and co-workers (38) have suggested that the increased osmolarity selectively blocks the initiation of polypeptide chains; neither chain elongation nor termination are affected. Interestingly, poliovirus RNA translation is resistant to the effects of hypertonic medium at levels which completely block cellular protein synthesis (39). (Viral protein synthesis is also blocked at very high osmolarity). This finding has provided a unique tool with which to study early events in the infectious cycle by permitting the unmasking of virus-specific polypeptide synthesis (39). It also suggested a few models to explain the shut-off phenomenon.

In their model Nuss and Koch (40) propose that after infection there is a non-specific reduction in the overall rate of polypeptide chain initiation. However, since each mRNA has its own intrinsic translational efficiency, not all mRNAs are affected to the same extent, and there is a differential reduction in the translation of individual host and viral mRNA species. In a somewhat different model, Carrasco and Smith (41, 42) propose that upon contact of the virus with the cell membrane, a viral coat protein associates with the membrane, changing the normal monovalent-ion gradient. During the course of viral RNA translation into progeny coat protein, these proteins are also inserted into the membrane, and the membrane alterations continue. Eventually, sodium leaks into the cell, there is an increase in the concentration of monovalent-ions inside the cells, and cellular but not viral protein chain initiation is inhibited.

Carrasco and Smith have tested their model by examining the effect of sodium on the cell-free translation of host and viral messages (42). Concentrations of sodium ions that inhibited globin

RNA and mouse cell poly(A)-containing RNA translation stimulated EMC viral RNA translation. They also measured the cellular Na^+/K^+ ATPase activity (the enzyme responsible for the maintenance of the monovalent ion gradient) at different times after infection by estimating the uptake into cells of the potassium ion analogue, $^{86}Rb^+$ (42). Up to 4 hours after infection no difference in the amount of $^{86}Rb^+$ uptake was observed. Yet, cellular protein synthesis inhibition takes place soon after infection and is nearly complete by 3 hours after infection. The uptake of $^{86}Rb^+$ decreased precipitously after 4 hr, coinciding with the peak of viral protein synthesis. This suggests that either the Na^+/K^+ ATPase activity is severly inhibited or that the plasma membrane becomes leaky to monovalent ions. The result of either is the breakdown of the monovalent ion gradient and an increase in the intracellular sodium ion concentration. However, this phenomenon occurs late in infection, several hours after the onset of cellular protein synthesis inhibition. Thus, it cannot explain the early inhibition of cellular protein synthesis.

Similarly, Egberts and co-workers (43) measured the intracellular concentration of K^+ and Na^+ after infection using flame spectrophotometry and found that the K^+ content slightly increased up to 3 hours after infection and then declined progressively, reaching a level of 50% of control by 5 hours after infection. The Na^+ content fell slightly after infection up to 3 hours, and then increased between 3 and 5 hours. The K^+/Na^+ ratio remained constant early in infection, at a time when cellular protein synthesis is declining most rapidly. Late in infection, there was a leakage into the medium of low-molecular weight substances, and it is possible that translation is affected at that time. Viral RNA translation, however, is also inhibited late in infection.

In summary, the available evidence in intact cells does not strongly support the theory that an increase in Na^+ ions into the cell is responsible for host cell protein synthesis inhibition at early times after infection. Late in infection there is an alteration in the membrane permeability and perhaps the effects on protein synthesis seen at that time result from this phenomenon.

C. Competition Between Cellular and Viral mRNAs

If viral and cellular RNAs are simultaneously added to a cell-free translation system, there is a supression of translation of the cellular RNA and a preferential translation of the viral message (26, 44). This competition among messages has been of interest for several years. There seems to be a hierarchy among messenger RNAs, some being able to utilize the components needed for protein synthesis more effectively than others (45, 46). Using a fractionated protein synthesis system from plasmacytoma cells, Golini _et al._ (46) showed that the initiating step of translation is the site of competition. Their data indicate that the competition is established at, or prior to, the formation of the first peptide bond. The viral and cellular mRNAs appear to be recognized equally well by 40S ribosomal subunits, suggesting that the components responsible for competition may be the initiation factors themselves. This suggestion was supported by evidence which showed that the addition of excess crude initiation factors from reticulocyte or plasmacytoma cells could relieve the competition, allowing the translation of cellular mRNA (10S RNA or globin mRNA) (47). To determine which of the initiation factors was responsible for the resumption of globin RNA synthesis, the purified initiation factors from reticulocytes were added one by one (47). These studies suggested that when initiation factor eIF-4B(IF-M3)was present in limiting quantities, viral RNA outcompetes cellular mRNA for translation. When the factor is present in excess, both messages are translated.

What is the relevance of these studies to the shut-off problem? For one, we know that shut-off occurs early after infection, before a substantial accumulation of viral RNA has occurred. Furthermore, shut-off occurs in the absence of RNA replication (14, 15). Finally, the competition effect is variable in cell-free extracts and depends to a great extent on the way in which the extract was prepared and on the molar ratios of one message to another (44). The concentration of initiation factors in whole cells is not known, and thus it is difficult to know if viral and cellular messages would behave in the same way in whole cells. Thus, competition does not seem to be the complete explanation of the shut-off phenomenon.

D. Synthesis of New, Viral Specific, Initiation Factors

Another theory to explain the shut-off, proposed by Cooper _et al._ (48) involves the synthesis of viral proteins which have an affinity for the small ribosomal subunit and the 5' end of viral mRNA. The viral protein would repress the synthesis of cellular proteins by combining with the 40S subunit, thereby blocking its link with host mRNA. By also binding to the 5' end of viral RNA, the proteins would facilitate the attachment of viral RNA to the 40S subunit and increase the translation of viral RNA. Viral proteins have been found to co-sediment with the 40S subunits of HeLa cells infected with poliovirus (49), of Ehrlich ascites tumor cells infected with EMC virus (50) and of L-cells infected with mengovirus (51). In poliovirus infected cells, the viral proteins co-sedimenting with ribosomes were identified as VPO, VP1 and VP3, all structural proteins (49). Both structural and non-structural proteins were found associated with ribosomes from EMC virus and mengovirus infected cells (50, 51). However, no one has yet been able to assign a role for these proteins in viral RNA translation or host cell protein synthesis inhibition.

In cell-free assays the native 40S subunits carrying the viral proteins appear to be as active, if not even slightly more active, than those from uninfected cells in terms of met-$tRNA_f$ binding (51). It is possible that the viral proteins exert their effect at some stage other than met-$tRNA_f$ binding, but this is not yet known.

E. Inactivation of Cellular Initiation Factors

Another way in which picornaviruses could inhibit cellular protein synthesis would be to inactivate an initiation factor needed for cellular, but not viral, mRNA translation. If this were the case one might expect to find a decreased capacity of extracts from infected cells to initiate translation of cellular mRNAs compared to extracts from uninfected cells. Such studies have yielded a variety of results. In some laboratories no differences in activity were detected between extracts from uninfected cells and from EMC infected plasmacytoma cells or mengovirus infected Ehrlich ascites tumor cells (26, 44). In one laboratory the ability of extracts from infected cells to translate exogenously added encephalomyocarditis (EMC) virus RNA and total Krebs II ascites cell mRNA was markedly diminished, but no evidence for the selective inhibition of translation of host cell mRNA was obtained (52).

In another case the initiation of translation on endogenous cellular mRNA in polysomes from poliovirus infected cells was reduced compared to that from uninfected cells (53). The ribosomal wash fraction prepared from the inhibited polysomes had reduced activity in stimulating translation of endogenous cell mRNA. However, the ability of the wash fraction to stimulate translation of viral mRNAs was not tested. Recently, Helentjaris and Ehrenfeld (54) have confirmed and extended these studies. Using cell-free systems prepared from uninfected and poliovirus-infected HeLa cells, they showed (Table 3) that crude preparations of initiation factors from late-infected cells do not stimulate the initiation of translation by polyribosomes containing endogenous host cell mRNA. However, they do stimulate the initiation of translation on polysomes containing endogenous viral mRNA nearly as well as initiation factors from uninfected cells. Factors from uninfected cells were able to stimulate translation of both cellular and viral mRNAs. In further experiments they found that the addition of increasing amounts of crude initiation factors from infected cells to an already initiated system caused an inhibition of initiation. This suggests that an inhibitor is present in the crude initiation

Table 3. Effect of ribosomal wash from uninfected and infected cells on initiation of cell and viral mRNA translation.[a]

Source of salt wash	F (^{35}S)met incorp. (cpm) by		
	Uninfected polysomes	Early-infected polysomes	Late infected polysomes
None	363	278	529
Uninfected	8,801	8,407	8,913
Early-infected	3,466	1,386	8,449
Late-infected	1,175	992	7,403

a. Fractionated S-10s were reconstituted with pH 5 fraction, polysomes, and crude salt wash and incubated for 30 min at 32° with (^{35}S)-fmet-tRNA as described by Helentjaris and Ehrenfeld (54). Modified from Helentjaris and Ehrenfeld (54).

factor preparation, which is responsible for the inactivation of an initiation factor.

Most recently Rose et al. (55) have identified an initiation factor that is inactivated after infection of HeLa cells with poliovirus. In their experiments Rose et al. (55) took advantage of the finding that translation of VSV mRNA, like host mRNA translation, is inhibited in cells superinfected with poliovirus (32, 36). They prepared extracts from poliovirus-infected and uninfected HeLa cells, and after a preincubation period and RNase treatment to eliminate endogenous mRNA translation, tested the ability of the extracts to translate exogenous poliovirus and VSV mRNA. Poliovirus mRNA was translated by both extracts, but VSV mRNA was translated only in the extracts from uninfected cells.

To determine the reason for the lack of VSV mRNA translation, purified initiation factors from reticulocytes were added one by one to the inactive extracts and the ability of the extract to support VSV mRNA translation was examined. Only one initiation factor, eIF-4B, could restore VSV mRNA translation, as shown in Figure 5. This is the same factor that Golini et al. (47) found could restore the translation of cellular mRNA when in competition with viral mRNA.

Their results further indicated that extracts from infected cells contain an activity that causes the slow inactivation of eIF-4B. Apparently, the translation of poliovirus mRNA in cell-free systems requires little or no eIF-4B, or can use an altered form of

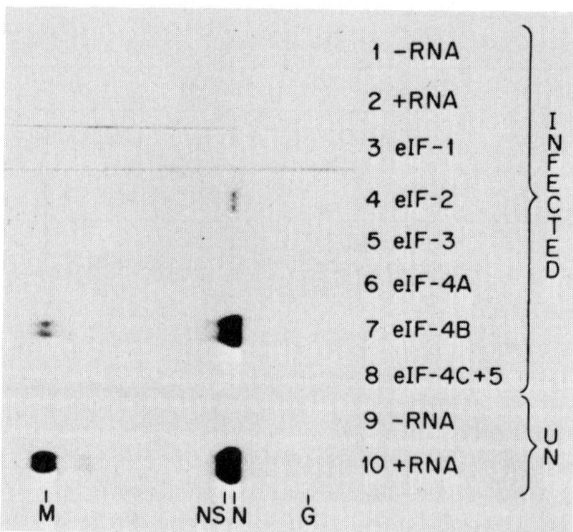

Figure 5. Effects of initiation factors on VSV mRNA translation. (^{35}S)- Methionine-labeled proteins synthesized <u>in vitro</u> were subjected to gel electrophoresis and detected by autoradiography as shown. Lanes 1-8 show products synthesized in a lysate from poliovirus-infected cells; lanes 9 and 10 show products synthesized in a mock-infected lysate. Lanes 1 and 9, no added RNA; lanes 2-8 and 10, plus VSV mRNA and the indicated initiation factors. Details are in publication by Rose <u>et al</u>. (55). From Rose <u>et al</u>. (55).

the factor. Factor eIF-4B has been found to interact with the "cap" on eukaryotic mRNA (56). The capped 5' end of VSV mRNA and of other mRNAs is important for ribosome recognition of mRNA (30). Since poliovirus mRNA lacks a capped 5' end, an inactivation of initiation factor eIF-4B would have little or no effect on its translation, but would severely limit the translation of "capped" mRNAs including VSV mRNA.

Hackett and co-workers (57, 58) have also observed that in a cell-free system from mengovirus infected Ehrlich ascites tumor cells, host mRNA was translated less well than in extracts from uninfected cells. Furthermore, at supraoptimal concentrations of Mg^{++} and K^{+}, the system from virus infected cells supported the translation of mengovirus RNA but not host mRNA, thereby demonstrating a selective translation of viral over host mRNA. A large part of the selectivity found in the translational system derived from infected cells was due to the initiation factor fraction.

In summary, the available evidence suggests that an initiation factor is inactivated after infection of HeLa cells by poliovirus, possibly by a virus-coded factor. The variety of results seen in comparative studies using extracts from infected and uninfected cells in terms of translation of cellular and viral messages is not understood, but may be a property of the virus or cell system under study.

F. Double-stranded RNA as an Inhibitor of Protein Synthesis

One way of searching for the presence of inhibitors of polypeptide initiation in infected cells was to add cytoplasmic fractions from virus infected cells to a cell-free system from rabbit reticulocytes. This system initiates the synthesis of new polypeptide chains at a very high rate. Cytoplasm from poliovirus infected HeLa cells, but not from uninfected cells, inhibited protein synthesis in the reticulocyte lysate (59). The inhibitor was isolated and identified as double-stranded (ds) RNA (60). To study the effect of ds RNA on host and viral protein synthesis, a cell-free system from HeLa cells was developed which initiated translation on endogenous cellular or viral mRNA. When added to this system, the ds RNA was found to inhibit the translation of both cellular and viral mRNAs (61). Furthermore, measurement of the amount of ds RNA present in cells early in infection (61, 62) revealed that an insufficient quantity was present to act as a direct agent of protein synthesis inhibition.

Interestingly, in reticulocyte lysates, the effect of ds RNA is similar to that resulting from heme deficiency (63), that is, polypeptide chain initiation is inhibited. (For an extensive discussion, see chapter 10 of this book). The ds RNA appears to induce a protein kinase which phosphorylates initiation factor eIF-2 (63). Experiments in our laboratory (M. Jaye, unpublished data) have failed to detect the presence of any phosphorylated components in the crude initiation factor preparation after virus infection corresponding to those elicited in the presence of ds RNA.

When added to whole cells, non-infectious bovine enterovirus ds RNA is cytotoxic (64). The cytotoxicity occurred in the presence of inhibitors of protein synthesis, suggesting that translation is not necessary for the effect. The ds RNA appeared to be selective, since it inhibited only cellular protein synthesis and not mengovirus protein synthesis (65). These findings are contrary to those found in extracts, where no selectivity of ds RNA was observed. No explanation for this difference is available at this time.

The general consensus of researchers in the field is that ds RNA is not involved in shut-off, since its effect in extracts showed no specificity between viral and cellular protein synthesis, as would be expected for such an inhibitor.

III. DISCUSSION

In this review I have outlined several theories that have been proposed to explain the mechanism by which picornaviruses inhibit cellular protein synthesis. Some theories seem less likely than others. Inhibition by ds RNA, for example, is no longer thought to be a likely possibility. In cell-free extracts ds RNA inhibits both cellular and viral mRNA translation (61). The inhibitor of cellular protein synthesis would be expected to be selective in its inhibitory activity. It is also apparent that picornavirus infection does not result in the degradation or alteration of cellular mRNA (9, 27, 29, 31). So, too, experiments demonstrating that protein synthesis inhibition takes place in the absence of significant viral RNA synthesis (14) tend to weaken the argument that protein synthesis inhibition results from direct competition of viral mRNA with cellular RNA for initiation factor eIF-4B (47). As mentioned earlier, superinfection with poliovirus of cells infected with VSV prevents VSV mRNA translation (32, 36). In lysates from uninfected HeLa cells, however, VSV mRNA translation is favored over poliovirus mRNA translation when both mRNA species are present in equimolar saturating concentrations (55). If competition were a major cause of cellular protein synthesis inhibition, one would have expected poliovirus mRNA to out-compete VSV mRNA in cell-free translation, not the contrary.

The theory that a virus-specific factor associates with 40S ribosomal subunits and the 5' end of viral mRNAs, thereby blocking the link of the 40S subunit with cellular mRNA, is difficult to test. Picornaviral proteins have been found to co-sediment with 40S subunits (49, 50, 51), but it is difficult to determine whether their presence is important or an artifact of isolation, particularly in view of the fact that no 40S alteration of activity has been found (51).

The model postulating an increase in the intracellular salt concentration due to an alteration of membrane permeability (41, 42) is also not strongly supported, at least not at early times after infection (42, 43). However, the possibility that the cell membrane has been altered resulting in changes in other intracellular components, which have a large effect on protein synthesis, has not been ruled out.

The theory receiving most support at this time involves the inactivation of an initiation factor necessary for cellular, but not viral, mRNA translation. The ability of initiation factor eIF-4B to restore the activity of extracts from infected cells in the translation of VSV mRNA is strong evidence that eIF-4B is the initiation factor inactivated after infection of HeLa cells by poliovirus (55). As mentioned it is reasonable to expect that inactivation of a factor which can differentiate between capped (cellular mRNAs) and uncapped (poliovirus) mRNAs would be the basis of poliovirus induced inhibition of protein synthesis. These results suggest that each virus must use a different strategy to inhibit cellular protein synthesis. Certainly, the inhibition of cellular protein synthesis by VSV cannot occur by inactivation of eIF-4B, since this factor is necessary for the translation of VSV mRNAs. Whatever the strategy that VSV uses, it does not disturb the translation of poliovirus mRNA, since poliovirus grows in VSV infected cells.

The idea that each virus uses its own strategy to inhibit cellular protein synthesis is further illustrated by herpes simplex virus (66) infection of Friend erythroleukemia cells. In this case cellular mRNA (globin) is degraded, in contrast to the infection of these same cells by VSV (66), in which no cellular mRNA degradation is seen.

Although the current evidence favors a mechanism involving the synthesis of a virus-induced inhibitor of an initiation factor, we cannot eliminate the possibility that the cell itself also contributes to shutting down the synthesis of its own macromolecules. The response of the cell to infection by cytotoxic viruses is not unlike its response to amino acid or glucose deprivation (67, 68), treatment with histidinol (69, 70), high cell density (71), exposure to hypertonic growth medium (37, 38), ATP depletion (72), exposure to elevated temperatures (73), and entrance into the mitotic phase of the cell cycle (74). In each case polysomes disaggregate, there is a concomitant accumulation of 80S ribosomes, rRNA and DNA synthesis are inhibited, and the cell seem to enter an early G1 or quiescent G_o state. The response of the cell to these perturbations, known as the negative pleiotypic response (75), has been studied for many years, but little is known about how it is mediated.

Interestingly, viral mRNA translation can occur in metaphase arrested cells (76), in cells arrested as a result of treatment with histidinol (70) and as mentioned, in cells treated with high salt (37, 38). Mengovirus also grows in a purine starved ade$^-$ mutant of Chinese hamster cells (77). Instead of only asking how the translation machinery can discriminate between viral and cellular mRNA, perhaps we should also be asking how the virus or viral mRNA can by-pass what seems to be a normal cellular regulatory mechanism. It is possible that virus-induced inhibition of protein synthesis is a two step process. In the first step, a translation product of the virus interacts with the cell membrane or other cellular component resulting in some sort of deficiency or stress condition, which in turn, sets off the negative pleiotypic response. The translation of those mRNAs molecules sensitive to this response will be inhibited.

Since viral mRNA translation is insensitive to this cellular control mechanism, it is readily translated. This theory differs from that of Carrasco (42) in that we do not restrict the perturbation in the cell to an increase in the intracellular salt concentration. Any condition that turns on the normal cellular control process will suffice. This theory also differs from that of Nuss and Koch (40), who advocate a non-specific inhibitor, in that the inhibitor is specific--it is the negative pleiotypic effector whose identity is unknown.

In the second step, the virus synthesizes an inhibitor that inactivates an initiation factor and by the mid-point of infection, most ribosomes are directed towards translating viral mRNA. This two step process could explain the differential effect of such perturbations as exposure to high salt concentrations and histidinol treatment on cellular and viral mRNA translation and the inactivation of initiation factor eIF-4B after infection of HeLa cells with poliovirus.

IV. CONCLUSIONS

The data available suggests that poliovirus induces the synthesis of some inhibitor which causes the slow inactivation of initiation factor eIF-4B. While it appears that there may be a break-through in the problem of protein synthesis shut-off, many questions still remain unanswered. For example, the nature of the eIF-4B inhibitor is unknown. Is it a viral product, or is it a cellular factor induced by the virus? Is the eIF-4B inactivated by proteolytic digestion or is it inactivated by modification?

Also, we do not yet know how viruses such as VSV, which have capped messengers, inhibit cellular protein synthesis. Whatever the strategy that VSV uses to turn off the host, poliovirus is apparently resistent to this step, since poliovirus grows in VSV infected cells. If VSV does inactivate an initiation factor other than eIF-4B, then this would suggest that poliovirus mRNA translation by-passes the requirement for several cellular initiation factors. This finding emphasizes our lack of knowledge of the exact initiation factor requirement for viral and cellular protein synthesis in cell-free systems.

ACKNOWLEDGMENTS

I am grateful to Dr. Robert Warrington for his many valuable suggestions in terms of formulating a hypothesis for shut-off and to Drs. E. Ehrenfeld and D. Baltimore for allowing me to use their published data in this review. Tables 1,2 and 3 and Figure 4 were reproduced by permission of the publishers.

REFERENCES

1. MARTIN, E.M. and KERR, I.M. Virus-induced changes in host-cell macromolecular synthesis. In The Molecular Biology of Viruses, Eighteenth Symposium of the Society for General Microbiology, eds., Crawford, L.V. and Stocker, M.G.P. (1968), pp. 15-56. London: Cambridge University Press.

2. METZ, D.H. Discrimination between viral and cellular macromolecular synthesis. In Control Processes in Virus Multiplication, Twenty-Fifth Symposium of the Society for General Microbiology, eds., Burke, D.C. and Russell W.C. (1975), pp. 323-353. London: Cambridge University Press.

3. ROIZMAN, B. and SPEAR, P.G. Macromolecular biosynthesis in animal cells infected with cytolytic viruses. Curr. Top. Develop. Biol. (1969), 4, 79-108.

4. BABLANIAN, R. Structural and functional alterations in cultured cells infected with cytocidal viruses. In Progress in Medical Virology, ed. Melnick, J.L. (1975), Vol.19 pp. 40-83. Basel: Karger.

5. MANAK, M.M. Protein synthesis inhibition by mengovirus. Studies on the role of native 40S ribosomal subunits. Ph. D. Thesis. (1976), University of Connecticut. pp. 1-154.

6. NAKAI, K. and LUCAS-LENARD, J. Processing of mengovirus precursor polypeptides in the presence of zinc ions and sulfydryl compounds. J. Virol. (1976), 18, 918-925.

7. PENMAN, S., SCHERRER, K., BECKER, Y. and DARNELL, J.E. Polysomes in normal and poliovirus-infected HeLa cells and their relationship to messenger-RNA. Proc. Natl. Acad. Sci. U.S.A. (1963), 49, 654-661.

8. DALGARNO, L., COX, R.A. and MARTIN, E.M. Polyribosomes in normal Krebs-2 Ascites tumor cells and in cells infected with encephalomyocarditis virus. Biochem. Biophys. Acta. (1967), 138, 316-328.

9. COLBY, D.S., FINNERTY, V. and LUCAS-LENARD, J. Fate of mRNA of L-cells infected with mengovirus. J. Virol. (1974), 13, 858-869.

10. SCHARFF, M.D., SHATKIN, A.J. and LEVINTOW, L. Association of newly formed viral protein with specific polyribosomes. Proc. Natl. Acad. Sci. U.S.A. (1963), 50, 686-694.

11. PENMAN, S. and SUMMERS, D. Effects on host cell metabolism following synchronous infection with poliovirus. Virology, (1965), 27, 614-620.

12. HELENTJARIS, T. and EHRENFELD, E. Inhibition of host protein synthesis by UV-inactivated poliovirus. J. Virol. (1977), 21, 259-267.

13. JACOBSON, M.F. and BALTIMORE, D. Polypeptide cleavages in the formation of poliovirus proteins. Proc. Natl. Acad. Sci. U.S.A. (1968), 61, 77-84.

14. STEINER-PRYOR, A. and COOPER, P.D. Temperature-sensitive poliovirus mutants defective in repression of host protein synthesis are also defective in structural protein. J. Gen. Virol. (1973), 21, 215-225.

15. HOLLAND, J. Inhibition of host cell macromolecular synthesis by high multiplicities of poliovirus under conditions preventing virus synthesis. J. Mol. Biol. (1963), 8, 574-581.

16. COLE, C.N. and BALTIMORE, D. Defective interfering particles of poliovirus. III. Nature of the defect. J. Mol. Biol. (1973), 76, 325-343.

17. FRANKLIN, R.M. and BALTIMORE, D. Patterns of macromolecular synthesis in normal and virus-infected mammalian cells. Cold Spring Harbour Symposium Quantitative Biology. (1962), 27, 175-198.

18. DARNELL, J.E., GIRARD, M., BALTIMORE, D. and MAIZEL, J.V. The synthesis and translation of poliovirus RNA. In The Molecular Biology of Viruses, eds., Colter, J.S. and Paranchych, W. (1967), pp. 375-401. New York: Academic Press.

19. WILLEMS, M. and PENMAN, S. The mechanism of host cell protein synthesis inhibition by poliovirus. Virology, (1966), 30, 355-367.

20. LEIBOWITZ, R. and PENMAN, S. Regulation of protein synthesis in HeLa cells. III. Inhibition during poliovirus infection. J. Virology. (1971), 8, 661-668.

21. SCHWARTZ, L.B., LAWRENCE, C., THACH, R.E. and ROEDER, R.G. Encephalomyocarditis virus infection of mose plasmacytoma cells. III. Effect on host RNA synthesis and RNA polymerases.

J. Virology. (1974), 14, 611-619.

22. APRILETTI, J.W. and PENHOET, E.E. Recovery of DNA-dependent RNA polymerase activities from L-cells after mengovirus infection. Virology, (1974), 61, 597-601.

23. ENSMINGER, W.D. and TAMM, I. The step in cellular DNA synthesis blocked by Newcastle disease virus or mengovirus infection. Virology, (1970), 40, 152-165.

24. HAND, R. and OBLIN, C. DNA synthesis in mengovirus-infected cells: Mechanism of inhibition. J. Gen.Virol. (1977), 37, 349-358.

25. BRAWERMAN, G. Eukaryotic messenger RNA. Ann. Rev. Biochem. (1974), 43, 621-642.

26. LAWRENCE, C. and THACH, R. Encephalomyocarditis virus infection of mouse plasmocytoma cells. I. Inhibition of cellular protein synthesis. J. Virol. (1974), 14, 598-610.

27. GALLWITZ, D., TRAUB, U. and TRAUB, P. Fate of histone messenger RNA in mengovirus-infected Ehrlich ascites tumor cells. Eur. J. Biochem. (1977), 81, 387-393.

28. SHEINESS, D. and DARNELL, J.E. Polyadenylic acid segment in mRNA becomes shorter with age. Nature New Biol. (1973), 241, 265-268.

29. KOSCHEL, K. Poliovirus infection and poly(A) sequences of cytoplasmic cellular RNA. J. Virol. (1974), 13, 1061-1066.

30. SHATKIN, A.J. Capping of eukaryotic mRNAs. Cell (1976), 9, 645-653.

31. FERNANDEZ-MUÑOZ, R. and DARNELL, J.E. Structural difference between the 5' termini of viral and cellular mRNA in poliovirus-infected cells: Possible basis for the inhibition of host protein synthesis. J. Virol. (1976), 18, 719-726.

32. DOYLE, S. and HOLLAND, J. Virus-induced interference in heterologously infected HeLa cells. J. Virol. (1972), 9, 22-28.

33. MUDD, J.A. and SUMMERS, D.F. Protein synthesis in vesicular stomatitis virus-infected HeLa cells. Virology,(1970), 18, 328-340.

34. WERTZ, G.W. and YOUNGNER, J.S. Inhibition of protein synthesis in L-cells infected with VSV. J. Virol. (1972), 9, 85-89.

35. MOYER, S.A., GRUBMAN, J.J., EHRENFELD, E. and BANERJEE, A.K. Studies on the in vivo messenger RNA species of vesicular stomatitis virus. Virology,(1975), 67, 463-473.

36. EHRENFELD, E. and LUND, H. Untranslated vesicular stomatitis virus messenger RNA after poliovirus infection. Virology,(1977), 80, 297-308.

37. WENGLER, G. and WENGLER, G. Medium hypertonicity and polyribosome structure in HeLa cells. The influence of hypertonicity of the growth medium on polyribosomes in HeLa cells. Eur. J. Biochem. (1972), 27, 162-173.

38. SABORIO, J.L., PONG, S.S. and KOCH, G. Selective and reversible inhibition of initiation of protein synthesis in mammalian cells. J. Mol. Biol. (1974), 85, 195-211.

39. NUSS, D.L., OPPERMAN, H. and KOCH, G. Selective blockage of initiation of host protein synthesis in RNA virus-infected cells. Proc. Natl. Acad. Sci. U.S.A. (1975), 72, 1258-1262.

40. NUSS, D.L., and KOCH, G. Translation of individual host mRNAs in MPC-11 cells is differentially suppressed after infection by vesicular stomatitis virus. J. Virol. (1976), 19, 272-278.

41. CARRASCO, L. The inhibition of cell functions after viral infection: A proposed general mechanism. FEBS Letters (1977), 76, 11-15.

42. CARRASCO, L. and SMITH, A.E. Sodium ions and the shut-off of host cell protein synthesis by picornaviruses. Nature (London), (1976), 264, 807-809.

43. EGBERTS, E., HACKETT, P.B. and TRAUB, P. Alteration of the intracellular energetic and ionic conditions by mengovirus infection of Ehrlich ascites tumor cells and its influence on protein synthesis in the midphase of infection. J. Virol. (1977), 22, 591-597.

44. ABREU, S. and LUCAS-LENARD, J. Cellular protein synthesis shut-off by mengovirus: Translation of nonviral and viral mRNAs in extracts from uninfected and infected Ehrlich ascites tumor cells. J. Virol. (1976), 18, 184-192.

45. LODISH, H.F. Model for the regulation of mRNA translation applied to haemoglobin synthesis. Nature (London), (1974), 251, 385-388.

46. GOLINI, F., THACH, S., LAWRENCE, C. and THACH, R. Regulation of protein synthesis in EMC virus-infected cells. In Animal

Virology, IV. ICN-UCLA Symposia on Molecular and Cellular Biology, eds., Baltimore, D., Huang, A.S. and Fox, C.F. (1976), pp. 717-734. New York: Academic Press.

47. GOLINI, F., THACH, S.S., BIRGE, C.H., SAFER, B., MERRICK, W.C. and THACH, R.E. Competition between cellular and viral mRNAs in vitro is regulated by a messenger discriminatory initiation factor. Proc. Natl. Acad. Sci. U.S.A. (1976), 73, 3040-3044.

48. COOPER, P.D., STEINER-PRYOR, A. and WRIGHT, P.J. A proposed regulator for poliovirus: The equestron. Intervirology,(1973), 1, 1-10.

49. WRIGHT, P.J. and COOPER, P.D. Poliovirus proteins associated with ribosomal structures in infected cells. Virology,(1974), 59, 1-20.

50. MEDVEDKINA, O.A., SCARLAT, I.V., KALININA, N.O. and AGOL, V.I. Virus-specific proteins associated with ribosomes of Krebs-II cells infected with encephalomyocarditis virus. FEBS Letters (1974), 39, 4-8.

51. MANAK, M., ABREU, S. and LUCAS-LENARD, J. Association of mengovirus proteins with host cell native 40S ribosomal subunits. INSERM (1975), 47, 387-396.

52. SVITKIN, Y.V., UGAROVA, T.Y., GINEVSKAYA, V.A., KALININA, N.O., SCARLAT, I.V. and AGOL, V.I. Efficiency of translation of viral and cellular mRNAs in extracts from cells infected with encephalomyocarditis virus. Intervirology,(1974), 4, 214-220.

53. KAUFMANN, Y., GOLDSTEIN, E. and PENMAN, S. Poliovirus-induced inhibition of polypeptide initiation in vitro on native polyribosomes. Proc. Natl. Acad. Sci. U.S.A. (1976), 73, 1834-1838.

54. HELENTJARIS, T. and EHRENFELD, E. Control of protein synthesis in extracts from poliovirus-infected cells. I. mRNA discrimination by crude initiation factors. J. Virol. (1978), 26, 510-521.

55. ROSE, J.K., TRACHSEL, H., LEONG, K. and BALTIMORE, D. Inhibition of translation by poliovirus: Inactivation of a specific initiation factor. Proc. Natl. Acad. Sci. U.S.A. (1978), 75, 2732-2736.

56. SHAFRITZ, D.A., WEINSTEIN, J.A., SAFER, B., MERRICK, W.C., WEBER, L.A., HICKEY, E.D. and BAGLIONI, E. Evidence for role of $m^7G^{5'}$- phosphate group in recognition of eukaryotic mRNA by initiation factor IF-M_3. Nature (London), (1976), 261, 291-294.

57. HACKETT, P.B., EGBERTS, E. and TRAUB, P. Translation of ascites and mengovirus RNA in fractionated cell-free systems from uninfected and mengovirus-infected Ehrlich-ascites tumor cells. Eur. J. Biochem. (1978), 83, 341-352.

58. HACKETT, P.B., EGBERTS, E. and TRAUB, P. Selective translation of mengovirus RNA over host mRNA in homologous, fractionated, cell-free translational systems from Ehrlich-ascites tumor cells. Eur. J. Biochem. (1978), 83, 353-361.

59. HUNT, T. and EHRENFELD, E. Cytoplasm from poliovirus-infected HeLa cells inhibits cell-free haemoglobin synthesis. Nature New Biology (1971), 230, 91-94.

60. EHRENFELD, E. and HUNT, T. Double-stranded poliovirus RNA inhibits initiation of protein synthesis by reticulocyte lysates. Proc. Natl. Acad. Sci. U.S.A. (1971), 68, 1075-1078.

61. CELMA, M.L. and EHRENFELD, E. Effect of poliovirus double-stranded RNA on viral and host cell protein synthesis. Proc. Natl. Acad. Sci. U.S.A. (1974), 71, 2440-2444.

62. COLLINS, F. and ROBERTS, W. Mechanism of mengovirus-induced cell injury in L-cells: Use of inhibitors of protein synthesis to dissociate virus-specific events. J. Virol. (1972), 10, 969-978.

63. LEVIN, D. and LONDON, I.M. Regulation of protein synthesis: Activation by double-stranded RNA of a protein kinase that phosphorylates eukaryotic initiation factor 2. Proc. Natl. Acad. Sci. U.S.A. (1978), 75, 1121-1125.

64. CORDELL-STEWART, B. and TAYLOR, M.W. Effect of viral double-stranded RNA on mammalian cells in culture: Cytotoxicity under conditions preventing viral replication and protein synthesis. J. Virol. (1973), 12, 360-366.

65. CORDELL-STEWART, B. and TAYLOR, M.W. Effect of viral double-stranded RNA on protein synthesis in intact cells. J. Virol. (1973), 11, 232-237.

66. NISHIOKA, Y. and SILVERSTEIN, S. Alterations in the protein synthetic apparatus of Friend erythroleukemia cells infected with vesicular stomatitis virus or Herpes simplex virus. J. Virol. (1978), 25, 422-426.

67. VAUGHAN, M.H., PAWLOWSKI, P.J. and FORCHHAMMER, J. Regulation of protein synthesis initiation in HeLa cells deprived of single essential amino acids. Proc. Natl. Acad. Sci. U.S.A. (1971), 68, 2057-2061.

68. VAN VENROOIJ, W.J.W., HENSHAW, E.C. and HIRSCH, C.A. Effects of deprival of glucose or individual amino acids on polyribosome distribution and rate of protein synthesis in cultured mammalian cells. Biochem. Biophys. Acta. (1972), 259, 127-137.

69. GRUMMT, F. and GRUMMT, I. Studies on the role of uncharged tRNA in pleiotypic response of animal cells. Eur. J. Biochem. (1976), 64, 307-312.

70. WARRINGTON, R.C., WRATTEN, N. and HECHTMAN, R. L-histidinol inhibits specifically and reversibly protein and ribosomal RNA synthesis in mouse L-cells. J. Biol. Chem. (1977), 252, 5251-5257.

71. LEVINE, E.M., BECKER, Y., BOONE, C.W. and EAGLE, H. Contact inhibition, macromolecular synthesis, and polyribosomes in cultured human diploid fibroblasts. Proc. Natl. Acad. Sci. U.S.A. (1965), 53, 350-356.

72. GILOH, H. and MAGER, J. Inhibition of peptide chain initiation in lysates from ATP-depleted cells. I. Stages in the evolution of the lesion and its reversal by thiol compounds, cyclic AMP or purine derivatives and phosphorylated sugars. Biochem. Biophys. Acta. (1975), 414, 293-308.

73. McCORMICK, W. and PENMAN, S. Regulation of protein synthesis in HeLa cells: Translation at elevated temperatures. J. Mol. Biol. (1969), 39, 315-333.

74. FAN, H. and PENMAN, S. Regulation of protein synthesis in mammalian cells. Inhibition of protein synthesis at the level of initiation during mitosis. J. Mol. Biol. (1970), 50, 655-670.

75. HERSHKO, A., MAMONT, P., SHIELDS, R. and TOMKINS, G.M. Pleiotypic response. Nature New Biology (1971), 232, 206-211.

76. MARCUS, P.I. and ROBINS, E. Viral inhibition in the metaphase-arrest cell. Proc. Natl. Acad. Sci. U.S.A. (1963), 50, 1156-1164.

77. SRIRAM, G. and TAYLOR, M.W. Purineless death: Ribosomal RNA turnover in a purine-starved ade$^-$ mutant of Chinese hamster cells. J. Biol. Chem. (1977), 252, 5350-5355.

THE ROLE OF INITIATION FACTORS IN THE SHUT-OFF OF PROTEIN SYNTHESIS

C. BAGLIONI, P.A. MARONEY and M. SIMILI

Department of Biological Sciences, State University of New York at Albany,
New York, 12222, U.S.A.

INTRODUCTION

The shut-off of protein synthesis in virus-infected cells is a phenomenon that has attracted the attention of several investigators. In an attempt to provide an explanation at the molecular level for the mechanism of shut-off, studies were initially carried out with virus-infected cells. Since the results of this kind of studies are extensively discussed in the preceeding chapter of this book, we will concentrate on a particular aspect, namely the role of initiation factors on the virus-induced inhibition of host protein synthesis.

The currently available evidence (1-5) suggested that the binding of host mRNAs to ribosomes was specifically inhibited in infected cells (2). Furthermore, viral mRNA had been shown to outcompete host mRNA at the initiation step of protein synthesis (5), and *in vitro* studies demontrated that picornavirus RNA was preferentially translated over host mRNA when both were present in saturating amounts (6-7). This competition can be relieved by the addition of a specific initiation factor, eIF-4B, suggesting that the amount of factor available and possibly the higher affinity of viral mRNA for this factor determine the preferential translation of viral templates (8).

The affinity of eIF-4B for a cellular mRNA, globin mRNA, and for encephalomyocarditis virus (EMC) RNA has been recently measured (9). The affinity of EMC RNA for eIF-4B was shown to be one order of magnitude higher than that of globin mRNA, as revealed by the dissociation constant of complex formation between the RNAs and the purified initiation factor. This result was confirmed by competition-binding experiments, where labelled and unlabelled mRNAs compete with each other for binding to eIF-4B (9).

The competition between viral and host mRNAs was, therefore, localized at the level of a specific protein and, probably, offered the most satisfactory explanation for the molecular mechanism of the shut-off. It was also suggested that eIF-4B was specifically inactivated in poliovirus-infected cells (10). This was not in contrast with the results of the experiments mentioned above on the effect of eIF-4B on *in vitro* translation and with the high affinity of EMC RNA for this initiation factor. However it has not definitely been established whether in infected cells eIF-4B is

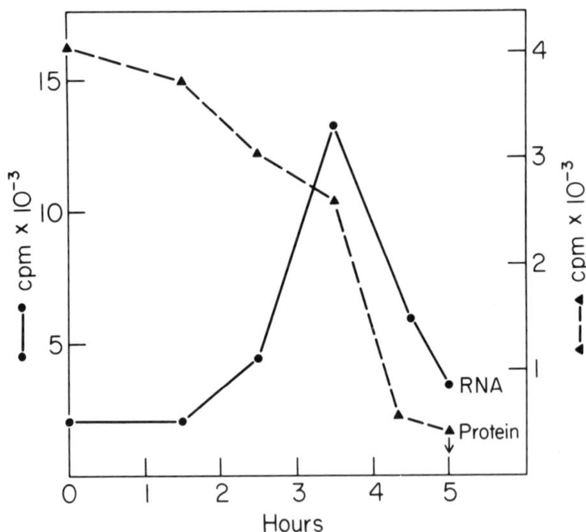

Figure 1. Time course of protein and viral RNA synthesis in HeLa cells infected with EMC. A culture at 4 x 10^6 cells/ml in MEM medium, 5% fetal calf serum, 2mM glutamine and 14 mM HEPES-KOH, pH 7.2, was infected with 300pfu/cell of EMC. To measure protein synthesis, 0.1 ml of the culture were added at the indicated times to 0.9 ml of methionine-free medium containing 10 μCi/ml of (^{35}S)-methionine and incubated 20 min. To measure RNA synthesis, 0.2 ml of the culture were added to 0.8 ml of medium containing 6.25 μg /ml of actinomycin D and incubated 5 min ; 10 μCi of (^{3}H)-uridine were then added for 15 min. The incubations were stopped by adding 10 ml of cold spinner salts solution and the cells collected by centrifugation. The pellet was dissolved in 1% Na-dodecyl sulphate and precipitated with 5% trichloroacetic acid for counting.

inactivated or sequestered by picornavirus RNA.

Cell extracts prepared at different times after infection of HeLa cells with EMC virus have been used to study the mechanism of shut-off. These studies were made possible by the observation of Weber _et al._ (11) that HeLa cell extracts supplemented with haemin and other components necessary for _in vitro_ protein synthesis were active in initiation.

I. PROTEIN SYNTHESIS WITH EXTRACTS OF EMC-INFECTED CELLS

Inhibition of protein synthesis in intact HeLa cells infected with EMC virus was observed at the same time that viral RNA synthesis reached its peak (Figure 1). The shut-off of host protein synthesis, however, preceded this general inhibition of protein synthesis (see below). When cell extracts were prepared at different times after infection and tested for endogenous protein synthesis, a progressive loss of activity with time of infection was observed. A similar result was obtained in L-cells infected with EMC (Figure 2).

Ehrenfeld and Hunt (12) showed that the inhibition of protein synthesis in reticulocyte lysates by extracts of polio-infected cells was due to the presence of double-stranded RNA (dsRNA) of viral origin. It was later established that low levels of dsRNA activate a protein kinase which blocks initiation of protein synthesis by phosphorylating a subunit of the initiation factor eIF-2 (13; for a detailed discussion, see chapters 10 and 11 of this book). The inactivation of eIF-2 seemed to be responsible for the inhibition of the binding Met-tRNA$_f$ (initiator tRNA) to the 40 S native ribosomal subunit (40 S^N) observed after the addition of synthetic (or viral) dsRNA to rabbit reticulocyte or HeLa cell extracts (13-14). A similar inhibition of Met-tRNA$_f$ binding to 40 S^N was observed in extracts prepared from EMC-infected cells (Figure 3), suggesting that eIF-2 was indeed inactivated in these extracts. However, protein synthesis in intact cells was only slightly inhibited at a corresponding time point in infection (compare Figures 1 and 3). Experiments designed to test the Met-tRNA$_f$ binding activity in intact cells (15-16) at different times after infection showed that the Met-tRNA$_f$ binding activity declined in infected cells at the same time that the total protein synthesis declined significantly (Figures 1 and 4). At 3 to 4 hours after infection protein synthesis was markedly inhibited and Met-tRNA$_f$ binding activity in intact cells also decreased. This was also shown by the reduction in polysomes and the increase in 80 S ribosomes (Figure 4). In extracts prepared from cells 2 hours after infection, the Met-tRNA$_f$ binding activity was practically abolished, whereas in intact cells there was no decrease in binding.

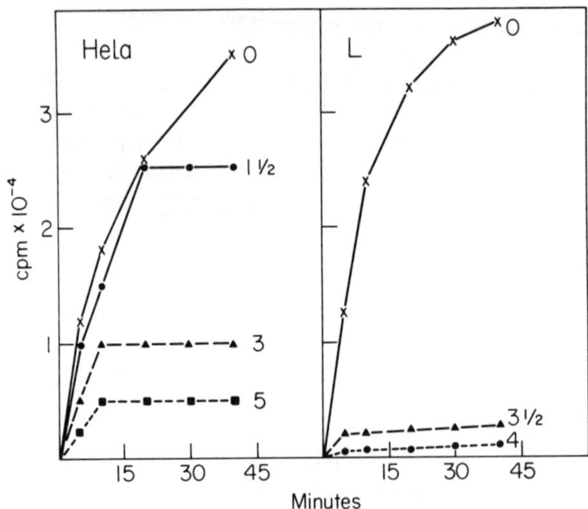

Figure 2. Protein synthesis with extracts prepared from HeLa and L-cells infected with EMC. The infection of cells is described in Figure 1. The preparations of cell extracts and the protein synthesis assay have previously been described (11). The incorporation of (^{3}H)-lysine per 5 µl of incubation is indicated. The time of preparation of extracts during infection is indicated: 0, mock-infected cells; 1½, 3, 3½, 4 and 5 hr after infection.

This suggested that the reduced initiation activity observed in the cell extracts did not reflect adequately the protein synthesizing activity in the intact cells. An explanation for these findings is that the activation of the inhibitory protein kinase took place in the cell extracts more readily than in the intact cells (see chapter 10).

The observed inhibition of protein synthesis in reticulocyte lysates by extracts of EMC-infected cells was, therefore, not due to the viral dsRNA itself but to the dsRNA-activated kinase present in the extract. It is not clear when the kinase is activated, but it can be reasonably argued that this happens during the homogenization of the infected cell and the centrifugation step during the preparation of the cell extract. The dsRNA-activated kinase, therefore, can account for the inhibition of Met-tRNA$_f$ binding observed in infected cell extracts (Figure 3).

II. THE ROLE OF INITIATION FACTORS IN THE SHUT-OFF

The inactivation of eIF-2 by the dsRNA-activated protein kinase prevented any study of active protein synthesis in extracts from

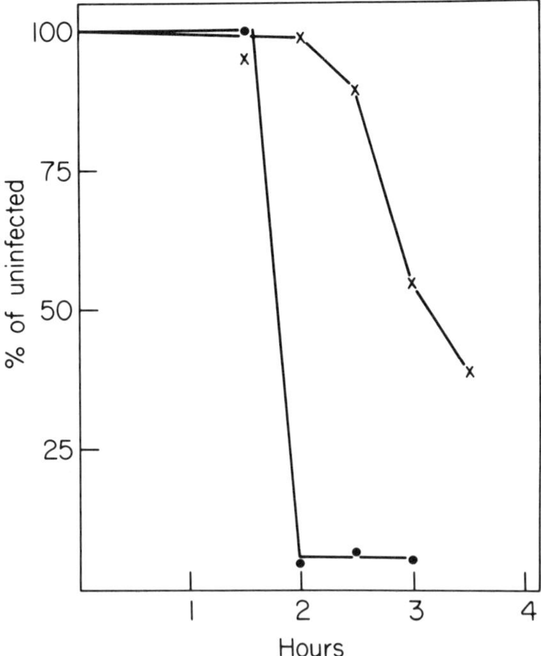

Figure 3. Binding of Met-tRNA$_f$ to 40 S ribosomal subunits in infected cells (x----x) and cell extracts (●----●). Each point represents the Met-tRNA$_f$ binding determined by sucrose gradient centrifugation (see Figure 4) in infected cells or in extracts prepared from these cells. Hours after infection are indicated in the abscissa. The binding of Met-tRNA$_f$ in extracts was measured as previously described using (^{35}S)-Met-tRNA$_f$ from rabbit liver (11). The in vivo binding is described in Figure 4. The binding activity is plotted as percentage of that of uninfected cells or of an extract prepared from these cells.

infected cells. The rate of initiation in these extracts was severely depressed by their low Met-tRNA$_f$ binding activity, though initiation was not completely inhibited. It was therefore necessary to supplement the extracts with eIF-2 or with a preparation of initiation factors (eIFs) containing eIF-2. We have supplemented incubations of EMC-infected cell extracts with rabbit reticulocyte eIFs and have observed a general stimulation of protein synthesis (Figure 5). The pattern of proteins synthesized was, however, altered by the addition of eIFs. Extracts prepared 2 hours after EMC infection synthesized predominantly host proteins, whereas extracts prepared at 3 hours synthesized predominantly viral proteins (Figure 5). Addition of eIFs resulted in an increased

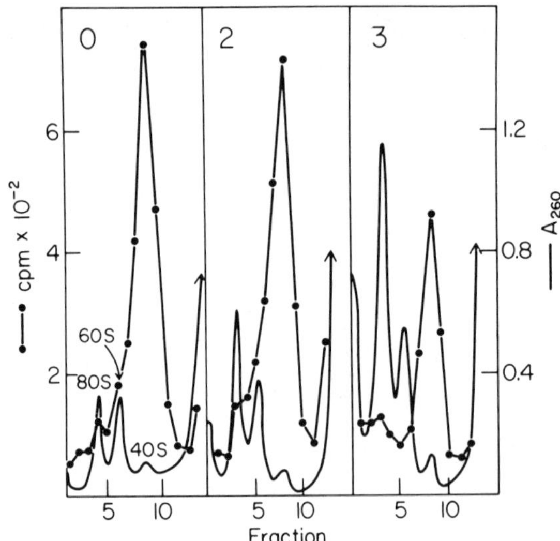

Figure 4. Binding of Met-$tRNA_f$ to 40 S ribosomal subunits in cells infected with EMC. At the times post-infection indicated in the top left corner of each panel, HeLa cells infected as described in Figure 1 were collected by centrifugation, resuspended in 1/4 vol of methionine-free medium, incubated 2 min with 0.1 mg/ml of emetine and then with 20 μCi/ml of (^{35}S)-methionine. The cells were washed and lysed to prepare a cell extract, which was fractionated by sucrose gradient centrifugation for 16 hr at 24,000 rpm (11). The position of ribosomes and subunits is indicated. Fractions were counted as previously described (11).

synthesis of host proteins in both samples. Viral polypeptides were identified by their synthesis in the presence of 7-methylguanosine-5'-monophosphate (pm^7G). This "cap" analog blocks the translation of cellular mRNAs, but has little (or no) effect on the translation of EMC RNA (17).

These experiments showed that eIFs relieved in part the competition between host and EMC RNA in extracts prepared from infected cells. This conclusion was in substantial agreement with the results obtained by Thach and collaborators (8) and by Rose et al. (10), who had shown that the addition of eIF-4B alone to a cell-free system suppressed the preferential translation of picornavirus RNA.

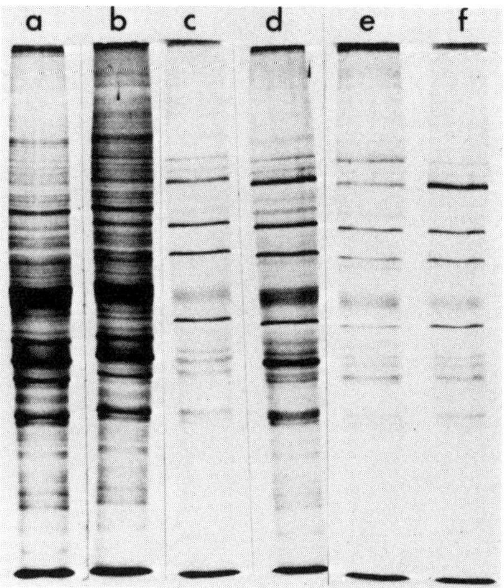

Figure 5. Protein synthesized by extracts of EMC-infected cells. Cell extracts were prepared from cells 2 hr (a, b and e) and 3 hr (c, d and f) after infection as described in Figure 2. Fifty μl incubations contained 50 μM amino acids-methionine, 120 mM KOAc, 2.5 mM Mg $(OAc)_2$, other components previously described, and no additions (a and c), 1 μl of rabbit reticulocyte initiation factors (b and d; 30 μg of protein), prepared as described (22), or 1 mM 7-methylguanosine-5'-monophosphate (e and f). The samples were preincubated 10 min and then 50 μCi of (^{35}S) methionine were added for 90 min at 30°C. The cpm incorporated per 2 μl were respectively: a= 45,200, b= 92,000, c= 19,800, d= 35,000, e= 17,000 and f= 22,000. Twenty μl of each incubation were fractionated on 12.5% polyacrylamide gels as previously described (14). The autoradiograph of the dried gel is shown.

Two different explanations have been proposed for this effect of eIF-4B:

i) different mRNAs may bind to eIF-4B with widely different affinity (8-9);

ii) in the course of picornavirus infection eIF-4B becomes gradually inactivated (10).

The first explanation was based on measurements of the relative affinity of EMC RNA, Vesicular Stomatitis Virus (VSV) mRNA, and globin mRNA for eIF-4B (9): EMC RNA outcompeted globin and VSV mRNAs for binding to eIF-4B (9). This finding was in agreement with the observation that in VSV-infected cells superinfected with poliovirus, a preferential translation of polio RNA took place: VSV mRNA was synthesized but not utilized for protein synthesis (18-19). It has not yet been established, however, whether the higher affinity of EMC RNA for eIF-4B results in the formation of a more stable complex between the initiation factor and EMC RNA, and it is not known how the "uncapped" picornavirus RNA binds to eIF-4B. This initiation factor interacts with the 5'-terminal $m^7G^{5'}$ p of eukaryotic mRNAs (17).

The second explanation considered above implied that either picornavirus RNA had a higher affinity for eIF-4B, so that when the factor became limiting it was preferentially translated, or that eIF-4B was not absolutely required for translation of picornavirus RNA. *In vitro* translation experiments with purified initiation factors suggested, however, that eIF-4B was required for translation of all mRNAs, even those which are not "capped" (20).

We have investigated the initiation factors of EMC-infected cells in an attempt to obtain evidence for the explanations discussed above. Since eIFs are found predominantly associated with the 40 S^N subunit (21), we isolated this ribosomal subunit from HeLa cells at different times after infection. The proteins associated with 40 S^N were analyzed by gel electrophoresis and some characteristic changes were observed (Figure 6). Two polypeptides of M_r 115,000 and 80,000 respectively were found in reduced amounts in the 40 S^N of infected cells. The molecular weights of these polypeptides correspond to that of a component of eIF-3 (21) and of eIF-4B respectively (17). The reason for the decrease in the amount of these factors associated with 40 S^N is not yet clear.

A search for these polypeptides in other cell fractions gave inconclusive results. Electrophoretic analysis of polysome-associated proteins,for example, showed a very complex pattern with many, poorly resolved bands. At the present time, it seems however possible that these polypeptides may be associated with other components of infected cells, namely with the viral RNA.

III. GENERAL CONCLUSIONS

Extracts prepared from infected cells may be a useful tool to investigate the mechanism of the shut-off of the host-cell protein synthesis. In the case of picornavirus-infected cells, however, the cell extracts are relatively inactive in protein synthesis unless supplemented with initiation factors. This inhibition of protein synthesis in cell extracts is probably caused by dsRNA of viral origin.

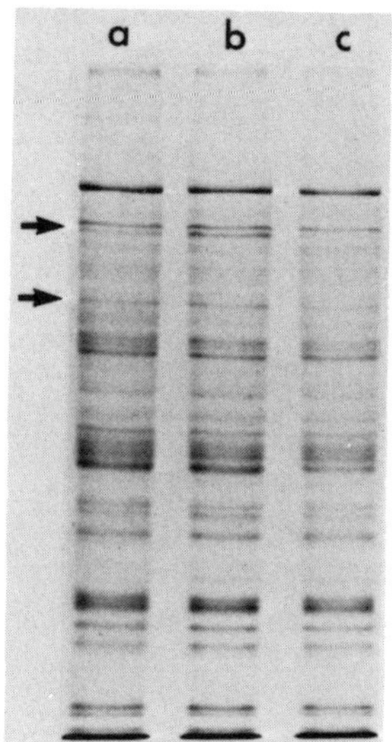

Figure 6. Electrophoretic analysis of proteins associated with native 40 S ribosomal subunits of control and EMC-infected HeLa cells. Extracts were prepared from mock-infected and infected cells as described in Figure 2; 0.5 ml of extract from control cells (b) or from cells infected 2 hr (a) or 3 hr (c) with EMC were layered on 15-30% sucrose gradients in 30 mM KCl, 2mM $Mg(OAc)_2$, 20 mM HEPES-KOH, pH 7.4, 1 mM dithiothreitol, and centrifuged 17 hr at 30,000 rpm. Fractions corresponding to the 40 S peak were combined, precipitated with 10% trichloroacetic acid, washed with acetone/ether (2/3) and ether, dissolved in sample buffer and fractionated on 12.5% polyacrylamide gels as previously described (14). The gels were stained with Coomassie Blue. Markers were run in parallel to assign M_r to the bands indicated by arrows.

Protein synthesis in extracts of cells infected with other RNA viruses is not inhibited to the same extent as in the case of picornaviruses. This has been clearly shown for VSV: extracts prepared from HeLa cells at different times after infection with VSV are very active in cell-free protein synthesis and synthesize only viral proteins (our unpublished observations). The replicative intermediate of VSV does not seem capable of activating the protein kinase of HeLa cells and of inhibiting protein synthesis.

The addition of eIFs to extracts of EMC-infected cells stimulates the synthesis of host proteins. This is probably due to the effect of eIF-4B. Two possible explanations for the action of this initiation factor have been discussed above. Further work should establish whether active eIF-4B can be recovered from picornavirus-infected cells. Extraction of eIF-4B with 0.5 M KCl, and purification and assay of the factor in a reconstituted cell-free system may be necessary to provide conclusive evidence for its presence in normal amounts or for its proposed inactivation in picornavirus-infected cells. For the time being, our preliminary studies on the electrophoretic pattern of eIFs in infected cells suggest only that changes at the level of initiation factors take place, but provide no information on the cause of these changes.

ACKNOWLEDGEMENT

The work described in this review was supported by Grants AI-11887 and HL-17710 from N.I.H. to C.B.

REFERENCES

1. BALTIMORE, D. In Biochemistry of Viruses (1969), Levy, H.B. ed. pp. 101-176, Marcel Dekker, New York.

2. LEIBOWITZ, R. and PENMAN, S. J. Virol. (1971), 8, 661-668.

3. FERNANDEZ-MUÑOZ, R. and DARNELL, J.E. J. Virol. (1976), 126, 719-726.

4. SABORIO, J.L., PONG, S.S. and KOCH, G. J. Mol. Biol. (1974), 85, 195-211.

5. NUSS, D.L., OPPERMAN, H. and KOCH, G. Proc. Natl. Acad. Sci. U.S.A. (1975), 72, 1258-1262.

6. LAWRENCE, C. and THACH, R.E. J. Virol. (1974), 14, 598-610.

7. ABREU, S. and LUCAS-LENARD, J. J. Virol. (1976), 18, 182-194.

8. GOLINI, F., THACH, S.S., BIRGE, C.H., SAFER, B., MERRICK, W.C. and THAHC, R.E. Proc. Natl. Acad. Sci. U.S.A. (1976), 73, 3040-3044.

9. BAGLIONI, C., SIMILI, M. and SHAFRITZ, D.A. Nature (London) (1978), 275, 240-243.

10. ROSE, J.K., TRACHSEL, H., LEONG, K. and BALTIMORE, D. Proc. Natl. Acad. Sci. U.S.A. (1978), 75, 2732-2736.

11. WEBER, L.A., FEMAN, E. and BAGLIONI, C. Biochemistry (1975), 14, 5315-5321.

12. EHRENFELD, E. and HUNT, T. Proc. Natl. Acad. Sci. U.S.A. (1971), 68, 1075-1078.

13. FARRELL, P.J., BALKOW, K., HUNT, T. and JACKSON, R.J. Cell, (1977), 11, 187-200.

14. LENZ, J.R. and BAGLIONI, C. J. Biol. Chem. (1978), 253, 4219-4223.

15. TARNOWKA, M.A. and BAGLIONI, C. J. Cell. Physiol. (1978), submitted for publication.

16. DARNBROUGH, C., LEGON, S., HUNT, T. and JACKSON, R.J. J. Mol. Biol. (1973), 76, 379-403.

17. SHAFRITZ, D.A., WEINSTEIN, J.A., SAFER, B., MERRICK, W.C., WEBER, L.A., HICKEY, E.D. and BAGLIONI, C. Nature (London) (1976), 261, 291-294.

18. DOYLE, S. and HOLLAND, J. J. Virol. (1972), 9, 22-28.

19. EHRENFELD, E. and LUND, H. Virology (1977) 80, 297-308.

20. STAEHELIN, T., TRACHSEL, H., ERNI, B., BOSCHETTI, A. and SHREIER, M.H. In Proc. 10th FEBS Meeting (1975), pp. 309-323, Elsevier Scientific Publishing Co., Amsterdam.

21. FREIESTEIN, C. and BLOBEL, G. Proc. Natl. Acad. Sci. U.S.A. (1975), 72, 3392-3396.

22. SHAFRITZ, D.A. and ANDERSON, W.F. J. Biol. Chem. (1970), 245, 5553-5559.

SECTION III:

SYNTHESIS AND PROCESSING OF VIRAL PROTEINS

SYNTHESIS AND PROCESSING OF PICORNAVIRAL POLYPROTEIN

R.R. RUECKERT, T.J. MATTHEWS[+], O.M. KEW, M. PALLANSCH, C. McLEAN and D. OMILIANOWSKI
Biophysics Laboratory, University of Wisconsin,

Madison, Wisconsin 53706, U.S.A.

INTRODUCTION

Under properly defined conditions picornaviruses interrupt host RNA and protein synthesis (1) and subvert the cellular machinery to production of viral protein and RNA. By feeding radiolabeled amino acids to virus-infected cells after cessation of host-protein synthesis, viral protein can be selectively labeled. In a pioneering study, which introduced the now widely used SDS-polyacrylamide gel electrophoresis technique, Summers et al. (2) identified some 14 different virus-specified polypeptides in extracts of poliovirus infected HeLa cells. The net mass of these polypeptides exceeded two-fold or more the known coding capacity of the viral genome. The explanation was later traced to cleavages which took place during and after synthesis of the viral protein.

A. Evidence that Picornaviral RNAs Contain a Single Initiation Site

In 1968, Jacobson and Baltimore (3) reported the accumulation, in poliovirus-infected cells treated with amino acid analogs, of a giant protein (NCVP 00) with an electrophoretic mobility in SDS-polyacrylamide gels just slightly less than that of myosin (MW 200,000). To explain this result they proposed that translation of polioviral RNA is initiated at a unique site near the 5'-end of the genome and that the ribosome then translated the RNA continuously to the other end. With this insight they proposed that one function of protein processing must be to separate the several gene functions

+ Current address: Tumor Virology Laboratory, Duke University Medical Center, Durham, N.C. 27710, U.S.A.

fused within the polyprotein; to rationalize this unusual mode of genetic expression they proposed that eukaryotic cells, unlike prokaryotic cells, may be unable to read "internal" initiation codons. This hypothesis has been exceedingly successful in explaining the function of segmentation in RNA viruses; it has further helped understand why viruses such as RNA tumor viruses with "silent" internal initiation sites elaborate special mRNAs for expressing envelope and transforming proteins. However recent work from some laboratories (see below and chapter 11) suggests that the "one cistron" model of polioviral translation needs careful re-examination.

B. Pactamycin as a Tool for Mapping Translational Order

Studies on effects of the drug pactamycin have, on the one hand, provided strong support for the idea that the picornaviral genome has a single initiation site and, on the other hand, afforded a method for deducing the order in which the polypeptides are encoded in the viral messenger RNA (see chapter 8). Using this tool the cistrons of poliovirus appeared to be ordered, in increasing distance from the initiation site: NCVP 1a, NCVP X and NCVP 1b (4). Confidence in the validity of the pactamycin technique, of the underlying assumption of a single initiation site and the specificity of the drug's action, mounted with the finding that the technique gave a consistent map location not only for the precursor polypeptides but also for their cleavage products (5-6). Moreover, the order of the capsid chains was consistent with results obtained from progressive labeling studies designed to order the chains in the absence of the drug (5) and with cyanogen bromide fragmentation studies (7), indicating that the VP4 chain lies on the amino-terminal end of the VPO precursor of ME virus.

That the gene order and cleavage pattern of a rhinovirus is similar to that of poliovirus and cardiovirus was first shown by McLean and Rueckert (8) who found that rhinoviruses 1A and 2, like poliovirus and EMC virus, redirect host-protein synthesis to that of the virus and produce a cleavage pattern homologous to that of EMC virus. These results were confirmed by Butterworth (9) who carried out a useful comparative study on all three of these viruses (Figure 1).

Stoichiometric analysis of chains in EMC virus-infected HeLa cells showed that, after correction for loss due to cleavage, the A, F and C regions are translated with equal frequency (10). This result, obtained on normally functioning virus in the absence of viral inhibitors or analogs, not only supported the idea of a single initiation site but also implied the operation of only one termination site, thus

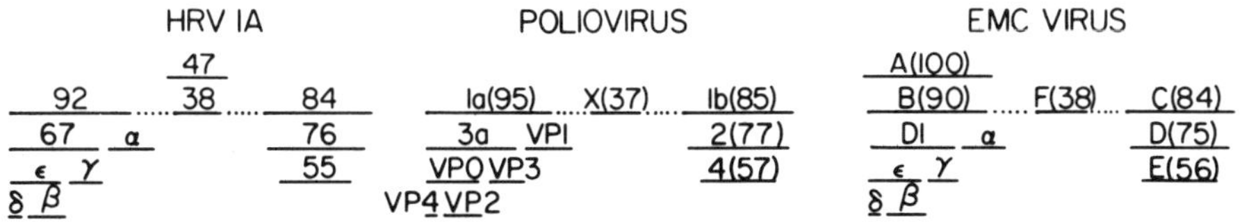

HRV IA | POLIOVIRUS | EMC VIRUS

47
92 ..... 38 ..... 84 | Ia(95) ..... X(37) ..... Ib(85) | A(100)
67 α 76 | 3a VPI 2(77) | B(90) ..... F(38) ..... C(84)
ε γ 55 | VP0 VP3 4(57) | DI α D(75)
δ β | VP4 VP2 | ε γ E(56)
δ β

Figure 1. Common features in the translational and processing maps of picornaviral protein: HRV-1A, human rhinovirus 1A; poliovirus; and EMC, encephalomyocarditis virus (after Butterworth, 9). Polypeptides with similar masses and pactamycin mapping positions are evident with all three viruses. The lateral position represents the relative location of the corresponding gene locus on the viral RNA (the 5'-end of the RNA is to the left). The vertical position represents precursor-product relationships or alternative cleavage forms. Line lengths are proportional to mass.

(initiate) P1-S-P2 (terminate)

where P1 and P2 are large precursor chains and S is a stable product which maps between them by the pactamycin method (see Figure 1).

C. Does Picornavirus RNA Contain a Second Termination Site?

In 1974 the existence of a second (internal) termination site at the end of the coat gene was proposed to account for a 2-3 fold "overproduction" of mengoviral coat protein (P1) relative to the S protein (6-11). To account for the observation that coat protein was overproduced only late in the infection cycle, the latter workers proposed that the internal termination signal is suppressed during the log phase of viral multiplication. McLean _et al_. (12) observed a similar overproduction of rhinoviral coat protein; they concluded however that asymmetric translation was an artifact due to undercounting the S protein and to selective degradation of the P2 proteins. The S protein is produced in two forms (p38 and p47); failure to score for both forms results in undercounting the number of translations. Neither of these possibilities was considered by the above workers and the case for a second termination site has been unresolved.

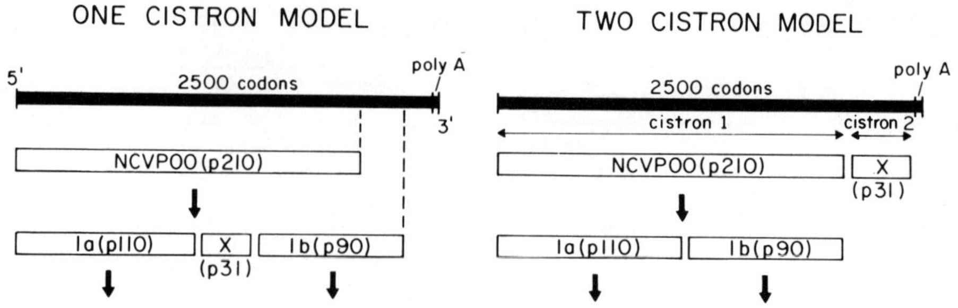

ONE CISTRON MODEL

TWO CISTRON MODEL

5' 2500 codons poly A 3'

NCVPOO(p210)

Ia(pII0) X (p31) Ib(p90)

2500 codons poly A

cistron 1 cistron 2

NCVPOO(p210) X (p31)

Ia(pII0) Ib(p90)

Figure 2. Two alternative models for translation of the picornaviral genome in virus-infected cells (14). The polyprotein (NCVP 00 , also called p210) is produced only under conditions designed to inhibit processing. Blocks indicate "primary" proteins which are named according to their apparent (electrophoretically determined) masses in kilodaltons. The translational order deduced for the one cistron model is based on the effect of pactamycin on relative labeling rates. The two cistron model was proposed to account for the reported absence of p31 sequences in p210 (13); hence the proposed cistron-2 might equally well lie nearest the 5'-end of the genome. Secondary processing products of p110 (coat protein) and p90 are not shown.

D. Are the Three Primary Products P1, S and P2 Located in One Translational Cistron or Two?

Abraham and Cooper (13) have reported that the tryptic peptides of the stable polioviral protein NCVP X (homologous to the stable rhinoviral protein 38) are not represented among the tryptic peptides of the 210,000 dalton polyprotein (NCVP 00). They correctly point out that the order deduced from the pactamycin studies is founded on the validity of the single cistron NCVP 00 (210 kD) and that of the primary products, NCVP 1a + X + NCVP 1b (231 kD). Cooper (14) suggested that the data were equally consistent with a two cistron genome (Figure 2). A report of two different initiation products from an _in vitro_ translation system programmed with polioviral RNA (15) reinforced this idea. Cooper's conclusion is hard to avoid if one accepts the crucial result that S peptides are not represented in the polyprotein. However we report below studies on the proteins of human rhinovirus 1A which favor the conventional one cistron model.

I. PROTEINS OF HUMAN RHINOVIRUS 1A

Matthews has isolated, from lysates of rhinovirus-infected HeLa cells, twenty two polypeptides radiolabeled with (^{3}H)- or (^{14}C)-leucine (Figure 3, previous page). Each was fingerprinted by chromatographing tryptic digests from mixed pairs of differentially labeled proteins (see below). The results confirmed many details of the translational and cleavage map proposed previously (Figure 1). Thus protein 92 was found to contain the sequences of capsid proteins (67, α, β, γ, δ, ε) and is therefore homologous to polioviral precursor la. Similarly p47 contained the sequences of protein 38; the latter is homologous to polioviral protein X. In addition it was shown that p146 contained the tryptic sequences of both p92 and p47 (Figure 4). This result implies that both proteins are synthesized under control of the same initiation site and bolsters the conventional one cistron model (Figure 2) for human rhinovirus 1A and presumably therefore also for poliovirus.

In other experiments (Figure 5) Matthews and Omilianowski showed that two pairs of proteins (p24, p55) and (r39a, r39b)

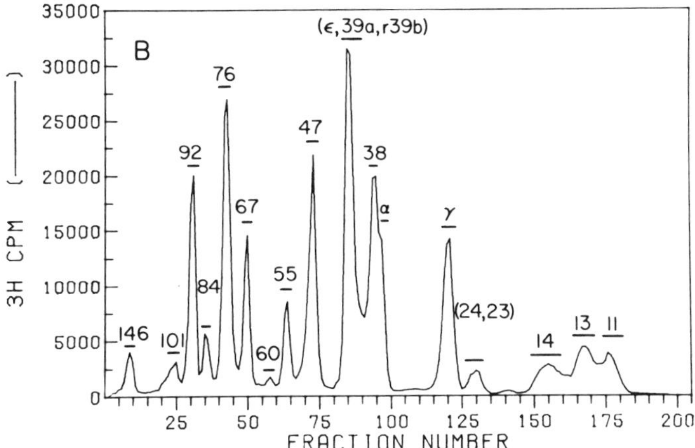

Figure 3. Radioactivity profile of rhinoviral 1A polypeptides on a preparative SDS-polyacrylamide gel. Bars indicate fractions pooled in preparing isolated proteins. In some cases further electrophoresis under altered conditions was necessary to resolve mixtures (ε, 39a, 39b), (38, α) and (24, 23).

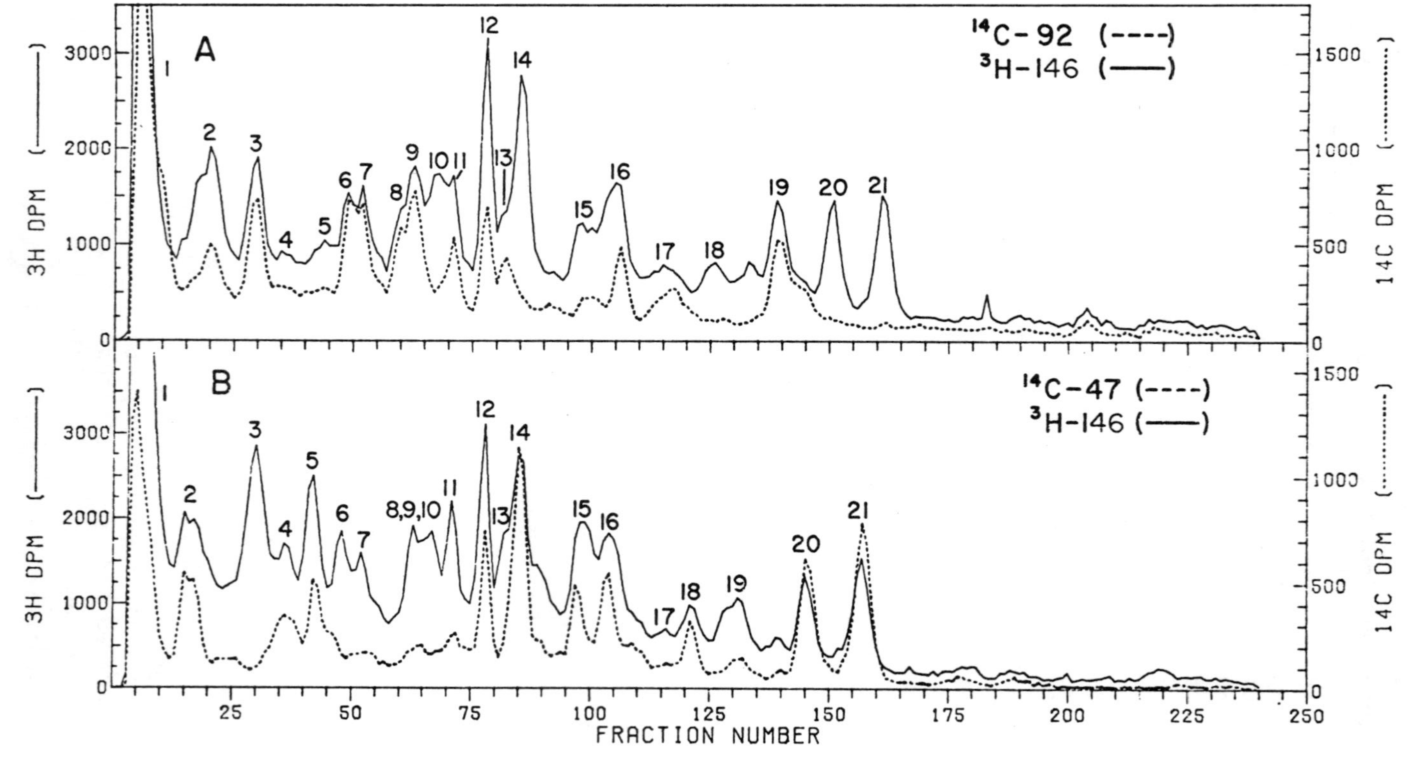

Figure 4. Identification of sequences from p92 and p47 in p146. About 10^5 cpm of electrophoretically repurified (^{3}H)-leucine p146 was digested with (A) 6.7 x 10^4 cpm (^{14}C)-leucine p92; or (B) 4.8 x 10^4 cpm (^{14}C)-leucine p47 and chromatographed.

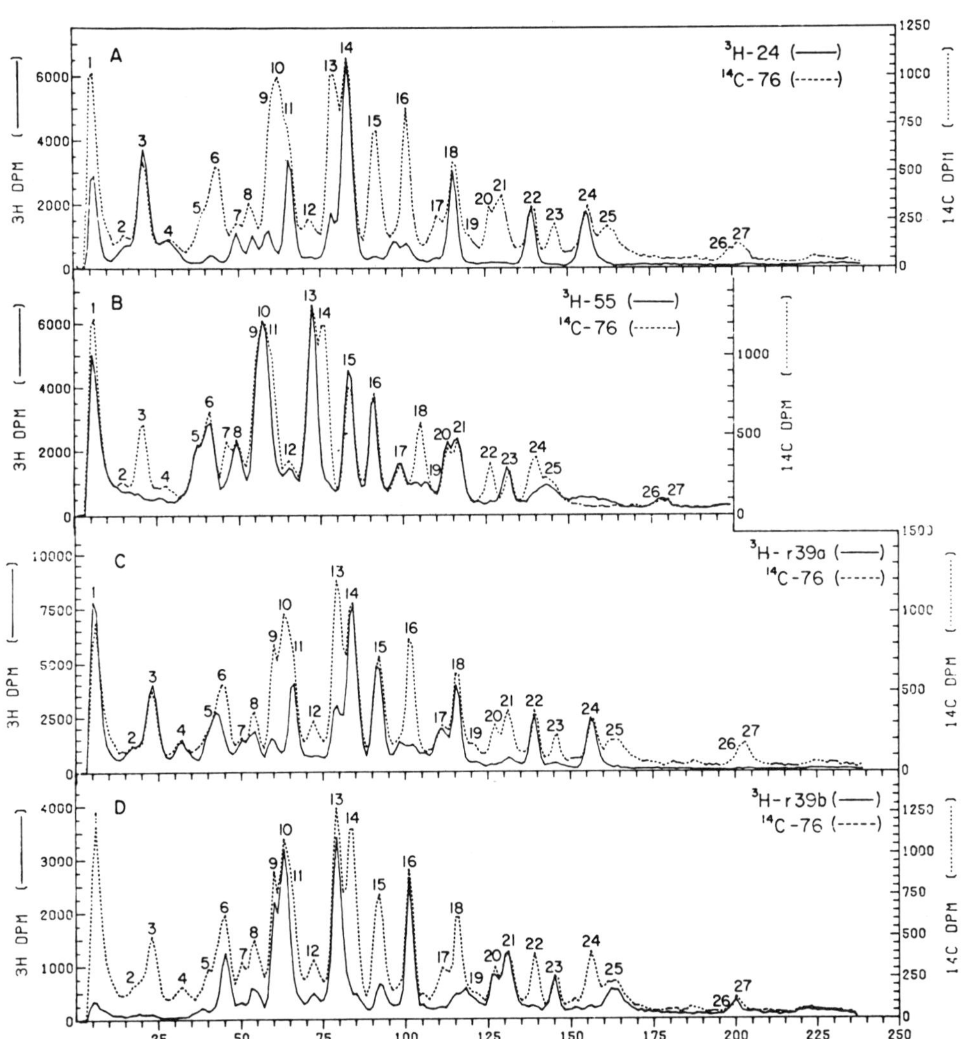

Figure 5. Evidence from tryptic patterns that p76 is cleaved to yield a (p24, p55) pair or a (r39a, r39b) pair). Comparisons were as described in Figure 4.

contain the sequences of p76. Moreover they showed with the aid of pactamycin that the sequences in p24 and r39a lie on the amino-terminal side of p76 (data not shown). These results confirm an earlier report (12) that p76 is cleaved in two different ways with the 76 ⟶ r39a + r39b pathway appearing to occur on ribosomes; this latter pathway is partially inhibited by pactamycin, an inhibitor of protein synthesis (12). Other results (Figure 6) showed that p14 is the cleavage complement of p76 and the p13 and p11, which contain the same sequences, are represented in none of the other three primary sequences (not shown).

The position of p13 on the pactamycin map (9) locates it near the midpoint of the polyprotein and argues against it being the product of a second short (30 kD or less) cistron whose product should map near the far left. Hence p13 is most likely located between p146 and p84 (Figure 6).

Assuming for the mRNA of rhinovirus 1A a mass of 2.6 million daltons this leaves some 25-30 kD, of the theoretical 270 kD protein-coding capacity, unaccounted for. With one significant

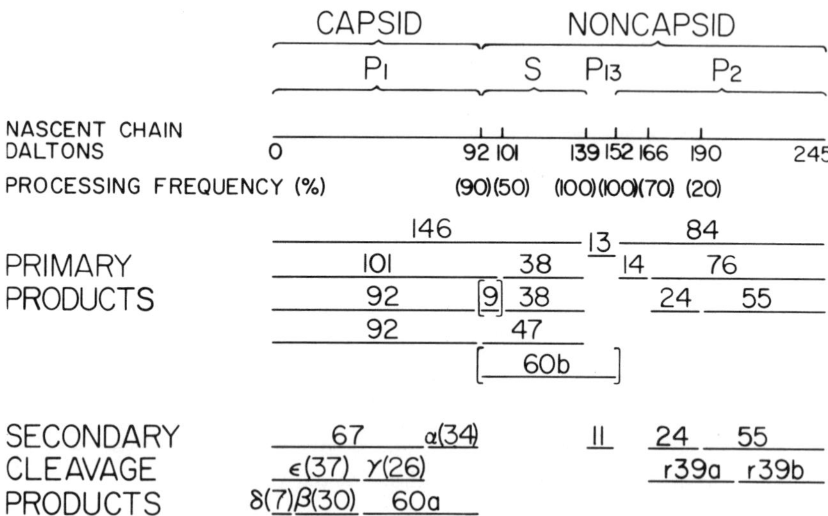

Figure 6. Protein processing map of human rhinovirus 1A. Identity of bracketed chains is tentative (Mathhews, Omilianowski and Rueckert, in preparation).

exception it is quite possible that all of the proteins encoded by the rhinovirus genome are now accounted for and that the discrepancy between theoretical and observed coding capacity is due to small errors in molecular weight of the RNA or proteins. On the other hand the existence of non-coding regions involved, for example, in replicase recognition at either end of the RNA molecule would not be surprising. In addition it is not yet possible to exclude the possibility of still undetected protein especially if it were the product of a cistron which is translated at a lower frequency than the major one. This returns us to an important omission in our picornaviral map, the position of the viral "capping" protein which is covalently bound to the 5'-end of the picornaviral RNAs (16-19) and which appears to be encoded by the virus (20).

II. THE PROCESSING MAP OF POLIOVIRAL POLYPROTEIN

The map of rhinovirus 1A raises the question as to which of the alternative cleavage pathways have functional significance and which are mere accidents. Evidence that the alternative cleavage modes are not mere pecularities of rhinovirus 1A, but are preserved in detail through evolution and therefore probably have functional significance,comes from comparative studies on the processing of polioviral protein.

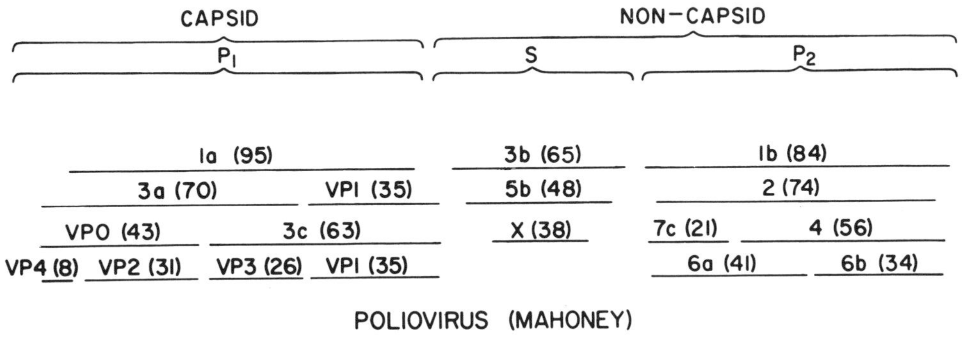

Figure 7. Proposed relationships among the major polypeptides produced in poliovirus-infected cells. The values enclosed in parentheses to the right of each polypeptide name refer to the apparent molecular weight of that polypeptide in the Mahoney strain. Not all poliovirus strains produce detectable amounts of polypeptides 3c, 6a and 6b (Kew and Rueckert, in preparation).

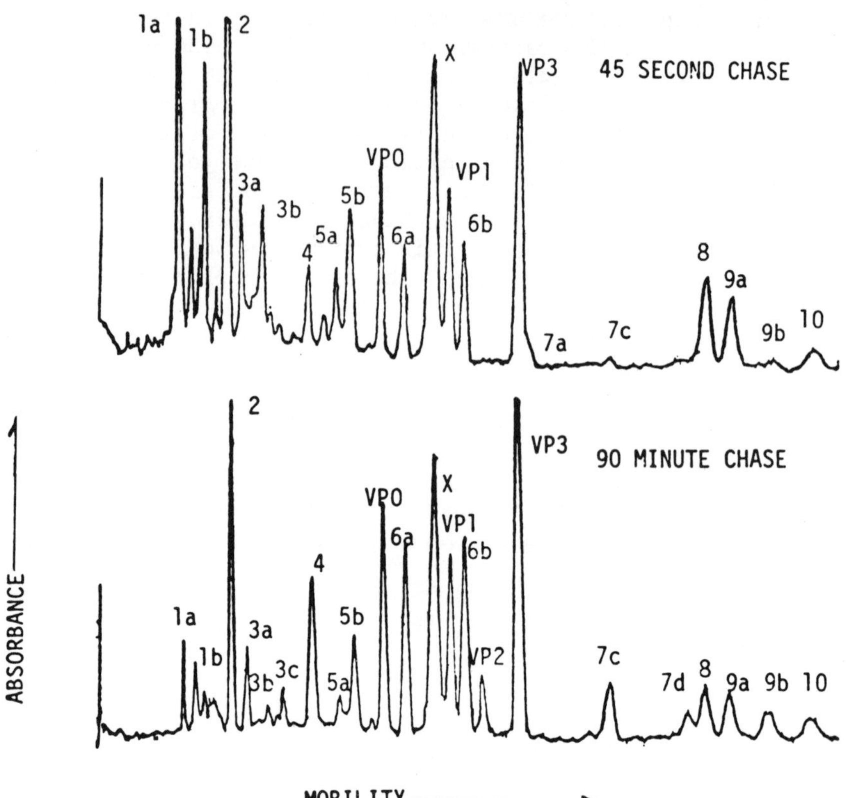

Figure 8. Densitometer tracing from autoradiograph of an electropherogram of poliovirus-infected HeLa cells. At 210 minutes post-infection the cells were pulsed 10 minutes with ^{35}S-methionine. The label was chased by adding excess unlabeled methionine and cells were lysed at the indicated chase periods and electrophoresed on an SDS-polyacrylamide slab gel.

Kew has compared 12 different strains of poliovirus, including neurovirulent and attenuated strains of all three serotypes; these strains display definite strain-dependent differences in cleavage pattern which provide useful insights into the processing of polioviral protein (Figure 7). For example some strains such as type 1 Mahoney and type 3 Saukett generate relatively large amounts of 6a and 6b, while others such as type 1, Brunhilde and type 3 FOX strains produce little if any. There is always a close correlation between the relative amounts of 7c and 4 and of 6a and 6b. For example if the amount of 4 is small, so is the amount of

7c; if the amount of 6b is large so is the amount of 6a (not shown). All strains of poliovirus also produce a protein 5b (MW 48,000) similar in size to that of rhinoviral protein 47. Tryptic analysis of Mahoney viral polypeptides has confirmed that 5b contains the sequences of X; so does 3b which is thus homologous to rhinoviral protein 60b. We have also confirmed that protein 2 contains the sequences of 4, 7c and 6a.

High resolution autoradiographic analysis of pulse-chase experiments on polyacrylamide gels suggest that 3b is a short-lived precursor of protein 5b, and possibly also of X (Figure 8); these patterns also show that 6a, 6b, 4, 7c, 7d and 9b are cleavage products of a precursor.

III. CONCLUSIONS

The data reviewed here are still compatible with the single initiation site-single termination site model for picornaviral translation. It is evident that the S-protein region of the polioviral polyprotein can be expressed in forms other than X, i.e., also to form 5b and 3b, and that the P2 proteins are cleaved to products not previously accounted for. Hence it is clear that until these facts are taken into account, apparent inequalities in family stoichiometry represent inadequate evidence for drawing conclusions about the number of translational initiation or termination sites operating in picornaviral RNA.

In searching for possible significance in the multiple processing modes of picornaviral protein we note the strong similarity between poliovirus and human rhinovirus processing. This argues against the idea that multiple processing is a mere accident, else why are the cleavage patterns so similar in two widely divergent viruses? On the other hand the processing pattern of cardioviruses (EMC, mengovirus) appears to be relatively simpler and the viruses multiply very well indeed. Yet here too it is likely that secondary processing modes emerge during the late stage of the growth cycle in mengo-infected L-cells (6-11). Analysis of these questions will undoubtedly lead to further useful insights into the role of protein processing in regulating viral functions such as nucleic acid synthesis, assembly, intracellular transport and cellular lysis.

REFERENCES

1. FRANKLIN, R. and BALTIMORE, D. Cold Spring Harbor Symp. Quant. Biol. (1962), 27, 175.

2. SUMMERS, D.F., MAIZEL, J.V. and DARNELL, J.E. Proc. Natl. Acad. Sci. U.S.A. (1965), 54, 505.

3. JACOBSON, M.F. and BALTIMORE, D. Proc. Natl. Acad. Sci. U.S.A. (1968), 61, 77.

4. TABER, R., REKOSH, D. and BALTIMORE, D. J. Virol. (1971), 8, 395.

5. BUTTERWORTH, B.E. and RUECKERT, R.R. J. Virol. (1972), 9, 823.

6. PAUCHA, E., SEEHAFER, J. and COLTER, J.S. Virology (1974), 61, 315.

7. STOLTZFUS, M. Ph. D. Thesis. (1971), University of Wisconsin, Madison.

8. McLEAN, C. and RUECKERT, R.R. J. Virol. (1973), 11, 341.

9. BUTTERWORTH, B.E. Virology (1973), 56, 439.

10. BUTTERWORTH, B.E., HALL, L., STOLTZFUS, C.M. and RUECKERT, R.R. Proc. Natl. Acad. Sci. U.S.A. (1971), 68, 3083.

11. LUCAS-LENARD, J. J. Virol. (1975), 14, 261.

12. McLEAN, C., MATTHEWS, T. and RUECKERT, R.R. J. Virol. (1976), 19, 903.

13. ABRAHAM, G. and COOPER, P.D. J. Gen. Virol. (1975), 29, 215.

14. COOPER, P.D. Genetics of picornaviruses. In Comprehensive Virology, Vol. 9 (1977), Fraenkel-Conrat, H. and Wagner, R. eds., pp.133-207, Plenum Press, New York.

15. CELMA, M. and EHRENFELD, E. J. Mol. Biol. (1975), 98, 761.

16. LEE, Y.F., NOMOTO, A., DETJEN, B.M. and WIMMER, E. Proc. Natl. Acad. Sci. U.S.A. (1977), 74, 59.

17. PETTERSSON, R.F., FLANEGAN, J.B., ROSE, J.K. and BALTIMORE, D. Nature (London) (1977), 268, 270.

18. HRUBY, D.E. and ROBERTS, W.K. J. Virol. (1978), 25, 413.

19. SANGAR, D.V., ROWLANDS, D.J., HARRIS, T.J.R. and BROWN, F. Nature (London) (1977), 268, 648.

20. GOLINI, F., NOMOTO, A. and WIMMER, E. Proc. Natl. Acad. Sci. U.S.A. (1978), in press.

VIRUS-DIRECTED PROTEIN SYNTHESIS

JEAN M. LUCAS-LENARD

Biochemistry and Biophysics Section, Biological Sciences Group, University of Connecticut,
Storrs, Connecticut 06268, U.S.A.

INTRODUCTION

Earlier in this conference (1) we discussed the fact that after infection of host cells by picornaviruses, there is a decrease in the rate of cellular protein synthesis. Whatever the mechanism by which these viruses inhibit cellular synthetic processes, the virus itself is unaffected by it. Within a short time after infection, viral messenger RNA molecules can be detected in polysomes and viral proteins are synthesized.

From the available evidence, it appears that viral protein synthesis in general proceeds in the same manner as cellular protein synthesis. The amino terminal amino acid of nascent viral peptides is methionine (2), donated presumably by unformylated met-$tRNA_{fmet}$ (3, 4), a specific initiatior molecule.

In poliovirus infected cells and presumably in cells infected by other picornaviruses, the polysome complex formed with viral mRNA is large. The average number of ribosomes in the largest polysomes (380S) is about 35 per viral mRNA (5), due at least partially to the large size of the viral mRNA (2.5×10^6 daltons (6)). The transit time for a ribosome to traverse a molecule of picornavirus mRNA is approximately 10 min (7, 8) early in infection. The rate of chain elongation is about 205 to 225 amino acids per min. This rate is considerably slower than the rate of synthesis of the α and β chains of hemoglobin, namely 600 and 400 amino acids per min (9), respectively.

Late in infection the average size of the polysome decreases (200S), and the number of ribosomes associated with the structures drops to about 20 per message (5). By this time after infection the rate of protein synthesis in general is decreased considerably.

Approximately 25% of the total polysomes in poliovirus-infected and uninfected HeLa cells are membrane-bound (10). The membrane-bound polysomes are about five times more active in translation per unit mass than free polysomes, as determined by incorporation of amino acids in cell-free extracts (11). The exact role of the membrane in this process has not yet been defined.

For a summary of the various aspects of the replicative cycle and structure of picornaviruses, the reader is directed to the review articles by Levintow (12) and Rueckert (13).

I. POLYPEPTIDE SYNTHESIS AND CLEAVAGE

If one examines the pattern of proteins synthesized in cells infected for three to four hours with a picornavirus, the only proteins synthesized are viral, for by this time the synthesis of most host proteins has ceased. In the earliest studies carried out by Summers *et al*. (14) in poliovirus-infected HeLa cells, about 14 different virus-specified polypeptides were identified by SDS-polyacrylamide gel electrophoresis. The net mass of these viral polypeptides, however, exceeded the coding capacity of the viral genome by a factor of 2. This finding was incompatible with what was then known about the translation of the small RNA bacteriophages such as Qβ and R17. In these cases, each of the three cistrons was found to have its own initiation and termination signals (15).

The observation that label, which had been incorporated into a specific polypeptide during a short pulse with radioactive amino acids, could be chased into different size products after incubation with unlabeled amino acids, suggested that some of the products were generated by the cleavage of others (16, 17, 18). These and related studies led to the concept that poliovirus proteins arise by the cleavage of larger primary products. This pattern of protein synthesis was later confirmed in cells infected with enteroviruses (18), cardioviruses (19, 20, 21, 22, 23) human rhinoviruses 1A and 2 (24, 25), and foot-and-mouth disease virus (26, 27). An example of post-translational processing in mengovirus-infected L-cells is shown in Figure 1. Where possible in this review, mengovirus will be used for illustrating viral protein synthesis, since this is the virus which is being studied in my laboratory and with which I am most familiar. The nomenclature of Butterworth *et al*. (19) for encephalomyocarditis (EMC) viral proteins will be used throughout this discussion.

The six panels show the patterns observed after a 12 min pulse of label followed by chases of 10, 20, 40 and 80 min. The

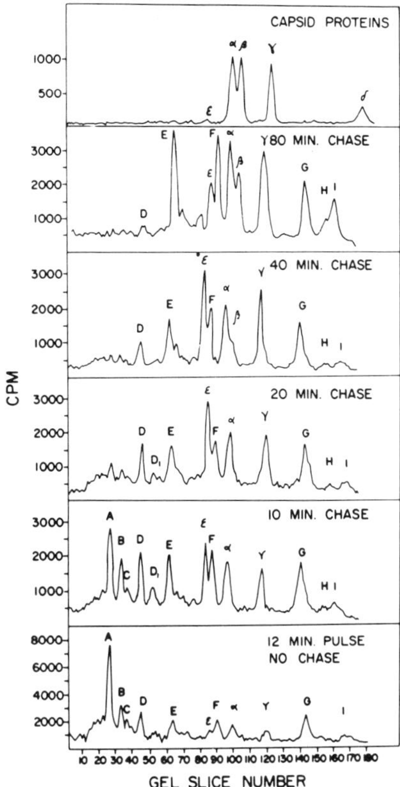

Figure 1. Pattern of labeling of mengovirus proteins during pulse-chase experiments. At 5 h after infection L-cell monolayer cultures were pulse-labeled for 12 min with a ^{3}(H)-amino acid mixture and chased with unlabeled amino acids for varying time periods as indicated in the figure. The samples were analyzed by SDS-polyacrylamide gel electrophoresis. See Lucas-Lenard (22) for details.

top panel demonstrates the capsid proteins of mengovirus, called α, β, γ and δ. As can be seen, there is a considerable amount of label in the protein designated A, and less in those designated B, C,D,E, F and G after a 12 min pulse period. After a 10 min chase period in medium containing unlabeled amino acids, there is a decrease in the amount of radioactivity in the A polypeptide and an increase in polypeptides ε, α and γ. After chases of 20, 40 and 80 min the radioactivity gradually shifts from proteins of higher molecular weight to proteins of lower molecular weight. By 80 min none of the large proteins are detectable in the polyacrylamide gel electrophoresis patterns and the amount of polypeptide ε, which has

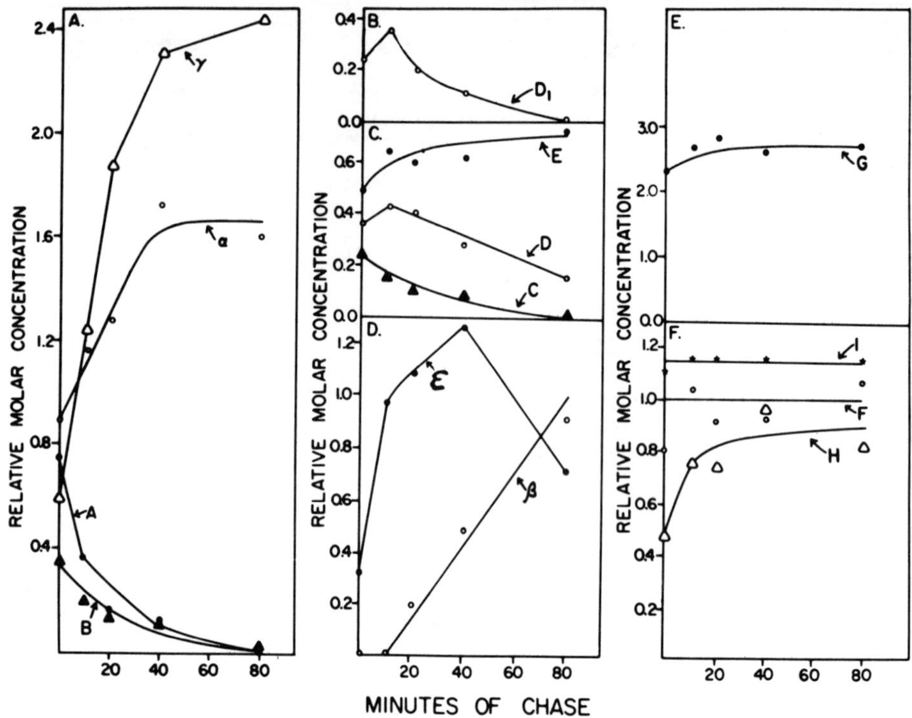

Figure 2. Kinetics of cleavage of mengovirus precursor polypeptides and of formation of stable viral proteins. The molar concentration of each component was calculated as described in the text. The data used for these calculations were taken from Figure 1. From Lucas-Lenard (22).

the slowest rate of cleavage, is also decreased.

The kinetic behavior of the various mengovirus proteins can be more clearly observed if the data in Figure 1 are plotted as shown in Figure 2. Here the relative molar concentration of each viral protein peak is expressed as a function of time. Operationally, the counts per minute in each peak from Figure 1 is first expressed as percent of total viral counts per minute. This number is then divided by the molecular weight of each viral peak. Finally, the relative molar concentration is standardized to a particular protein which presumably is stable throughout the course of the pulse-chase period. In the example shown in Figure 2, the data have been standardized to the F protein, a presumed primary cleavage product. The curves in Figure 2 demonstrate the precursor nature of polypeptides A, B, C, D, D1 and ε, and the flow of label into components ε, α, β, γ and E.

The half life of at least some of the precursors can be estimated by plotting the logarithm of the percent composition of the precursor in question against time of chase period. If the decay of precursors A and C is thus plotted, as shown in Figure 3, a straight line is obtained, suggesting that each is cleaved with first order kinetics. The half life of the A protein was thus found to be 7 min and that of the C polypeptide, 9.5 min.

On the basis of pulse-chase experiments such as shown in Figures 1 and 2, it has been possible to determine the number of polypeptides specified by the particular virus under study and to estimate their size, stability and relative molar concentration in infected cells. As shown in Table 1, in mengovirus infected L-cells

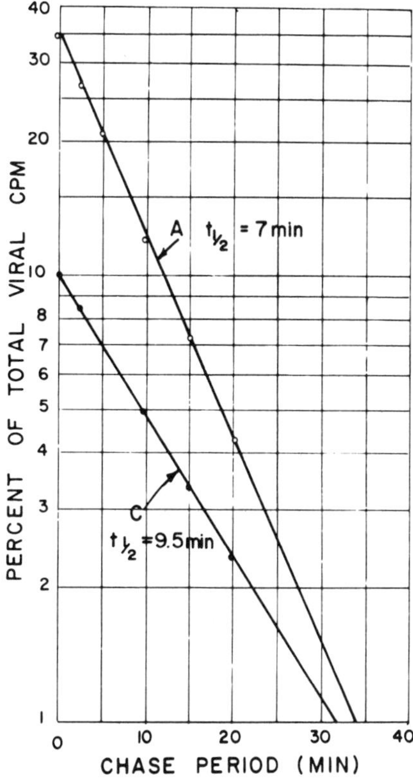

Figure 3. Rate of cleavage of mengovirus A and C polypeptides. Five h after infection L-cells were pulsed for 7 min with a mixture of ^{3}(H)-amino acids and chased as indicated in the figure. See Lucas-Lenard (22) for details.

Table 1. Size and stability of viral polypeptides and molar concentration of end-product polypeptides

Polypeptide	Apparent mol wt $\times 10^{-3}$	Stability	% of total[a] viral counts/min	Relative molar concn.
A	110 ± 4	u		
B	99 ± 5	u		
C	95 ± 6	u		
D	80 ± 4	u	3	0.14
D_1	70 ± 2	u	3	0.14
E	60 ± 3	s	12	0.64
ε	41 ± 2	u	7	0.61
F	39 ± 1	s	12	1.00
α	34 ± 1	s	14	1.34
β	31 ± 1	s	7	0.79
γ	24 ± 1	s	15	2.07
G	17 ± 1	s	12	2.39
H	12 ± 1	s	4	1.14
I	11 ± 1	s	9	2.53
δ	< 10	s	?	?

a. This represents the total viral counts per minute after a 12-min pulse and an 80-min chase (see Fig. 1). From Lucas-Lenard (22).

about 6 unstable and 9 stable viral polypeptides can be detected. "Stable" polypeptides are defined as those that do not show a decrease in counts during a chase with unlabeled amino acids, and "unstable" polypeptides are those that do lose counts during a chase period. The total molecular weight of the stable polypeptides is 238,000 daltons, which approximates the coding capacity of the virus, based on a molecular weight of 2.4×10^6 daltons for its genome (28). The molecular weights of the various polypeptides range from approximately 110,000 to less than 10,000 daltons. The relative molar concentration of each viral polypeptide as defined earlier in this review is also given in Table 1.

II. PRECURSOR-PRODUCT RELATIONSHIPS

Although the rapid rate of disappearance of label from certain polypeptides resembled the rate of appearance of others, the exact relationship among the precursors and products could not be unequivocally determined by this method alone. Cyanogen bromide fragmentation (19) and two-dimensional tryptic peptide mapping (29) proved to be very useful in establishing precursor-product relationships. Through a combination of kinetic studies and cyanogen bromide fragmentation and tryptic mapping techniques, several relationships were clearly established. First it was demonstrated that polypeptide A is the precursor of all stable virus capsid chains (19, 29). It was also shown that peptides C, D and E are related, since their cyanogen bromide fragmentation patterns or tryptic fingerprints were similar (19, 29). From kinetic studies and such mapping experiments the cleavage sequence C → D → E was established (19). Kinetic studies suggested that polypeptide F is neither a precursor nor a product and was thus considered to be a primary product (19). The fact that the cyanogen bromide fragmentation profile of polypeptide F showed little correlation with the profiles of polypeptides A or C provided further indirect evidence that it is a primary product and not derived from, or related to, A or C (19).

Further comparative studies on the cyanogen bromide fragmentation patterns of the ε chain showed similarities to patterns from the β and δ chains (19), thus lending support to kinetic studies which suggested that the β and δ polypeptides are produced by the cleavage of the ε chain. Results of both cyanogen bromide fragmentation and tryptic mapping techniques indicated that low molecular weight proteins, G, H and I are a mixture of different polypeptides of similar molecular weight. The pattern of polypeptide G differed greatly from that of polypeptide A or C and thus was not considered to be a cleavage product of A or C, but rather to be another primary product like polypeptide F. Studies on the precursor-product relationships of poliovirus proteins led to similar groupings of related polypeptides (30, 31).

The realization that picornaviral proteins arise from the cleavage of large precursors led Jacobson and Baltimore (17) to examine the effect of incorporating the amino acid analogue, p-fluorophenylalanine, into poliovirus proteins. In their studies they reasoned that the analogue might affect the configuration of the precursors, preventing recognition by the cleavage enzymes. The cleavage of NCVP-1, precursor of the capsid proteins and equivalent to polypeptide A in the cardiovirus terminology, was indeed prevented in the presence of the analogue. They then tested the effect on cleavage of using four amino acid analogues, p-fluorophenylalanine, ethionine, canavanine, and azetidine-2-carboxylic acid. Under these conditions at least two precursors larger than NCVP-1 accumulated (17, 30). One of these, NCVP-00, had an estimated molecular weight of greater than 200,000 daltons and was speculated to contain all the information encoded within the viral genome.

Similar results were obtained with mengovirus. In the presence of p-fluorophenylalanine, canavanine and azetidine-2-carboxylic acid, two precursors larger than polypeptide A were detected (22). If these analogues plus ethionine were present, then a small amount of a polypeptide equivalent in size to poliovirus NCVP-00 was observed (23). Figure 4 shows the pattern of labeling obtained from mengovirus-infected L-cells (upper panel) and poliovirus-infected HeLa cells (lower panel) in the presence of the analogues.

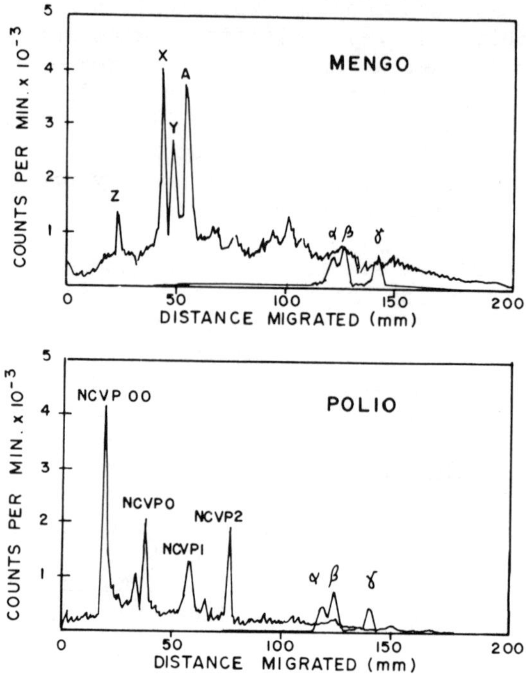

Figure 4. Profiles obtained from electrophoresis in 5% SDS-polyacrylamide gels of lysates of L and HeLa cells pulse labeled with ^{3}H-amino acids in the presence of the amino acid analogues mentioned in the text. The positions of mengovirus capsid proteins are also shown. Upper panel: mengovirus infected L-cells pulse labeled for 30 min at 6 hr post infection. Lower panel: poliovirus infected HeLa cells pulse labeled for 30 min at 3 hr post infection. Modified from Paucha _et al_. (23).

The finding of a viral polypeptide equivalent in size to the coding capacity of the viral genome suggested that all of the viral proteins are generated by proteolytic cleavage of a single polypeptide. The fact that no polypeptide greater than the capsid precursor (NCVP-1 or A) is seen in the absence of amino acid analogues suggests that cleavage takes place during growth of the polypeptide on ribosomes. To explain this unusual processing mechanism, it was hypothesized (17) that the polycistronic mRNA of poliovirus and other picornaviruses may lack "internal" initiation codons such as those found in the polycistronic mRNA of bacteria and their phages. To compensate for the lack of internal initiation of polypeptide synthesis, a system has evolved in which a giant precursor polyprotein is synthesized which is then cleaved, probably by a combination of cellular and viral enzymes, to give the separate functional proteins encoded within the polycistronic viral mRNA (see chapter 8).

III. GENE ORDER OF PICORNAVIRUS PROTEINS

Based on the hypothesis that there is only one site of initiation of protein synthesis, an attempt (32, 33) was made to map the poliovirus genome using the polypeptide chain initiation inhibitor, pactamycin (34). If this inhibitor is added to virus infected cells followed immediately or within a few minutes by a pulse of labeled amino acids, those proteins that have already been initiated are completed, but no new chains are initiated. Since proteins are synthesized from the amino to the carboxyl terminus (35) of the polypeptide chain in the 5' to 3' direction of the mRNA (36), this would mean that the carboxyl end of the protein (corresponding to the 3' end of the mRNA) would have the highest radioactivity and the amino end (corresponding to the 5' end of the mRNA) would have the least radioactivity.(See Figure 5 for a schematic diagram of the process). The fraction of total radioactivity incorporated into a particular polypeptide during the run off period in the presence of pactamycin relative to the fraction incorporated into the same polypeptide in the absence of the drug is a measure of the relative map distance of that protein along the mRNA. In this way the poliovirus gene order was found to be (5' → 3'), NCVP-1, NCVP-X, NCVP-2 (32, 33).

Figure 6 shows the effect of pactamycin on the distribution of radioactivity incorporated into EMC-viral proteins. As can be seen, the radioactivity in polypeptide A has decreased in comparison to the control, whereas the radioactivity in polypeptide C, another primary

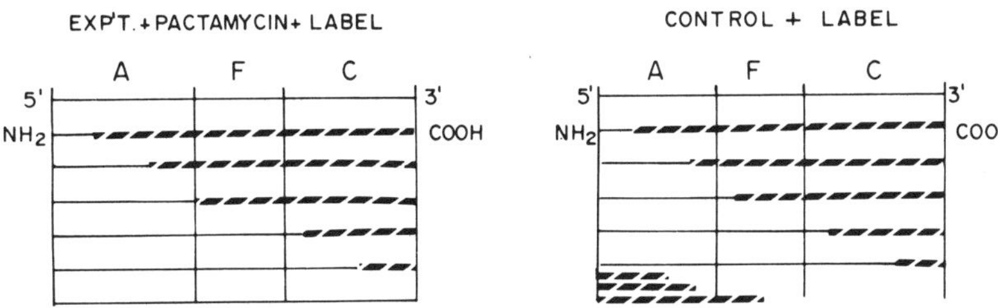

Figure 5. Schematic diagram of labeling of polypeptides in the presence and absence of pactamycin. The straight line represents unlabeled polypeptide chain. The hatched lines represent polypeptide chains labeled with radioactive amino acids.

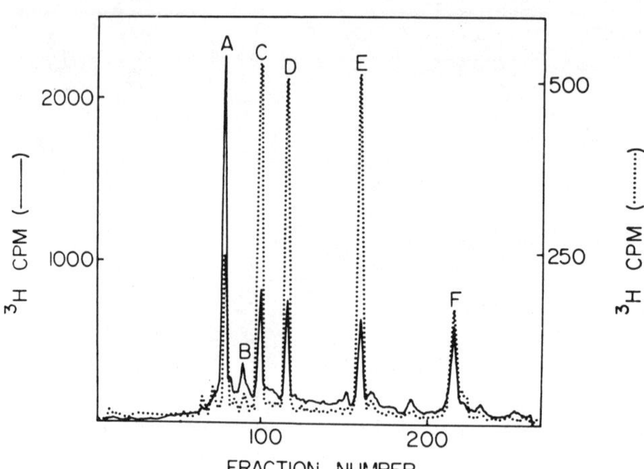

Figure 6. Effect of pactamycin on the distribution of radioactivity incorporated into the primary products of EMC-virus directed protein synthesis. At 3 hr 49 min post infection, an infected cell suspension was exposed to 10^{-7}M pactamycin. Six min thereafter, samples were removed, pulse labeled with a ^{3}(H)-amino acid mixture and electrophoresed on SDS-polyacrylamide gels (see ref. 37). Control, 6 min pulse (solid line); pactamycin, 6 min delay, 6 min pulse (dotted line). From Butterworth and Rueckert (37).

product, has increased relative to that of the control. The radioactivity of polypeptide F has remained essentially the same. These results indicate that the order of the primary products on the EMC-virus genome is 5' → 3' A-F-C (37).

The order of the capsid proteins within the viral genome can also be established in essentially the same manner, except that the addition of the radioactive label is followed by a chase period with unlabeled amino acids. As shown in Figure 7, the gene sequence obtained for the stable viral polypeptides is 5'→ 3', δ-β-γ-α-G-I-F-H-E. The precursor ε is still present and maps to the left of γ.

In this way the gene order of the capsid proteins of poliovirus (8), EMC-virus (37), rhinovirus 1A (24) and mengovirus (23) was determined. In keeping with our use of mengovirus as a model picornavirus, the gene order of mengovirus polypeptides as determined by Paucha et al. (23) is shown in Figure 8, including the pactamycin to control ratios. The gene sequence of mengovirus polypeptides closely resembles that of EMC-virus (37), the only difference being the inversion of the positions of polypeptides H and I (23).

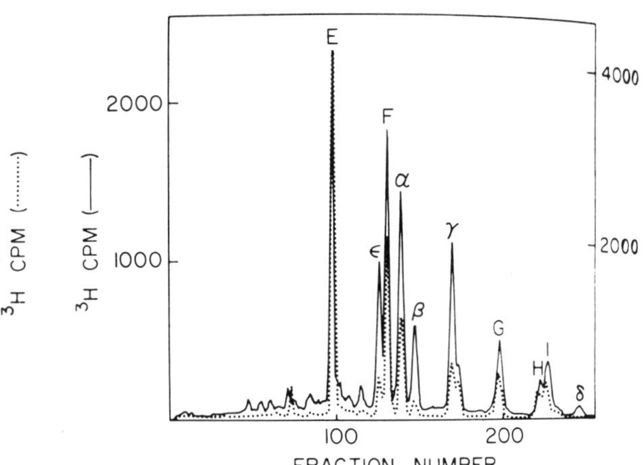

Figure 7. Effect of pactamycin on the distribution of radioactivity incorporated into the stable EMC virus polypeptides. At 3 hr 39 min post infection, an infected cell suspension was exposed to 10^{-7}M pactamycin and 80 μC of a ^{3}(H) amino acid mixture per ml for 20 min. Following the incubation period, the cells were chased for 80 min in medium free of pactamycin and isotope. Control, 20 min pulse, 80 min chase (solid line); pactamycin, 20 min pulse, 80 min chase (dotted line). From Butterworth and Rueckert (37). See ref. 37 for details.

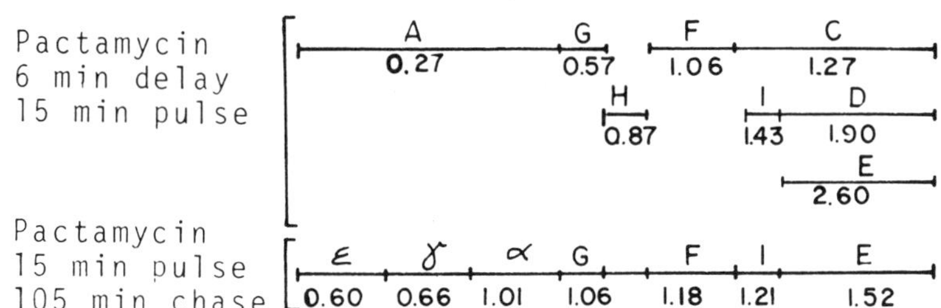

Figure 8. Gene sequence of mengovirus polypeptides as determined by the pactamycin mapping technique. The numerical values shown in the figure are pactamycin: control ratios, calculated as outlined in the text. Pactamycin concentration was 5 x 10^{-7}M. (From Paucha et al. (23)).

The order of the capsid chains was also determined for polio and EMC virus using progressive labeling studies in the absence of pactamycin (8, 38). In each case the gene order was the same as that determined by the pactamycin mapping technique, thereby confirming the validity of the pactamycin technique. The fact that a linear gene order was obtained by the pactamycin technique provides strong support for the concept of a single site of initiation of protein synthesis.

The determination of the precursor-product relationships by cyanogen bromide fragmentation and tryptic peptide mapping (19, 29) and the elucidation of the gene order of several picornavirus genomes have allowed the construction of a cleavage scheme for several picornavirus genomes. That for mengovirus constructed by Paucha et al. (23) is shown in Figure 9.

In summary, during the process of translation of a picornaviral mRNA, a ribosome attaches to the initiation site on the mRNA and translation begins. During elongation of the polyprotein, cleavage takes place, resulting in the formation of three large primary polypeptides, A, F and C, probably two smaller ones, G and H. Polypeptide C, corresponding to the 3' end of the viral RNA, undergoes further processing to produce the unstable non-capsid polypeptide D, and stable polypeptide I. Polypeptide D is further processed to give the stable polypeptide E. Product A, which corresponds to the 5' end of the viral RNA, is cleaved to unstable polypeptide B, which is then cleaved to produce unstable polypeptide D2. Capsid proteins ε, γ and α are eventually produced from the cleavage of D2 and B. Polypeptide ε is the last to be cleaved and results in the formation of stable capsid proteins δ and β .

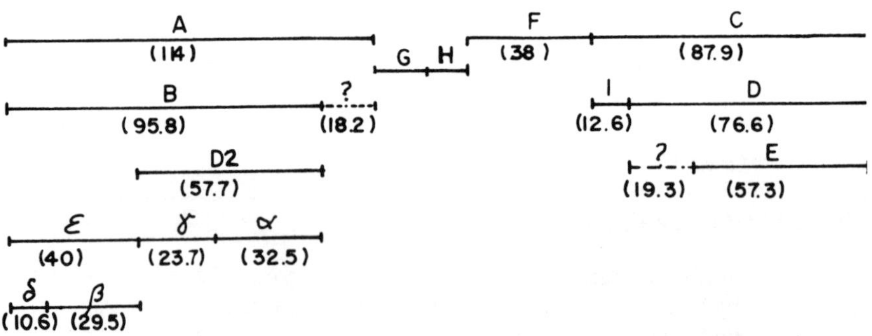

Figure 9. Proposed scheme for cleavage of mengovirus polypeptides. (From Paucha et al. (23)).

IV. MOLAR RATIOS OF VIRAL POLYPEPTIDES

If the entire picornavirus genome is translated into a single giant polyprotein which is cleaved to form the easily detectable primary proteins, A, F and C, then these proteins and their cleavage products should appear in a molar ratio of 1:1:1. If one calculates the molar ratios of the primary products as mentioned earlier, the results shown in Table 2 are obtained for mengovirus (22, 23), EMC virus (25) human rhinovirus 1A (25) and poliovirus (39). The A:F:C ratio for EMC virus (25) was found to be close to the theoretical value of 1:1:1. That for HRV-1A was also close to the theoretical value, 0.85:1.00:0.44, though the value for C was low. In contrast, the ratios for mengovirus (determined in two independent laboratories) and poliovirus deviated from the expected values. The capsid proteins appeared to be overproduced nearly by a factor of 2.

Table 2. Relative Molar Ratios of Viral Polypeptides

Viral polypeptide	Relative molar ratio[a]				
	Mengo[d]	Mengo[e]	EMC[f]	HRV-1A[f]	Polio[g]
A(total)[b]	1.60	1.93	1.03	0.85	1.96
F	1.00	1.00	1.00	1.00	1.00
C(total)[c]	0.78	0.86	0.76	0.44	0.70

a. All of these results are from pulse-chase experiments.

b. A(total) = $[\varepsilon + \beta + \alpha + \gamma / 3]$

c. C(total) = C + D + E

d. Obtained from Lucas-Lenard (22)

e. Obtained from Paucha et al (23)

f. Obtained from Butterworth (25)

g. Obtained from Paucha and Colter (38)

Paucha and Colter (39) also observed that from early to late log phase of virus production, there was a progressive increase in the ratio of capsid to non-capsid polypeptides synthesized both in mengovirus and poliovirus infected cells (see Figure 10). They suggested that the synthesis of polio and mengovirus proteins may be under some sort of translational control.

Several possible explanations for these unusual molar ratios have been proposed. (1) There is more than one site of initiation of protein synthesis, one at the beginning of the gene coding for polypeptide A and one internally. Steric hindrance at the internal initiation site might affect the efficiency with which it is initiated, resulting in an abnormal A:F:C ratio . From our

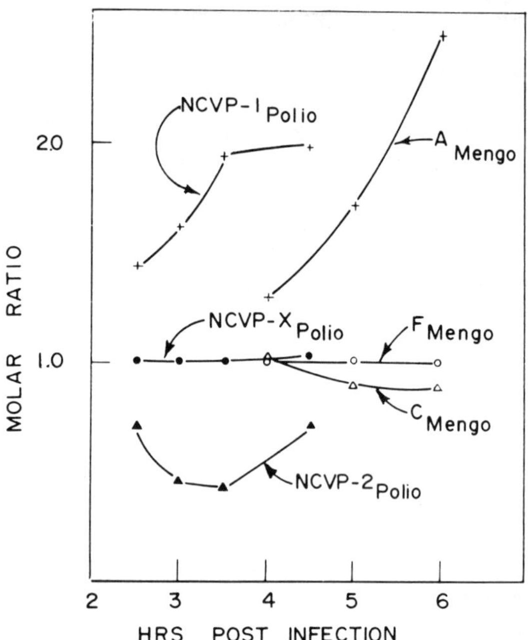

Figure 10. Relative molar ratios of mengo- and polio-virus specific polypeptides at various times after infection. At the time points indicated in the figure, infected cells were pulse-labeled with a mixture of ^{3}(H)-amino acids for 15 min and then chased for 105 min with medium containing unlabeled amino acids. The results are a modified version of those reported by Paucha and Colter (39).

discussion of picornavirus protein synthesis, this possibility is unlikely. One would not expect to see a giant polyprotein containing the entire informational content of the viral genome in the presence of amino acid analogues if there were more than one site of initiation of protein synthesis. Also, the pactamycin mapping data would not have indicated a linear map if there were internal initiation sites. On the other hand, this possibility cannot be set aside completely until the significance of the two initiation sites detected during the cell-free translation of poliovirus RNA is clarified (40).

(2) Another possible explanation for the unusual ratios is that polypeptide F, which is used to standardize the molar ratios, is unstable in the sense that it is non-specifically degraded during the course of infection. A low value for F would tend to increase the A:F ratio. Thus far, no-one has reported that polypeptide F is unstable and no change in its molar concentration was detected even when the viral proteins were isolated in the

presence of proteolytic enzyme inhibitors such as phenylmethylsulfonylfluoride (23).

(3) Multiple cleavage modes is another possible explanation for the observed high molar ratios (41). In their studies on the synthesis, processing and mapping of human rhinovirus (HRV) 1A proteins, McLean _et al_. (41) came upon some differences in the production of HRV proteins in comparison to the production of EMC virus proteins. One difference was the fact that the HRV equivalent of F protein did not quite behave like a stable primary product. On the basis of its kinetic behavior it was proposed that it consists of a mixture of two polypeptides, 38a, a major product, and 38b, a minor product. (The peaks are named after their estimated molecular weights). They also observed two other proteins which behaved like primary products by kinetic analysis. One of these, peak 47, appeared to have sequence homology with peak 38a, equivalent to protein F, suggesting the proteins 38a and 47 are alternative products of the F region. They also found that the HRV equivalent of the C region was unstable, and in 2 hr about one-half of the family was lost by degradation to molecules smaller than 10,000 daltons in molecular weight. Also, the C equivalent precursor was found to have more than one cleavage pathway, generating at least four discrete products.

McLean _et al_. (41) pointed out that if the molar ratio P1:S:P2 (the equivalent of A:F:C) is calculated by normalization of each to the amount of peak 47, the values 2.2:1:2 are obtained in a pulse-chase experiment. However, if the ratio is normalized to the sum of peaks 47 and 38a, the ratio 1.2:1:0.98 is obtained, which is close to the theoretical 1:1:1.

The possibility of multiple cleavage modes of polypeptide F has not been thoroughly investigated with either mengovirus or poliovirus. Thus, this explanation cannot be excluded with any certainty at this point.

(4) The unequal ratios could also be explained if near the middle of the viral RNA there is a weak termination signal which is read with increasing frequency towards the end of the infectious cycle (23). This would result in an overproduction of the capsid proteins with respect to the F and C proteins. This hypothesis was tested by Paucha and Colter (39) in the following way. At various times after infection of L-cells with mengovirus or HeLa cells with poliovirus, the cells were labeled with ^{3}(H)-amino acids for 30 min in the presence of amino acid analogues (39). This treatment results in the accumulation of the poliovirus polyproteins NCVP-00, NCVP-0a, NCVP-0b and NCVP-1 (23), and mengovirus polyproteins X, Y and Z and A (23). If the relative molar ratios are normalized with respect to the capsid precursor NCVP-1 in the case of poliovirus or polypeptide A in the case of mengovirus, the molar ratio of NCVP-00 (Figure 11) and NCVP-0b decreases with time

post-infection while that of NCVP-0a remains approximately the same. Similar results are obtained with mengovirus infected L-cells, except that the decrease in polyproteins Z and X is not as dramatic, since there is not much of these species present to begin with.

These findings are consistent with the hypothesis that the inequalities in capsid protein production result from a premature termination of translation somewhere at a site located near the midpoint of the viral RNA. If mengovirus polyprotein Y represents the region translated from the 5' end of the viral RNA and polyprotein X, the 3' end, as shown in the upper part of Figure 11, then one would not have expected protein Y to decrease but would have expected protein X to decrease, as was found. The placement of proteins X and Y in the positions indicated is tenuous. It is

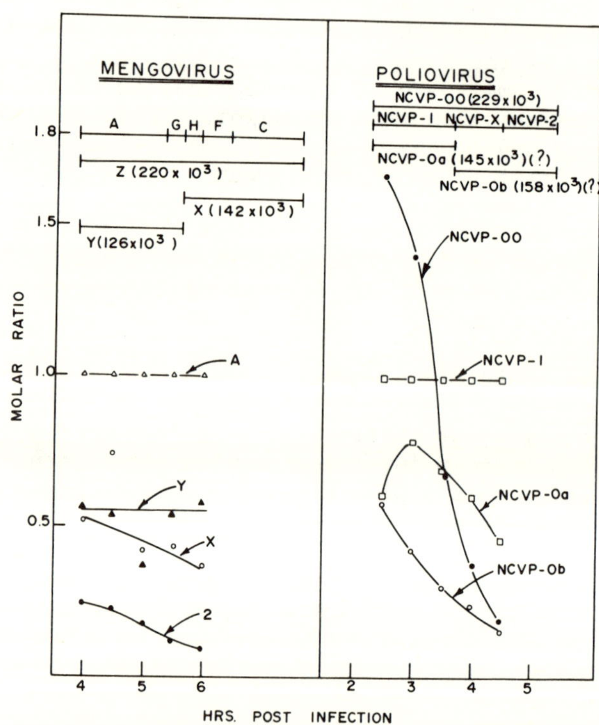

Figure 11. Relative molar ratios of mengovirus and poliovirus specific polypeptides produced in the presence of amino acid analogues. At the times indicated mengo- or polio-virus infected cells were labeled with ^{3}(H) amino acids for 30 min and analyzed by polyacrylamide gel electrophoresis as described (23, 39). Modified from Paucha and Colter (39).

based on the finding of Paucha and Colter (39) that antiserum to mengovirus capsid proteins precipitates both A and Y, but not X. Along these lines of reasoning, one would have expected giant polyprotein Z to decrease with time, as was found. The same arguments apply to poliovirus, except that the placement of polyproteins NCVP-0a and NCVP-0b on the map shown in Figure 11 are completely arbitrary.

Another interpretation of the data shown in Figure 10 is that with time after infection, there is an increased rate of degradation of the giant polypeptide synthesized in the presence of amino acid analogues. Products at the carboxyl-terminus would be degraded at a faster rate than those at the amino end of the giant precursor. This is not an unlikely possibility, since it has been shown that the products of the "C" region of human rhinovirus 1A (41) are quite unstable.

In conclusion, at this time there is no definite explanation available for the increased production of capsid proteins in polio- and mengovirus infected cells. Certainly, the possibility of multiple cleavage modes has not been eliminated and premature termination has not been proven. Thus, the question of whether the inequalities in capsid protein production represents some kind of translational control is rather speculative at this time.

V. FUNCTION OF PICORNAVIRUS PROTEINS

Little is known about the function of the 9 or so stable proteins synthesized by the picornaviruses. The stable polypeptides of mengovirus account for about 238,000 daltons (see Table 1). Of these the capsid proteins constitute about 99,000 daltons.

Using temperature sensitive mutants of poliovirus, Cooper *et al.* (42) devised a map based on the recombination frequencies obtained in crosses between viruses carrying different temperature sensitive markers. This map corresponds well to the map obtained using pactamycin (8, 32). The capsid protein mutants mapped near one end and the RNA polymerase mutants near the other end. Since poliovirus is similar in its map to other picornaviruses, this suggests that mengovirus polypeptide C, D, or E may be involved in RNA synthesis. Polypeptide E has been shown to be a constituent of a 250S mengovirus-induced RNA polymerase structure (43), but its role in viral RNA synthesis has not been elucidated.

The functions of polypeptides F, G, H and I are also not known. There are a number of reactions such as protein and rRNA synthesis inhibition and cleavage in which viral proteins appear to play a role. However, further study is needed to associate functions with the various picornavirus polypeptides.

ACKNOWLEDGMENTS

I should like to thank Drs. R. Rueckert and J. Colter for allowing me to use their published data in this review. Figures 1-4 and 6-11 and Table 1 were reproduced by permission of the publishers.

REFERENCES

1. LUCAS-LENARD, Jean M. Inhibition of cellular protein synthesis after virus infection: Fate of host cell mRNA. In The Molecular Biology of Picornaviruses. ed. Perez-Bercoff, R. (1979), chapter 4. Plenum Publishing Co. Ltd., New York and London.

2. CHATTERJEE, N.K., KOCH, G. and WEISSBACH, H. Initiation of protein synthesis *in vivo* in poliovirus-infected HeLa cells. Arch. Biochem. Biophys. (1973), 154, 431-437.

3. OBERG, B.F. and SHATKIN, A.J. Initiation of picornavirus protein synthesis in ascites cell extracts. Proc. Natl. Acad. Sci. U.S.A. (1972), 69, 3589-3593.

4. SMITH, A.E. The initiation of protein synthesis directed by the RNA from encephalomyocarditis virus. Eur. J. Biochem. (1973), 33, 301-313.

5. SUMMERS, D.F., MAIZEL, J.V. Jr., and DARNELL, J.E. Jr., The decrease in size and synthetic activity of poliovirus polysomes late in the infectious cycle. Virology (1967), 31, 427-435.

6. GRANBOULAN, N. and GIRARD, M. Molecular weight of poliovirus ribonucleic acid. J. Virol. (1969), 4, 475-479.

7. HUANG, A.S. and BALTIMORE, D. Initiation of polyribosomes formation in poliovirus-infected HeLa cells. J. Mol. Biol. (1970), 47, 275-291.

8. REKOSH, D. Gene order of the poliovirus capsid proteins. J. Virol. (1972), 9, 479-487.

9. HUNT, T., HUNTER, T. and MUNRO, A. Control of haemoglobin synthesis: Rate of translation of the messenger RNA for the α and β chains. J. Mol. Biol. (1969), 43, 123-133.

10. ROUMIANTZEFF, M., MAIZEL, J.V. Jr., and SUMMERS, D.F. Comparison of polysomal structures of uninfected and poliovirus infected HeLa cells. Virology (1971), 44, 239-248.

11. ROUMIANTZEFF, M., SUMMERS, D.F. and MAIZEL, J.V. Jr., In vitro protein synthetic activity of membrane-bound poliovirus polyribosomes. Virology (1971), 44, 249-258.

12. LEVINTOW, L. The reproduction of picornaviruses. In Comprehensive Virology, Vol. 2. ed. Fraenkel-Conrat, H. and Wagner, R.R. (1974), pp.109-169. Plenum Press, New York.

13. RUECKERT, R.R. On the structure and morphogenesis of picornaviruses. In Comprehensive Virology, Vol.6 ed. Fraenkel-Conrat, H. and Wagner, R.R. (1976), pp.131-213. Plenum Press, New York.

14. SUMMERS, D.F., MAIZEL, J.V. Jr., and DARNELL, J.E. Evidence for virus-specific non-capsid proteins in poliovirus-infected HeLa cells. Proc. Natl. Acad. Sci. U.S.A. (1965), 54, 505-513.

15. KOZAK, M. and NATHANS, D. Translation of the genome of a ribonucleic acid bacteriophage. Bacteriological Reviews (1972), 36, 109-134.

16. SUMMERS, D.F. and MAIZEL, J.V. Evidence for large precursor proteins in poliovirus synthesis. Proc. Natl. Acad. Sci. U.S.A. (1968), 59, 966-971.

17. JACOBSON, M.F. and BALTIMORE, D. Polypeptide cleavages in the formation of poliovirus proteins. Proc. Natl. Acad. Sci. U.S.A. (1968), 61, 77-84.

18. HOLLAND, J.J. and KIEHN, E.D. Specific cleavage of viral proteins as steps in the synthesis and maturation of enteroviruses. Proc. Natl. Acad. Sci. U.S.A. (1968), 60, 1015-1022.

19. BUTTERWORTH, B.E., HALL, L., STOLTZFUS, C.M. and RUECKERT, R.R. Virus-specific proteins synthesized in encephalomyocarditis virus-infected HeLa cells. Proc. Natl. Acad. Sci. U.S.A. (1971), 68, 3083-3087.

20. DOBOS, P. and MARTIN, E.M. Virus-specific polypeptides in ascites cells infected with encephalomyocarditis virus. J. Gen. Virol. (1972), 17, 197-212.

21. GINEVSKAYA, V.A., SCARLAT, I.V., KALININA, N.O. and AGOL, V.I. Synthesis and cleavage of virus-specific proteins in Krebs II carcinoma cells infected with encephalomyocarditis virus. Arch. Ges. Virusforsch. (1972), 39, 98-107.

22. LUCAS-LENARD, J. Cleavage of mengovirus polyproteins in vivo. J. Virol. (1974), 14, 261-269.

23. PAUCHA, E., SEEHAFER, J. and COLTER, J.S. Synthesis of viral-specific polypeptides in mengovirus-infected L-cells: Evidence for asymmetric translation of the viral genome. Virology (1974), 61, 315-326.

24. McLEAN, C. and RUECKERT, R.R. Picornaviral gene order: Comparison of a rhinovirus and a cardiovirus. J. Virol. (1973), 11, 341-344.

25. BUTTERWORTH, B.E. A comparison of the virus-specific polypeptides of encephalomyocarditis virus, human rhinovirus 1A, and poliovirus. Virology (1973), 56, 439-453.

26. LAPORTE, J. and LENOIR, G. Structural proteins of foot-and-mouth disease virus. J. Gen. Virol. (1973), 20, 161-168.

27. BLACK, D.N. Proteins induced in BHK cells by infection with foot-and-mouth disease virus. J. Gen. Virol. (1975), 26, 109-120.

28. ZIOLA, B.R. and SCRABA, D.G. Structure of the mengo virion. I. Polypeptide and ribonucleate components of the virus particle. Virology (1974), 57, 531-542.

29. DOBOS, P. and PLOURDE, J.Y. Precursor-product relationship of encephalomyocarditis virus-specific polypeptides. Comparisons by tryptic peptide mapping. Eur. J. Biochem. (1973), 39, 463-469.

30. JACOBSON, M.F., ASSO, J. and BALTIMORE, D. Further evidence on the formation of poliovirus proteins. J. Mol. Biol. (1970), 49, 657-669.

31. ABRAHAM, G. and COOPER, P.D. Relations between poliovirus polypeptides as shown by tryptic peptide analysis. J. Gen. Virol. (1975), 29, 215-221.

32. TABER, R., REKOSH, D. and BALTIMORE, D. Effect of pactamycin on synthesis of poliovirus proteins: A method for genetic mapping. J. Virol. (1971), 8, 395-401.

33. SUMMERS, D.F. and MAIZEL, J.V. Jr., Determination of the gene sequence of poliovirus with pactamycin. Proc. Natl. Acad. Sci. U.S.A. (1971), 68, 2852-2856.

34. COHEN, L.B., GOLDBERG, I.H. and HERNER, A.E. Inhibition by pactamycin of the initiation of protein synthesis. Effect on the 30S ribosomal subunit. Biochemistry (1969), 8, 1327-1335.

35. DINTZIS, H.M. Assembly of the peptide chains of hemoglobin.

Proc. Natl. Acad. Sci. U.S.A. (1961), 47, 247-261.

36. KOSSEL, H., MORGAN, A.R. and KHORANA, H.G. Studies on polynucleotides. LX-XIII. Direction of reading of messenger RNA. J. Mol. Biol. (1967), 26, 449-475.

37. BUTTERWORTH, B.E. and RUECKERT, R.R. Gene order of encephalomyocarditis virus as determined by studies with pactamycin. J. Virol. (1972), 9, 823-828.

38. BUTTERWORTH, B.E. and RUECKERT, R.R. Kinetics of synthesis and cleavage of encephalomyocarditis virus-specific proteins. Virology (1972), 50, 535-549.

39. PAUCHA, E. and COLTER, J.S. Evidence for control of translation of the viral genome during replication of mengovirus and poliovirus. Virology (1975), 67, 300-305.

40. CELMA, M. and EHRENFELD, E. Translation of poliovirus RNA in vitro: Detection of two different initiation sites. J. Mol. Biol. (1975), 98, 761-780.

41. McLEAN, C., MATTHEWS, T.J. and RUECKERT, R.R. Evidence of ambiguous processing and selective degradation in the non-capsid proteins of rhinovirus 1A. J. Virol. (1976), 19, 903-914.

42. COOPER, P.D., GEISSLER, D., SCOTTI, P.D. and TANNOCK, G.A. Further characterization of the genetic map of poliovirus temperature-sensitive mutants. In CIBA Symposium: The Strategy of the Viral Genome. eds., Wolstenholm and O'Connor. (1971), pp.75-100. Churchill Livingston, London.

43. LOESCH, W.T. Jr., and ARLINGHAUS, R.B. Stable polypeptides associated with the 250S mengovirus-induced RNA polymerase structure. Archives of Virol. (1975), 47, 201-215.

ROLE OF CELLULAR AND VIRAL PROTEASES IN THE PROCESSING OF PICORNAVIRUS PROTEINS

BRUCE D. KORANT

Central Research & Development Department, E.I. du Pont de Nemours and Company, Experimental Station
Wilmington, Delaware 19898, U.S.A.

INTRODUCTION

The replication of picornaviruses requires the action of proteolytic enzymes on viral precursor proteins. The protein cleavages are often crucial to the virus replication process, so that a complete understanding of their role could lead to development of new types of antiviral agents. On a more fundamental level, the characterization of the proteolytic reactions in virus infections is leading to a fuller appreciation of regulation of function by limited proteolysis. The picornavirus is a model in which proteolytic modifications yield macromolecular complexes with altered conformational and antigenic properties, and altered affinities for nucleic acids and receptors on cell membranes. Obviously, there is a general significance of these phenomena in understanding similar events in cellular interactions and in the numerous proteolytic events occurring in the tissues and fluids of higher organisms. For recent reviews of protein cleavage in picornavirus replication, see ref. 1-3.

It is now known that many diverse groups of viruses, with little or no genetic relationship to one another, display the phenomenon of proteolytic processing. Many of these viruses have proteins cleaved during assembly of subunits into larger aggregates, or in the maturation process when the viral genome is finally combined with capsid proteins to yield an infectious virion. Although proteolysis accompanying maturation is the most common theme, there are other variations in the requirements for proteolytic processing during early stages of infection, or the replication process itself.

I. POLYPROTEINS OF PICORNAVIRUSES: PROTEOLYTIC REGULATION OF ALL VIRAL FUNCTIONS

The picornaviruses are the prototype for studies on animal virus protein processing. A large number of laboratories have confirmed and extended the initial reports (4, 5, 6), and a substantial, although still incomplete picture has emerged of these viruses whose proteins are virtually all produced by proteolysis. A schematic is shown in Figure 1 which summarizes the important stages in the processing reactions.

The viral messenger RNA codes maximally for approximately 250 kD of protein. In vivo, at the midpoint of infection, there appears to be one initiation site for translation, apparently of a sequence adjacent to the amino terminus of the coat protein gene (6, 7, 8). As the ribosome leaves the region of the 95 kD structural gene, a cleavage of the nascent chain occurs.

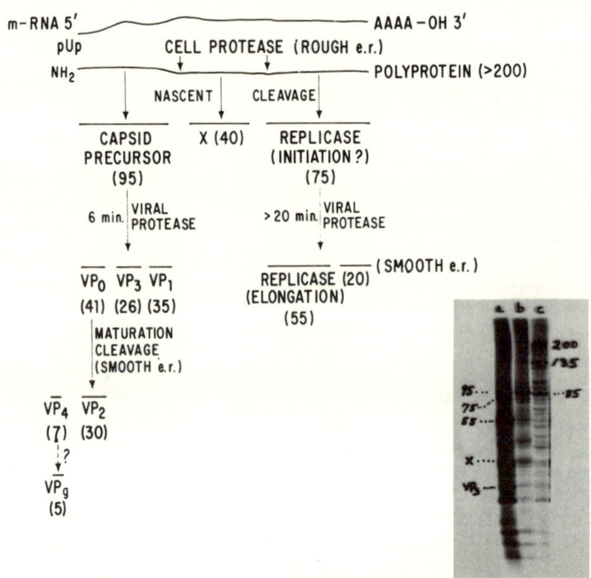

Figure 1. Processing of picornavirus polyprotein by proteolytic enzymes. The numbers in brackets represent molecular weights in thousands. Insert: Sodium dodecyl sulfate/polyacrylamide gel electrophoresis of poliovirus polypeptides, labelled with ^{35}S-methionine in infected HeLa cells. a. Infected, labelled at 2.5 hours post infection. b. Labelled in presence of 0.5 mM iodoacetamide. c. Labelled in presence of 0.1 mM tosyl lysine chloromethyl ketone.

Nevertheless the ribosome proceeds and translates the next region, coding for a polypeptide of about 40 kD. A second nascent cleavage occurs, but again the ribosome proceeds and completes the 75 kD replicase gene product (7, 9-11). Since these cleavages occur during translation, little or no polyprotein is normally observed. Special inhibitory treatments, such as addition of amino acid analogs or protease inhibitors, are required to clearly demonstrate the species larger than the coat precursor of 95 kD (see gel patterns a, b and c in Figure 1). The transit time of the ribosome through the structural gene region is short and the cleavage signal is probably a simple one built into the primary sequence of the polyprotein. In essence, the cleavages of the nascent picornavirus chains resemble those removing precursor regions of prohormones and other secretory proteins (12, 13), and the viruses may well be utilizing parts of the same processing system.

Additional proteolytic reactions produce the stable structural polypeptides. The secondary cleavages take several minutes to occur and lead to extensive refolding of the structural precursor as it assembles (14). An interesting aspect of these reactions is that the proteases responsible are not active in uninfected cells and the enzymes are probably virus-coded (see below). Processing of the RNA replicase polypeptides takes much longer (15 min or more at the midcycle of infection), and may provide a control over the action of viral RNA synthesis. Processing reactions leading to production of stable picornavirus proteins may be slowed by treatments which do not delay the nascent cleavages (Figure 1, gel pattern b). This implies that the protease or cleavage sites are different than those required for the nascent cleavages.

As shown in Figure 1, gel a, there are numerous proteins present in infected cells which are not noted in the diagram of Figure 1. These have been reported, and represent short-lived intermediates, products of ambiguous cleavage, and artifacts of incomplete dissociation (15-17). There have been several examinations of cell-free translation using picornavirus RNA. The more recent studies have been successful in translating the entire genome, and demonstrating the primary nascent cleavages. Three reports found clear evidence of production of stable capsid proteins, implying that secondary cleavage took place (18-20). There is general support from these studies of a single "strong" initiation site, but there are also indications of a second, rarely used site (21-24), see chapters 7 and 11 of this book). Additional studies may help to resolve the question. Peptide mapping techniques produced a result implying two initiation sites exist _in vivo_ (24), but this has not been confirmed. In fact, other results indicate that the three major primary translation products are present in larger, precursor forms (Figure 2 and ref. 25).

II. ROLE OF CELLULAR PROTEASES IN PICORNAVIRUS INFECTIONS

There are very clear evidences of a role of host proteases in the replication of certain RNA viruses, particularly the myxo- and paramyxoviruses (see ref. 1 for a review). By comparison, the evidence for participation of cellular proteases in picornavirus replication is less direct. The cleavage of nascent picornavirus polyproteins is probably carried out by cellular enzymes which are associated with membrane-bound polyribosomes. This was concluded from both <u>in vivo</u> (26) and cell-free studies (22, 25, 26). With poliovirus polyprotein as substrate, uninfected HeLa cell extracts were able to produce primary cleavage products with the molecular weights and antigens of intermediates in the normal cleavage process. Little or no production of capsid polypeptides was reported (25).

There have been no detailed descriptions of the cellular proteases which are utilized in the cleavage of viral proteins. In this connection, it is noteworthy that some cellular proteins, including hormones and other secretory polypeptides, are produced by limited cleavage of precursors. The proteases involved are not fully characterized, but seem to be located on membrane-bound structures (13, 27, 28), have optimum activities at neutral to slightly alkaline pH, and are sensitive to phosphofluoridate or other serine esterase inhibitors (12, 27, 29).

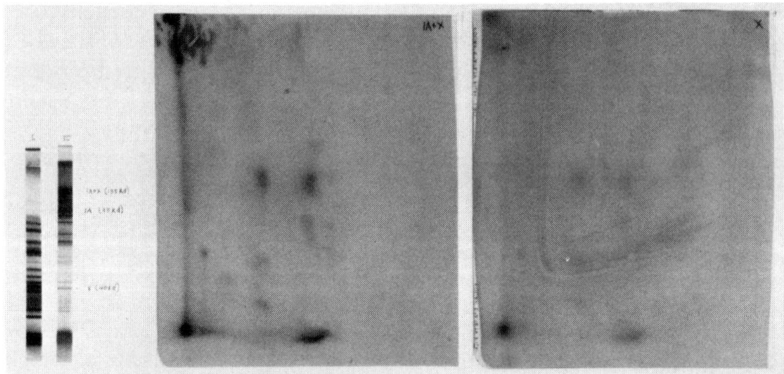

Figure 2. Tryptide analyses of poliovirus NCVPx, and a 140 kD apparent precursor to NCVPx. <u>Bona fide</u> poliovirus (^{35}S) NCVPx (gel I) was isolated, and its tryptic map compared with that of the 140 kD protein isolated in zinc (0.5 mM)-treated infected cells (gel II) by two-dimensional electrophoresis (pH 2.1) and chromatography (butanol: acetic acid: water, 3:1:1) on cellulose thin layer plates. Other tryptic fragments in the 140 kD protein are present in poliovirus coat proteins (not shown).

Recent evidence points to the presence of protease activity associated with polysomes and ribosomes when extracts of uninfected cells are assayed (refs. 27-32, Figure 3). Characteristic of infection of cells by poliovirus is drastic, rapid inhibition of protein synthesis. Poliovirus infection also depresses the ribosomal protease activity (27, 29, 33). Ribosomes from uninfected cells have been reported to possess an autoproteolytic activity (31, 32), and this has been confirmed by two-dimensional gel analysis (Figure 4). Poliovirus infection of HeLa cells reduces the autoproteolysis of isolated 80S ribosomes markedly (not shown). The inhibition of HeLa cell ribosomal protease activity requires protein synthesis, but proceeds in the presence of guanidine (33). The inhibition will not proceed if virus protein processing is prevented,e.g., by addition of zinc ion. The inhibition by poliovirus of HeLa cell ribosomal protease activity was studied further using iodine-labelled Trasylol, a potent protease inhibitor. As shown in Figure 5, Trasylol is a homogeneous protein of molecular weight approximately 6,500. The protein was labelled with ^{125}iodine, and separated from unbound iodine. Ribosomes were prepared from uninfected HeLa cell cultures, and from cells infected for three hours with poliovirus in the presence of 2 mM guanidine. The

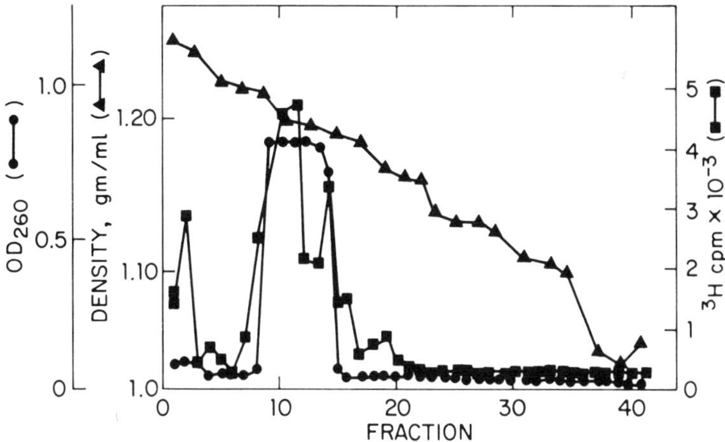

Figure 3. Polyribosomal protease activity. Polyribosomes were prepared from uninfected HeLa cells (3×10^7 cells) and rebanded in a discontinuous sucrose gradient for 17 hours at 25,000 rpm. The polysomes were initially adjusted to 25% sucrose (w/v), but banded in approximately 45% sucrose. Protease activity was assayed on ^{3}H labelled HeLa cell protein extracts (shown) and on ^{125}I-labelled hemoglobin (not shown), with similar results (29).

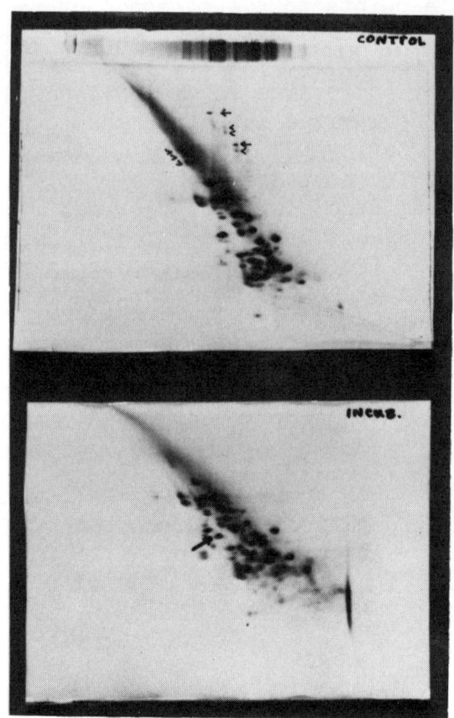

Figure 4. Autoproteolysis of rat liver ribosomal proteins, assayed by two-dimensional gel electrophoresis. Incubated ribosomes were held at 37° for 17 hours. First dimension was pH 4.3, 8 M urea, 5% polyacrylamide; second dimension was 10-22% polyacrylamide, 0.1% SDS. Ribosomal protein preparations were provided by J. Langner, Martin-Luther-University, Halle, DDR.

purified monosomes were incubated with the labelled Trasylol and chromatographed to separate unbound Trasylol from the ribosomes. As shown in Figure 5, much less of the protease inhibitor is bound to the ribosomes from infected cells which correlates with their reduced proteolytic activity. It is not clear whether the lack of binding is due to specific loss of the protease polypeptide, or to a more general effect of the infection on ribosome structure (29).

In a sensitive, solid-phase assay for endo-proteases (27, 34), the activity of polysomes and monosomes were compared. Shown in Table 1 is the loss of activity of 80S ribosomes from infected cells, but the retention of activity on polysomes. This data suggests continuing participation of a polysomal protease activity in viral

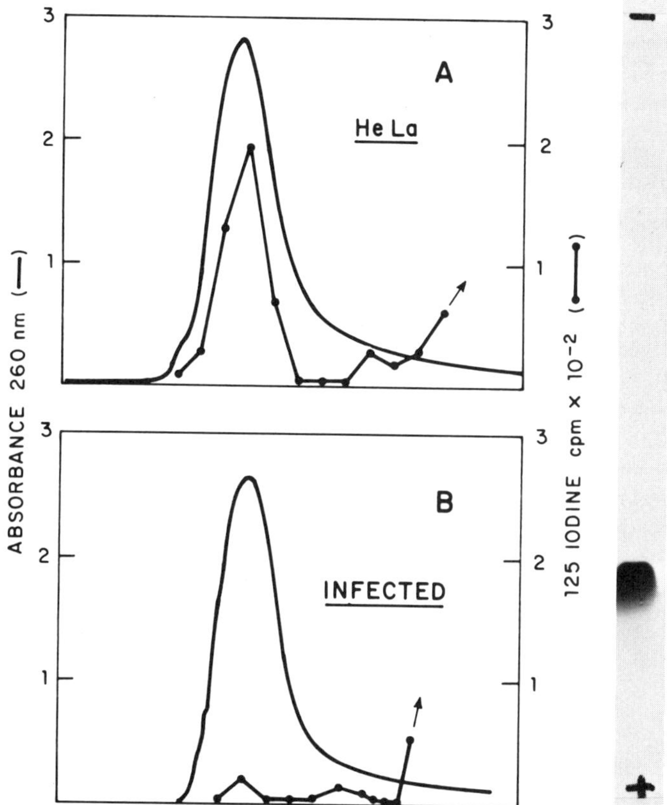

Figure 5. Binding of ^{125}I Trasylol to ribosomes. Ribosomes were purified by zonal sedimentation, following detergent treatment of uninfected or polio-infected HeLa cells. The ribosomes were incubated for 30 min at 25° with Trasylol labelled by the Bolton-Hunter reagent. The mixture was then passed through a Biogel P150 column, buffered at pH 7.2 and containing 1 mM $MgCl_2$ and 0.01 M NaCl. The monosomes were recovered in the void volume, and unbound Trasylol was retained. ^{125}I was determined by liquid scintillation counting. Panel A. Uninfected cells; Panel B. Poliovirus-infected. Insert: SDS gel of Trasylol. Autoradiography showed only one radioactive species (29).

protein synthesis and processing, throughout infection. A polysome-associated protease suggests a requirement for a proteolytic activity in synthesis of cellular proteins, which activity the picornavirus message may utilize initially.

Table 1

Proteolytic Activity of HeLa Cell Polysomes and Ribosomes

Sample	(^{35}S) CPM released/mg protein*
I. Polysomes	
a. uninfected	957
b. infected (4 hrs)	912
II. 80S ribosomes	
a. uninfected	405
b. infected	81

* Poliovirus TLCK radioactive polyprotein as substrate

Subsequently, the virus supplements it with another enzyme, coded by the viral genome.

III. EVIDENCE FOR A PICORNAVIRUS-SPECIFIC PROTEASE

In an attempt to comprehend the complex process of viral replication, it is crucial to know the origin of the enzymes which carry out the rate-limiting reactions. With viral protein cleavage, the enzyme(s) which produces the capsid polypeptides is of great interest. The simplest possibility is that host proteases are used to cleave the virus-coded substrates. However, this reduces the ability of the virus to control the reactions, and in fact, there are several pieces of direct evidence which support the existence of a viral-coded enzyme.

Virus-infected cells contain in their cytoplasm greater quantities of protease activity than do uninfected cells (26), and the induced activity parallels the time course of infection (33, 35-37) and the amount of virus used to initiate infection (33). Extracts of infected cells are able to carry out cleavages of viral precursors leading to production of capsid polypeptides or fragments of them, while extracts of uninfected cells lack these activities (26, 36-38).

The time course of induction of virus-specific protease activity in poliovirus and EMC virus-infected cells has been examined (33, 37). The rate of new enzyme production and the amount is positively correlated with the quantity of infecting virus (33). A characteristic of the new protease is a less alkaline optimum, compared to the major cellular protease, and the optimum pH of approximately 7 is unlike that of the major lysosomal enzymes (33, 37). Inhibitor studies suggest the enzyme requires an unblocked sulfhydryl group (20, 36).

Production of the new protease in poliovirus infection requires viral RNA synthesis, since the activity is not present in guanidine-treated cells (33). Protein synthesis is also necessary for production of the protease, because cycloheximide blocks its appearance. When amino acid analogs are added to poliovirus-infected cells at mid-cycle of replication, virus yields are greatly diminished; by comparison there is a virtually normal yield of protease. However, the protease produced is heat sensitive, compared to the enzyme from untreated cells (Figure 7). This indicates the

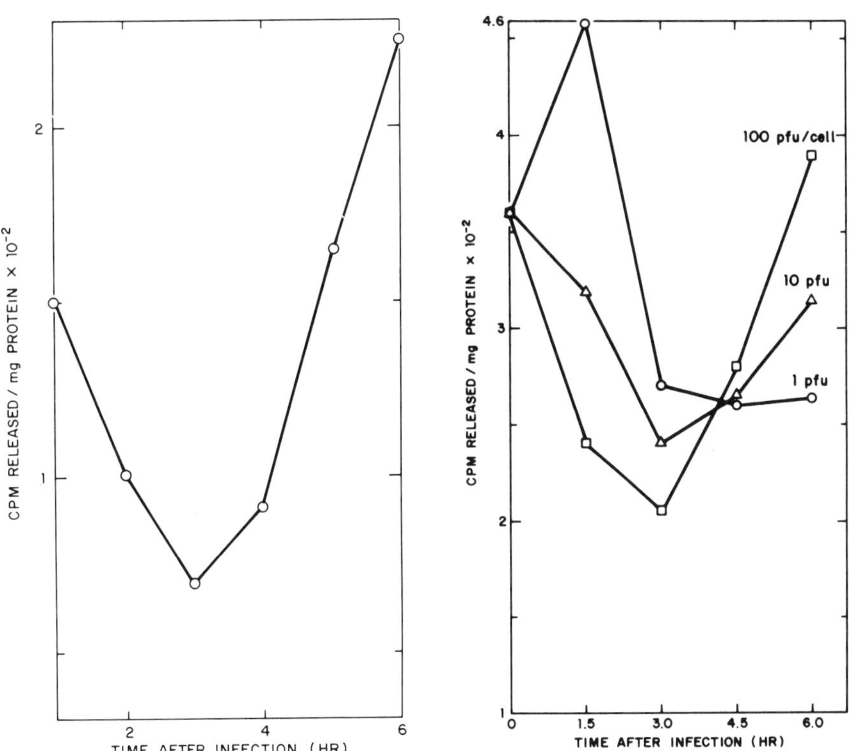

Figure 6. Left panel: Protease activities in poliovirus-infected HeLa cells at various times after infection. Substrate was solid phase-linked poliovirus precursor polypeptides. Right panel: Effect of multiplicity of infecting virus on protease activity in poliovirus-infected cells (33).

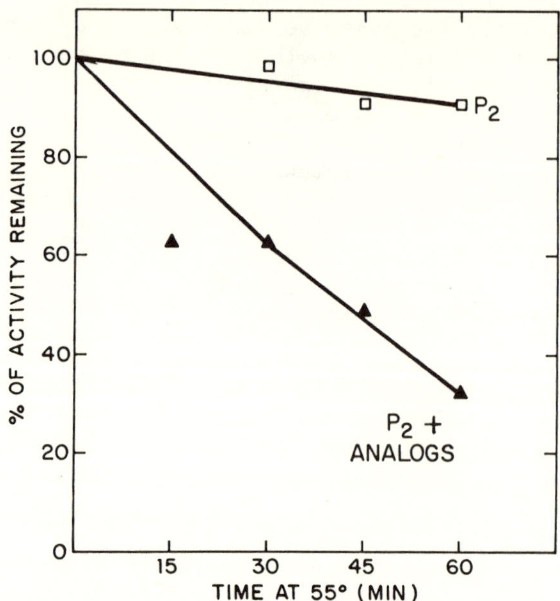

Figure 7. Stability of poliovirus protease in heated cell extracts. Protease extracted from amino acid analog-treated cells or from control cultures was heated at 55° for up to one hour, then chilled to 0°, and assayed at 35°. Untreated (—— □ ——); plus 30 μg/ml of each analog (—▲—). Analogs were fluorophenylalanine, fluorotryptophan and azaleucine (29).

enzyme contains analogs, and therefore has been synthesized at a time after infection when host protein synthesis is abolished.

There is a remarkable lack of effect of the virus-induced protease on cellular protein stability. This was measured by two-dimensional gel analysis of cellular proteins, labelled prior to infection. Samples were monitored up to the development of the cytopathic effect (Figure 8). The rapid cleavage of viral proteins is not observed at all with the cellular polypeptides, indicating a high degree of specificity of the protease, or its sequestration from cellular structures.

The specificity of poliovirus protease may be inferred from end group analyses of processed virus capsid polypeptides. The enzyme cleaves peptide bonds whose carboxyl group is donated by leucine or glutamic acid residues. This permitted use of a synthetic protease inhibitor terminating in a leucyl chloromethyl

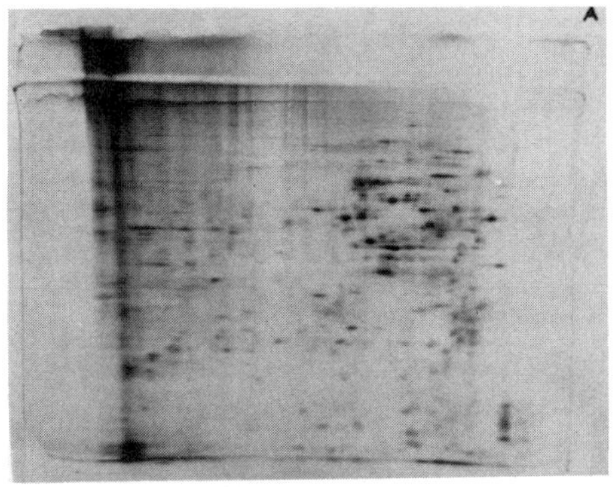

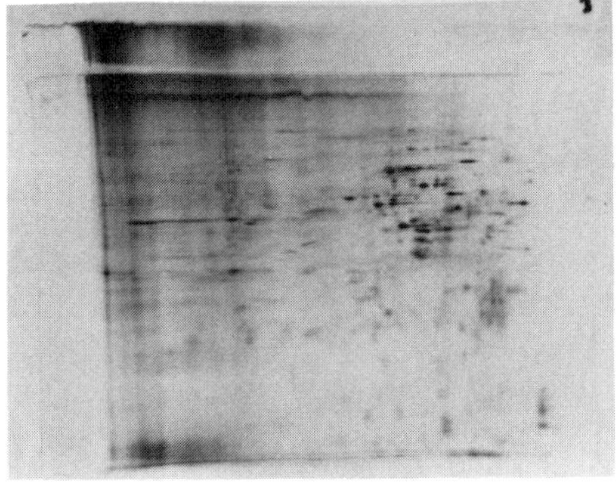

Figure 8. Two-dimensional analysis of proteins in infected HeLa cells. HeLa cells were labelled for 24 hours with (^{14}C) amino acids, washed thoroughly, then infected with poliovirus type 1 Mahoney. Samples were taken at sixty min intervals, lysed with 8 M urea, and analyzed by two-dimensional electrophoresis. The first gel was 5% polyacrylamide, 8 M urea, 27% ampholine, pH 5-8. Samples were separated at 300 volts, overnight at 6-7°. The gels were sliced, and placed on SDS-polyacrylamide gradient gels, and electrophoresed for 4.5 hours at 20 mA/gel. Gels were stained, dried, and autoradiographed. Gel A: uninfected cells; Gel B: infected with poliovirus for five hours.

ketone residue. The effect of such an inhibitor on poliovirus protein processing is shown in Figure 9. Figure 9-2, in which the infected cells are treated with 0.1 mM of the compound, shows that normal processing of virus coat precursor polypeptides is rapidly blocked, and high molecular weight polypeptides accumulate.

Although chloroketone protease inhibitors react irreversibly,

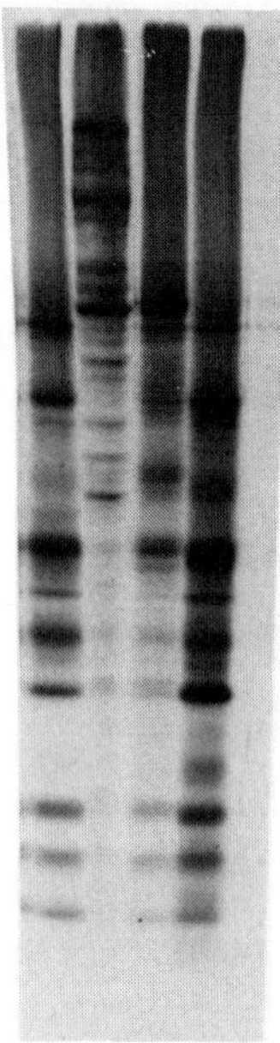

Figure 9. Polyacrylamide SDS gel analyses of poliovirus cytoplasmic polypeptides. Infected HeLa cells were labelled with (^{35}S) methionine, in the presence or absence of carbobenzoxy leucyl chloromethyl ketone. After an appropriate chase period, cells were lysed and proteins resolved. Gel sample identification : 1-virus control; 2-addition of carbobenzoxy leucyl chloromethyl ketone (0.1 mM); 3-remove excess chloromethyl ketone, add 100 μg/ml cycloheximide, chase 60 min; 4-same as sample 3, but no cycloheximide present. The chloromethyl ketone was supplied by J. Powers, Georgia Institute of Technology, Atlanta, Georgia.

viral-specific cleavage will restart if excess compound is removed by sufficient washing (Figure 9-3). However, reversal of the effect requires protein synthesis (Figure 9-4). This indicates the enzyme is continuously synthesized in infected cells when host protein synthesis is abolished.

There have been several reports of protease activities associated with purified virions of animal viruses, including picornaviruses (39). One study suggested individual capsid polypeptides of a cardiovirus possesses intrinsic protease activities on a viral substrate (37). In the study, neither purified substrate nor homogeneous enzyme were used, so that the presence of a host

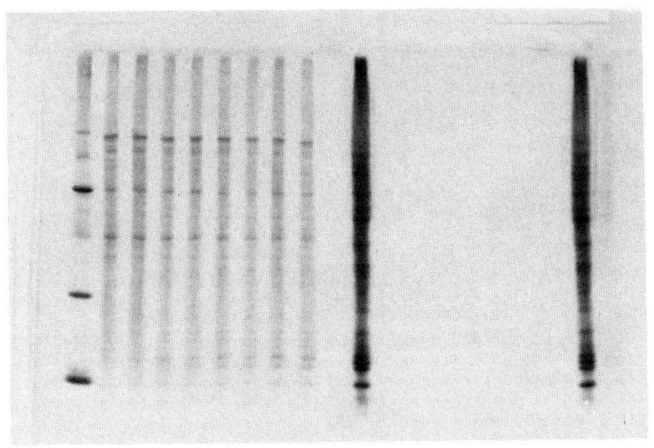

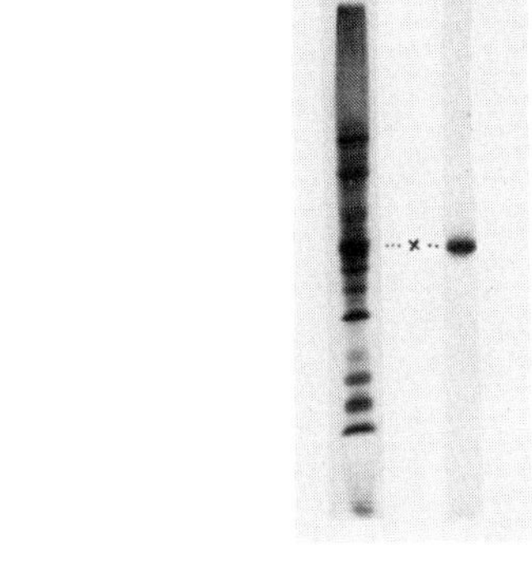

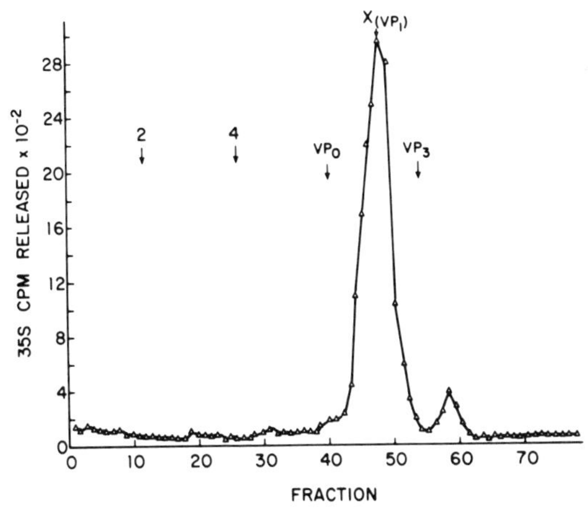

Figure 10. Preparation of poliovirus protease. Infected cell homogenates were chromatographed on DEAE-cellulose, then on hydroxylapatite, and analyzed on SDS-polyacrylamide gels. Upper panel: Left side indicates protein content of protease fraction after DEAE cellulose; right side shows removal of host proteins by hydroxylapatite (Coomassie Blue staining). Middle

panel: Virus protein contribution to the purified protease fraction is NCVPx, by autoradiography. Bottom panel: Recovery of protease activity from preparative polyacrylamide gel of poliovirus proteins.

enzyme cannot be completely eliminated as a contributing factor.

The protease was prepared from poliovirus-infected cells by homogenization, followed by chromatography on DEAE-Sephadex, and a second purification on an SDS-hydroxylapatite column (B. Korant, M. Lively and J. Powers, in preparation). The protease-containing fractions were pooled, and electrophoresed in a polyacrylamide gel containing SDS. As shown in Figure 10, the protease activity has a mobility corresponding to molecular weight of approximately 40,000. Autoradiography of the protease fraction to establish the virus protein contribution indicated that of the poliovirus proteins, only NCVPx was present (29). The possibility that the NCVPx preparation was significantly contaminated with VP1, the largest virus capsid polypeptide, was examined by assaying protease levels in cells infected with either of two poliovirus mutants, both of which produce no significant levels of capsid polypeptides (40). With both mutants, compared to a non-defective virus, there was production of a normal level of specific protease activity (Figure 11). This supports the view that there is no contribution of a virus coat polypeptide to the virus-specific protease.

A poliovirus mutant was previously described(1, 29, 33), which displayed a lack of regulation of viral RNA metabolism. Extracts of cells infected with this virus display lower amounts of protease than do those of wild type virus. Moreover, the protease produced by the mutant virus is more temperature-sensitive than that of its parent (Figure 12). These results indicate that the protease itself is coded for by the viral genome.

Pelham (20) has provided independent evidence for a viral contribution to the proteolysis of EMC virus precursor proteins. A reticulocyte cell-free protein synthesizing system, programmed with EMC RNA, produced the full complement of viral polypeptides. In addition, a proteolytic activity was detected; apparently synthesized in the cell-free system, which cleaved the coat precursor polypeptide of the virus into fragments matching in size the virus coat proteins produced in infected cells. In agreement with the results discussed for poliovirus, the results tend to rule out an autoproteolytic activity associated with the coat proteins. This is in conflict with an earlier study on EMC virus proteolysis (37), in which coat polypeptide gamma co-purified with an endo proteolytic activity.

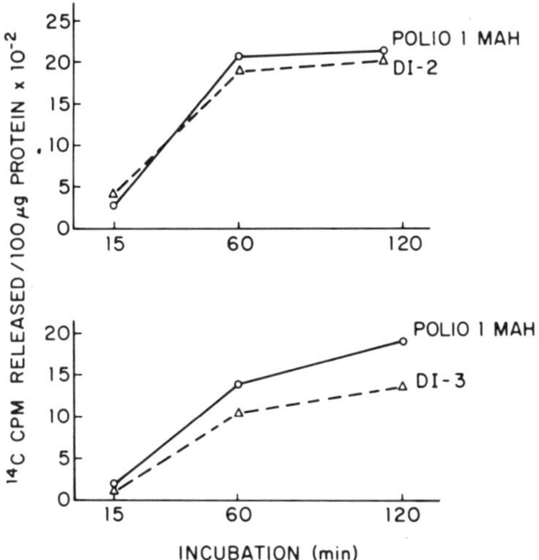

Figure 11. Protease activities detected in extracts of HeLa cells infected with either poliovirus type 1 (Mahoney) or defective-interfering mutants, the latter supplied by A. Nomoto and E. Wimmer, SUNY, Stony Brook, New York.

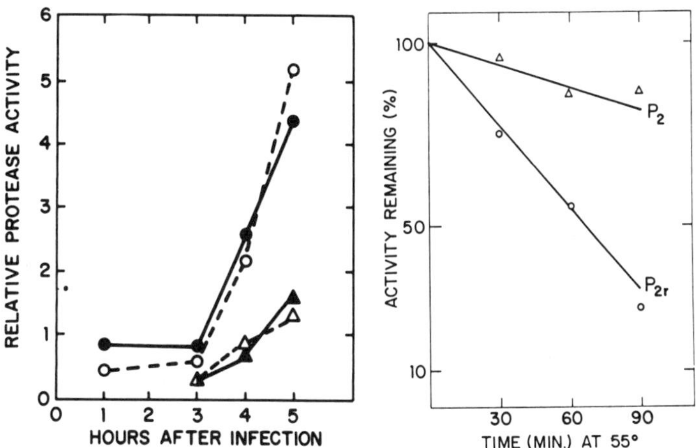

Figure 12. Protease activities in cells infected by parental or mutant polioviruses. Left panel: Infected HeLa cells were extracted at the indicated times after infection, and assayed on substrates of either the parental (P2) or mutant virus (P2r). Parent virus enzyme, parental substrate (o-----o); parent virus enzyme, mutant substrate (●——●); mutant virus enzyme, parental

substrate (▲———▲); mutant virus enzyme, mutant substrate (△-----△). Right panel: Heat inactivation of protease in extracts of P2 or P2r-infected HeLa cells (29).

In summary, evidence for a poliovirus gene product participating in the proteolytic cleavage of virus coat protein includes: a. kinetics of production; b. characteristic pH optimum; c. abolition of activity by inhibitors of virus RNA and protein synthesis; d. labilization to heat inactivation by amino acid analogs after cellular protein synthesis was abolished; e. in similar fashion, recovery of protease activity, after inhibition of virus proteolysis by a leucyl chloromethyl ketone, required protein synthesis; f. finally, protease activity co-purified with virus non-structural polypeptide X.

IV. CLEAVAGE SITES IN PICORNAVIRUS PRECURSOR PROTEINS

Complete sequence analyses of the picornavirus precursor proteins are not available. However, there have been determinations of amino and carboxyl terminal sequences of capsid polypeptides, which permit educated guesses of the structure of the cleavage region. Obviously, the information obtained is only valid if exopeptidases have not removed terminal residues after the primary peptide bond scission (this latter phenomenon has been described with proinsulin).

Although the number of examples are limited, there is striking repetitiveness of carboxyl terminal leucine and glutamine, or glutamic acid (41, 42). Since these terminal residues will fit the common binding pocket of certain proteolytic enzymes (42), it may be that the cleavage enzymes of the different picornaviruses have a common origin. Alternatively, the presence of adjacent residues may provide structural uniqueness of the region to be cleaved in the precursor.

From an inspection of the accumulated data, Bachrach has proposed that cleavage sites occur between α-helix maker-breaker pairs (B-bends) at exposed sites in the precursor (42). In support of this hypothesis are the presence of proline, glycine, serine and asparagine residues at the new amino termini; these are all "strong" helix breakers. The new carboxyl terminal leucine and glutamine (glutamic acid) are helix formers, and might be expected to fit spacially into a single binding pocket, although clearly their chemical properties are quite distinct. The viral cleavage sites would thus be explained in a manner similar to the activation of other protein zymogens.

This model can be tested by additional sequencing of other viral structural and non-structural polypeptides. In addition, there have been reports of ambiguous cleavages of picornavirus proteins (16, 17). Once sequences of these altered sites are known, the requirements of the processing reactions will be better understood.

V. INHIBITORS OF VIRAL PROTEOLYSIS

There are two tactics that have been widely used to inhibit viral proteolysis; these are alteration of primary structure and/or conformation of the precursors, and the inactivation of the cleaving enzymes. Some of the inhibitory treatments are particularly useful in demonstrating that proteolysis is actually occurring, and in accumulating precursor proteins for comparative peptide analyses with stable end products.

A. Alteration of the Substrate

The device most often used in the *in vivo* modification of the substrate is by the addition of unusual analogs of L-amino acids to the tissue culture medium. At moderate analog concentrations, viral protein synthesis continues, but the products contain the altered residues. This probably alters configuration of the precursors. In one early study, combinations of fluorophenylalanine, canavanine (arginine analog), ethionine and azetidine-2-carboxylic acid (proline analog) were used to accumulate very large poliovirus polyproteins (7). There are reports of proteolytic processing of analog-containing proteins. These reactions proceed more slowly than normal, and may signify random degradation by lysosomal enzymes. There is another interpretation of the effect of analogs which is possible. In light of evidence for virus-induced proteases with picornaviruses (see above) it cannot be discounted that the analogs are altering the enzymes directly.

Abnormally high temperatures are known to block viral protein cleavage with numerous viruses (1, 3, 43). Usually, a few minutes at 39-41 °C is sufficient to stop the cleavage reactions, which is a temperature range well below the point of heat inactivation of the proteases. Most likely the subtrates are being denatured, or reversibly altered, although effects on the host cells are possibly involved.

Zinc ion is an inhibitor of virus production (44) and blocks protein cleavage of rhinovirus, enterovirus, and cardiovirus precursors (25). Compared to other methods of inhibition, zinc treatment is reasonably gentle, so that there is minimal cytotoxicity, and its effect is reversible by washing the treated cells in zinc-

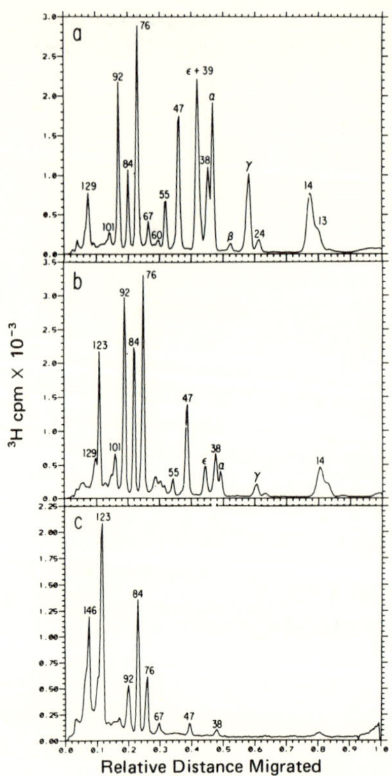

Figure 13. Effect of zinc ions on rhinovirus type 1A protein cleavage in HeLa cells. Panel a. No treatment, b. 0.1 mM $ZnCl_2$, c. 0.8 mM $ZnCl_2$ (44).

free medium (45). Accumulation of large quantities of viral precursors is possible with zinc treatment, and these may be used in sequence analysis or as antigens (Figure 13).

A plausible interpretation is that zinc acts to alter the viral substrate (45, 46). The conversion of capsid precursor is more sensitive than the other cleavages, however, increasing concentrations will accumulate progressively larger precursors; these all contain capsid sequences (25). With coxsackievirus and rhinovirus, zinc rapidly blocks viral maturation, and zinc will bind to the rhinovirus capsids, and prevent their crystallization (45, 47). Rhinoviruses selected for zinc resistance display altered capsid antigens, consistent with a binding site for zinc in the capsid sequence (45). After mengovirus capsid precursor synthesis is completed, its processing is resistant to added zinc ion (46).

A function of the picornaviruses which is poorly understood is their ability to halt host protein synthesis. A model which has received recent support is the simple out-competing of the cellular mRNAs for initiation factors by the viral RNA (48). There is good evidence that this occurs in vitro (49); this however is not sufficient to explain many features of shut-off in infected cells (for a review see chapter 4 of this book and ref. 50). More likely is the additional requirement for a viral protein in the inhibitory role. The coat protein of poliovirus has been proposed as an "equestron", modulating host ribosome function (51). If a stable viral protein is required for inhibition, then a specific blockade of viral protein processing should allow host protein synthesis to continue. Taking advantage of the sensitivity of rhinovirus protein processing to the presence of zinc ions, the requirement for cleavage of viral precursors was assessed. HeLa cells were infected with 100 plaque-forming units per cell of rhinovirus type 1A. Evaluation of host protein synthesis was made by zonal sedimentation of cellular polysomes and by incorporation of radioactive amino acids into cellular proteins. If zinc ions (0.3 mM) were present at the start of infection, the virus could do little harm to cellular translation (Figure 14 and Table 2). In the absence of zinc, protein synthesis was abolished (Figure 14). Zinc alone had no effect on HeLa cell protein synthesis at the concentration used. Viruses which are relatively insensitive to zinc ion, poliovirus and zinc-resistant rhinovirus 1A (44, 45) were able to stop host protein synthesis in up to 0.3 mM zinc ion. The conclusion to be drawn from this experiment is that although saturating levels (10^4-10^5 particles/cell) of virus were used and viral mRNA was introduced into the cells, host protein synthesis was spared if viral proteins were not synthesized and normally processed.

B. Chemical Protease Inhibitors

Inhibitors of well known metallo-, sulfhydryl and serine proteases block viral proteolysis in infected cells. The use of iodoacetamide (IAM) to stabilize poliovirus and foot-and-mouth virus precursors has been described (36, 52). The precursors were subsequently cleaved in vitro to yield normal-sized products (36), and polio RNA replicase activity was stabilized in cell-free extracts of IAM-treated cells (53). N-ethyl maleimide inactivates an EMC virus-specific protease in reticulocyte lysates (20). These results could signify that a sulfhydryl-type protease, e.g., cathepsin B, is required for some of the cleavage reactions. However, when such compounds are added to cells, numerous proteins react, and there is substantial inhibition of protein synthesis. Energy metabolism in such cells is depressed, and ATP-requiring proteolytic reactions blocked.

Table 2

Inhibition of HeLa Cell Protein Synthesis by Rhinovirus Type 1A

Treatment	^3H amino acids cpm incorporated	% inhibition
none	2.8×10^5	-
HRV 1A	0.6×10^5	79
HRV 1A + 0.3 mM $ZnCl_2$	2.6×10^5	7

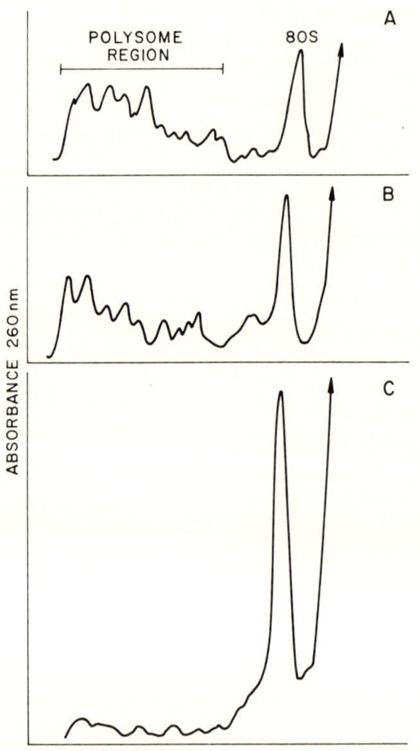

Figure 14. Inhibition of cellular protein synthesis by rhinovirus type 1A in the presence of zinc ion. Zonal sedimentation of cell extracts in 5-46% sucrose gradients was done to display the polysomes. Extracts were made 0.5% in deoxycholate and Triton X-100 before centrifugation. Sedimentation from right to left. Panel a. cell control. Panel b. HRV 1A-infected plus 0.3 mM Zn^{2+}. Panel c. HRV 1A-infected.

The protease/esterase inhibitor diisopropyl phosphofluoridate (DFP) was shown to block the cleavage of poliovirus polyprotein (7). This implicates a protease with a serine-containing active site.

The most interesting class of protease inhibitors in viral studies has been the chloromethyl ketone derivatives of amino acids (54, 55). These were designed as affinity labels of serine-type proteases, and react irreversibly with histidine and serine residues in the active sites of proteases. There is a basis for selectivity of the chloromethyl ketones: phenylalanyl and lysyl derivatives were synthesized, which had specificity for chymotrypsin and trypsin, respectively (54). This specificity led to studies on inhibition of poliovirus protein cleavage, with positive results (25, 26).

There have been conflicting reports on the efficacy of particular chloromethyl ketones. With picornaviruses, this can perhaps be explained by the various cell types used supplying distinct cleavage enzymes (25), at least for the nascent cleavages. However, there are several troublesome facts to be fit into any model of chloroketone action. For example, cleavage of poliovirus precursors was blocked by a D-rotatory chloroketone of phenylalanine (56). If the enzyme were actually chymotrypsin-like, the D-inhibitor should have been ineffective. The various chloroketones, particularly tosyl phenylalanine chloromethyl ketone (TPCK) block protein synthesis (57, 58); this has been attributed to reactivity with glutathione, or to induction by TPCK of an inhibitory substance. The property of non-specific alkylation of sulfhydryl groups and non-active site histidines and serine is difficult to deal with at present, but may be of relevance. For example, it was proposed that TPCK directly alters the antigenic specificity of a papova virus precursor (59), implying a chemical or conformation change, which conceivably could affect normal cleavage.

In summary, chemical inhibitors of cleavage are available, but one must be found which is absolutely directed against either a viral substrate or a virus-specific protease. Additional experimentation will be required, if eventually an inhibitor of proteolysis is to develop into a clinically useful antiviral.

ACKNOWLEDGEMENTS

My thanks go to several colleagues who have participated in the studies described. These include Drs. B. Butterworth, N. Chow, J. Kauer, E. Knight, Jr., M. Lively, A. Nomoto, J. Powers and E. Wimmer. Excellent assistance was provided by M. Parise and A. Vilda.

REFERENCES

1. KORANT, B.D. In Proteases and Biological Control (1975), Reich, E., Rifkin, D. and Shaw, E. eds., p. 621, Cold Spring Harbor Laboratory.

2. HERSHKO, A. and FRY, M. Ann. Rev. Biochem.(1975), 44, 775.

3. BUTTERWORTH, B. In Current Topics in Microb. and Immunol. (1977), Vol.77, p.1, Springer-Verlag, New York.

4. SUMMERS, D. and MAIZEL, J. Proc. Nat. Acad. Sci. U.S.A. (1968), 59, 966.

5. HOLLAND, J. and KIEHN, E. Proc. Nat. Acad. Sci. U.S.A. (1968), 60, 1015.

6. JACOBSON, M. and BALTIMORE, D. Proc. Nat. Acad. Sci. U.S.A. (1968), 61, 77.

7. JACOBSON, M., ASSO, J. and BALTIMORE, D. J. Mol . Biol. (1970), 49, 657.

8. SHATKIN, A. Ann Rev. Biochem. (1974), 43, 643.

9. BUTTERWORTH, B., HALL, L., STOLTZFUS, C. and RUECKERT, R. Proc. Nat. Acad. Sci. U.S.A. (1971), 68, 3083.

10. TABER, R., REKOSH, D. and BALTIMORE, D. J. Virol. (1971), 8, 395.

11. SUMMERS, D. and MAIZEL, J. Proc. Nat. Acad. Sci. U.S.A. (1971), 68, 2852.

12. STEINER, D., KEMMLER, W., TAGER, H., RUBENSTEIN, A., LERNMARK, A. and ZUHLKE, H. In Proteases and Biological Control. (1975), Reich, E., Rifkin, D. and Shaw, E. eds., p.531, Cold Spring Harbor Laboratory.

13. DEVILLERS-THIERY, A., KINDT, T., SCHEELE, G. and BLOBEL, G. Proc. Nat. Acad. Sci. U.S.A. (1975), 72, 5016.

14. BUTTERWORTH, B. and RUECKERT, R. Virology (1972), 50, 535.

15. ABRAHAM, G. and COOPER, P. J. Gen. Virol. (1975), 29, 199.

16. McLEAN, C., MATHEWS, T. and RUECKERT, R. J. Virol. (1976), 19, 903.

17. BECKMAN, L., CALIGUIRI, L. and LILLY, L. Virology (1976),

73, 216.

18. CHATTARJEE, N., POLATNIK, J. and BACHRACH, H. J. Gen. Virol. (1976), 32, 383.

19. SVITKIN, Y. and AGOL, V. FEBS Letters (1978), 87, 7.

20. PELHAM, H. Eur. J. Biochem. (1978), 85, 457.

21. SMITH, A. Eur. J. Biochem. (1973), 33, 301.

22. VILLA-KOMAROFF, L., GUTTMAN, N., BALTIMORE, D. and LODISH, H. Proc. Nat. Acad. Sci. U.S.A. (1975), 72, 4157.

23. CELMA, M. and EHRENFELD,E. J. Mol . Biol. (1975), 98, 761.

24. ABRAHAM, G. and COOPER, P. J.Gen. Virol. (1975), 29, 215.

25. BUTTERWORTH, B. and KORANT, B. J. Virol. (1974), 14, 282.

26. KORANT, B. J. Virol. (1972), 10, 751.

27. KORANT, B. Bioch. Biophys. Res. Commun. (1977), 74, 926.

28. DOBBERSTEIN, B., BLOBEL, G. and CHUA, N. Proc. Nat. Acad. Sci. U.S.A. (1977), 74, 1082.

29. KORANT, B. Proc. 3rd. Int., Symp. Intracell. Prot. Catal. In Acta Biol. Med. Germ. (1977), 36, 1565.

30. LEVJANT, M., BILINKIRA, V., TRUDOLJUBOWA, M. and OREKHOVICH, V. Molek. Biol. (1976), 10, 770.

31. LANGNER, J., ANSORGE, S., BOHLEY, P., WELFE, H. and BIELKE, H. Acta. Biol. Med. Germ. (1977), 36, 1729.

32. LEWIS, J. and SABATINI, D. Bioch. Biophys. Acta. (1977), 478, 331.

33. KORANT, B. In In Vitro Transcription and Translation of Viral Genomes. Haenni, A. and Beaud, G. eds.,(1975), p.273 INSERM, Paris.

34. SEVIER, E. Analyt. Bioch. (1976), 74, 592.

35. KIEHN, E. and HOLLAND, J. J. Virol. (1970), 5, 358.

36. KORANT, B. J. Virol. (1973), 12, 556.

37. LAWRENCE, C. and THACH, R. J. Virol. (1975),15, 918.

38. ESTEBAN, M. and KERR, I. Eur. J. Bioch. (1974), 45, 567.

39. HOLLAND, J., DOYLE, M., PERRAULT, J., KINGSBURY, D. and ETCHISON, J. Bioch. Biophys. Res. Commun. (1972), 46, 634.

40. COLE, D. and BALTIMORE, D. J. Mol . Biol. (1973), 76, 325.

41. ZIOLA, B. and SCRABA, D. Virology (1976), 71, 111.

42. BACHRACH, H. In Virology in Agriculture (1977), Romberger, J. ed., p.3,Allanheld, Osmun, Montclair.

43. GARFINKLE, B. and TERSHAK, D. J. Mol. Biol. (1971), 59, 537.

44. KORANT, B., KAUER, J. and BUTTERWORTH, B. Nature (London) (1974), 248, 588.

45. KORANT, B. and BUTTERWORTH, B. J. Virol. (1976), 18, 298.

46. NAKAI, K. and LUCAS-LENARD, J. J. Virol. (1976), 18, 918.

47. GEVAUDAU, P., PIERONI, G., GEVAUDAU, M. and CHARREL, J. Ann. Inst. Pasteur (1972), 122, 115.

48. LAWRENCE, C. and THACH, R. J. Virol. (1974), 14, 598.

49. GOLINI, F., THACH, S., BIRGE, C., SAFER, B., MERRICK, W. and THACH, R. Proc. Nat. Acad. Sci. U.S.A. (1976), 73, 3040.

50. CARRASCO, L. and SMITH, A. Nature (London) (1976), 264, 807.

51. COOPER, P., STEINER-PRYOR, A. and WRIGHT, P. Intervirol. (1973), 1, 1.

52. SANGAR, D., BLACK, D., ROWLANDS, D. and BROWN F. J. Gen. Virol. (1977), 35, 281.

53. RODER, A. and KOSCHEL, K. J. Virol. (1974), 14, 846.

54. SHAW, E. In Proteases and Biological Control (1975), Reich, E, Rifkin, D. and Shaw, E. eds., p.455. Cold Spring Harbor Laboratory.

55. POWERS, J. In Chemistry and Biochemistry of Amino Acids, Peptides and Proteins (1977), Weinstein, B. ed. Vol. 4, p.65, Marcel Dekker, New York.

56. SUMMERS, D., SHAW, E., STEWART, M. and MAIZEL, J. J. Virol. (1972), 10, 880.

57. PONG, S., NUSS, D. and KOCH, G. J. Biol. Chem. (1975), 250, 240.

58. FREEDMAN, M., FRIEDBERG, D., MUCHA, J. and TROLL, W. Bioch. Pharm. (1973), 22, 2441.

59. CARROLL, R. and SMITH, A. Proc. Nat. Acad. Sci. U.S.A. (1976), 73, 2254.

THE GENOME-LINKED PROTEIN OF PICORNAVIRUSES: DISCOVERY, PROPERTIES AND POSSIBLE FUNCTIONS

ECKARD WIMMER

Department of Microbiology, School of Basic Health Sciences, State University of New York at Stony Brook Stony Brook, L.I., N.Y. 11794, U.S.A.

I. DISCOVERY OF VPg AND ITS LOCATION IN VIRION RNA

The discovery that eukaryotic mRNAs were capped at the 5' end (for a review, see ref.1) prompted us and others to examine the 5' end of poliovirus mRNA. It had been shown before that virion RNA does not contain any nucleoside triphosphates (2). Consequently, virion RNA cannot be capped with $m^7G(5')ppp(5')Np$.

Poliovirus mRNA has previously been considered to be identical to virion RNA because mRNA and virion RNA have the same sedimentation coefficient of 35S, have the same base composition (3, 4) and yield indistinguishable fingerprints of RNase T1-resistant oligonucleotides (5). It was possible, however, that the infecting viral genome was modified by the formation of a capping group after entry into the cytoplasm. This possible modification of virion RNA and the capping of all other newly synthesized viral mRNA would not be evident in progeny virion RNA because viral mRNA is not encapsidated (for a review, see ref.6).

The 5' terminus of an RNA molecule may be identified by characterization of the products obtained following complete hydrolysis to mononucleotides with ribonuclease. The only products expected from RNase T2 digestion of RNA are (i) the four nucleoside 3'-monophosphates derived from internal positions in the ribonucleotide chain, and (ii) whatever structure arises from the 5' terminus. The capping group, pppNp, ppNp or pNp can be separated from internal Np and identified. Poliovirus ^{32}P-labeled mRNA was therefore isolated from polyribosomes and digested to completion with RNase T2. The products were separated by DEAE-cellulose chromatography at pH 5, using a gradient of the volatile salt triethylammonium acetate to elute the nucleotides (5). Fractions were collected into small scintillation vials and monitored for ^{32}P-label by detection of Cerenkov radiation. Materials of interest were conveniently recovered by lyophilization of fractions. Figure 1B shows a DEAE-cellulose separation of products resulting from RNase T2 digestion of polio (^{32}P)mRNA. No radioactive material with

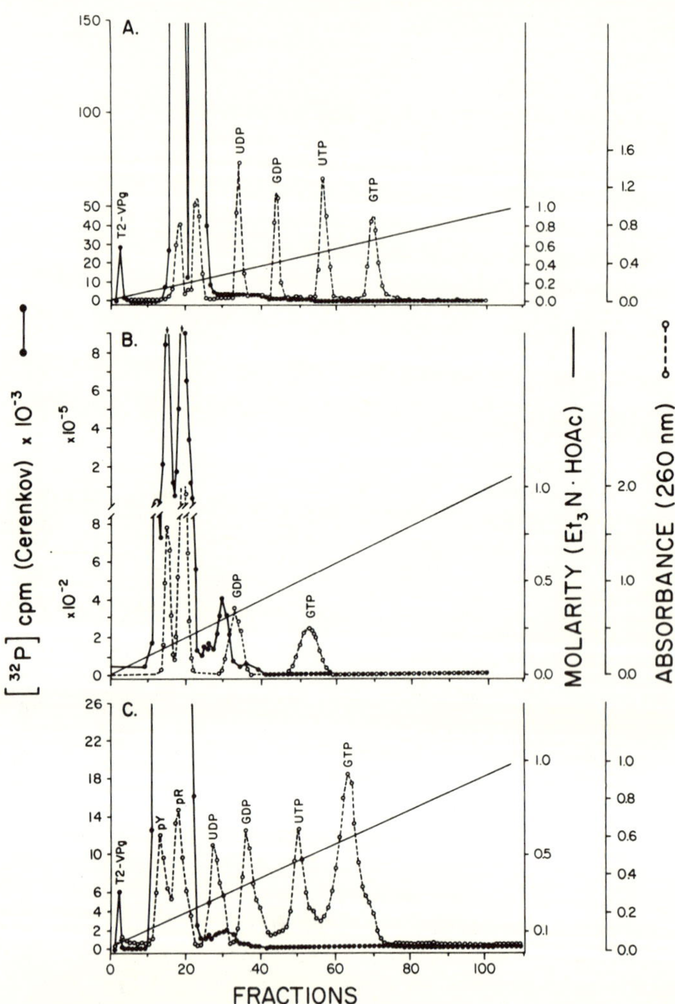

Figure 1. Separation of RNase T2 digests of ^{32}P-labeled polio virion RNA (A); mRNA (B); and replicative intermediate RNA (C); on DEAE-cellulose in the presence of marker nucleotides. Columns were in 5 ml plastic pipettes (Falcon, Serological); elution with triethylammonium acetate, pH 5, was as described (5, 11, 12).

a negative charge greater than that of GDP eluted from the column (the capping group elutes under these conditions together with GTP; ref. 5). Owing to the specificity of RNase T2, pUp can only originate from the 5' end of the mRNA. Poliovirus mRNA is therefore uncapped and except for mitochondrial mRNA (Grohmann _et al._ 1978) it still remains as the only known example of an uncapped mammalian mRNA. It is likely, however, that all picornavirus mRNAs carry a free pUp 5' end (1, 8). Hewlett _et al._ (9) and Fernandez-Mūnoz and Darnell (10) have come to the same conclusion using different methods of analysis.

As a general rule, _de novo_ initiation of RNA synthesis _in vivo_ occurs with purine rather than pyrimidine triphosphates. pUp is therefore an unlikely terminus for nascent viral RNA. This consideration led us to suggest that the pUp terminus of mRNA arose from the 5' end of viral RNA by processing (e.g. cleavage of a primer) as the newly made RNA is released from the replication complex (5). Evidence that virion RNA has a 5' terminus distinct from that of mRNA has been obtained as follows:

When virion(^{32}P)RNA was digested with RNase T2 and subjected to DEAE-cellulose column chromatography the elution profile shown in Figure 1A was obtained. No labeled compounds eluted after the nucleoside 3'-monophosphates (5, 11) indicating that virion RNA is neither terminated with a capping group, pppNp, ppNp, nor surprisingly, with pNp. It remained possible that the virion RNA had a free hydroxyl group at the 5' end, since, after RNase T2 digestion this would yield Np indistinguishable from internal nucleotides. This possibility, however, was ruled out because virion RNA could not be phosphorylated with (γ-^{32}P)ATP by polynucleotide kinase (2).

In contrast to the digests of (^{32}P)mRNA (Figure 1B) we consistently observed in the digests of virion(^{32}P)RNA, labeled material (0.015% of the total radioactive RNA) eluting in the void volume of the chromatography column (Figure 1A; ref. 5, 11, 12). This material was identified as the protein VPg (i) by labeling with (^{3}H)amino acids (for example, lys, arg, leu, or tyr); (ii) by its sensitivity to pronase, protease K, or amino peptidase (iii) by its electrophoretic behavior on paper or in polyacrylamide gels (Figure 2); and (iv) by the fact that it is insoluble in chloroform/methanol or acetone, and only partially soluble in 5-20% trichloroacetic acid (for details, see ref. 8, 11, 12, 13). Attempts to remove the protein from the RNA prior to RNase T2 digestion by various physical methods failed (11). Unambiguous proof that VPg was covalently linked to virion RNA came from the isolation of a nucleotidyl-protein (VPg-pUp) and a nonanucleotidyl-protein (VPg-pU-U-A-A-A-A-C-A-Gp) after digestion of virion RNA with RNase T2 and RNase T1, respectively (11, 13, 14).

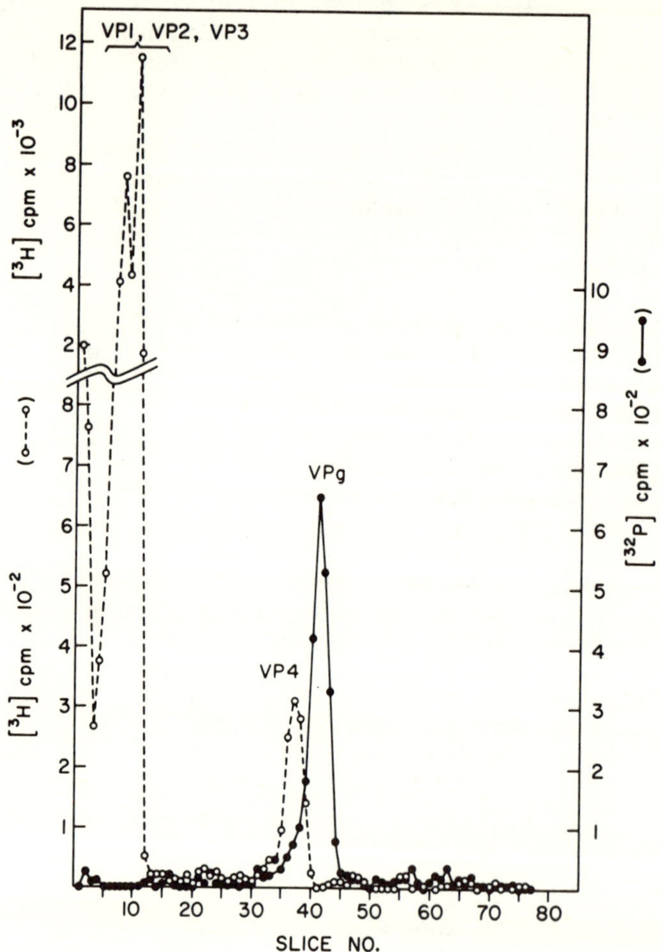

Figure 2. Polyacrylamide gel electrophoresis of ^{32}P-labeled VPg-pUp (denoted here simply as VPg) as it was released from virion (^{32}P)RNA with RNase T2 (Figure 1A). Coelectrophoresed in the same gel were the ^3H-labeled poliovirus capsid proteins (for details, see ref. 11).

The evidence that VPg is linked to the 5' end of virion RNA rather than internally can be summarized as follows: (i) Digestion of virion RNA with RNase T2 yields VPg-pUp containing two phosphates, whereas digestion of virion RNA with nuclease P1 yields VPg-pU containing only one phosphate. This observation is compatible only with VPg being linked to a 5'-terminal phosphate of uridine and not to an internal nucleotide (13); (ii) Visualization of the protein at the 5' end of virion RNA by electron microscopy (15), and (iii) Sequence analysis of the protein-linked nonanucleotide of virion RNA (VPg-pU-U-A-A-A-A-C-A-Gp) which was found to be identical to the 5'-terminal nonanucleotide of mRNA (pU-U-A-A-A-A-C-A-Gp; ref. 16, 17).

II. PROPERTIES OF VPg

From the very first analyses of VPg it was clear that the protein possessed unusual properties. When a portion of fraction 2 of the column chromatogram shown in Figure 1A was analyzed for ^{32}P-labeled material by liquid scintillation it was found to be radioisotope-free. The reason was that the (^{32}P)VPg-pUp, collected in fraction 2, had been quantitatively adsorbed to the glass vial from which it could be removed with 0.1% $NaDodSO_4$ but not with methanol/chloroform (11, 12). Adsorption to the surface of test tubes or vials occurred whether the glass had been pretreated with dimethyldichlorosilane and/or with a solution of bovine serum albumin and adsorption to plastic surfaces also occurred, although at a slower rate. Once solubilized in 0.1% $NaDodSO_4$ VPg does not precipitate with 10% acetic acid or 5 to 20% trichloroacetic acid even in the presence of carrier protein. This property probably explains why VPg elutes from polyacrylamide slab gels during staining procedures (11, 12). On the other hand, VPg can be quantitatively precipitated with acetone (8).

The solubility of VPg in acid suggested that VPg is basic and small. The positive charge of VPg was confirmed by the following observations:

(i) during electrophoresis on Whatman 3MM paper at pH 3.5 or 7.5, (^{32}P)VPg-pUp migrates toward the cathode, that is, it migrates in the opposite direction to nucleotides (11, 12, 14); (ii) The 5'-terminal protein-linked nonanucleotide, VPg-pU-U-A-A-A-A-C-A-Gp elutes from DEAE-cellulose (in 7 M urea at pH 7.5) together with trinucleotides, NpNpNp, an observation indicating that it has a net charge of -4 (13). Since a nonanucleotide has a net charge of approximately -11 under these conditions, the protein must have a net charge of about +7.

The small size of VPg was demonstrated by polyacrylamide gel electrophoresis; (^{32}P) VPg-pUp migrated slightly faster than capsid protein VP4 (Figure 2; ref.11). (^{32}P) VPg-pU behaves exactly like (^{32}P)VPg-pUp on polyacrylamide gels, showing that the loss of the 2 negative charges does not influence migration (Detjen, Nomoto and Wimmer, unpublished results). The mobility of VPg-pUp on the highly cross-linked polyacrylamide gels (18) relative to cytochrome-c and CNBr cleavage products of cytochrome-c suggests that it has a molecular weight of about 5,000 (19). Based on gel filtration studies under denaturing conditions, on the other hand, Ambros and Baltimore (20) have concluded that the molecular weight of VPg is 12,000. Since VPg is both hydrophobic and charged, deviation from normal behavior in gel electrophoresis as well as in gel filtration is possible. Thus, the exact molecular weight of VPg remains to be determined.

The amino acid composition has not been established for any picornavirus VPg. It is noteworthy, however, that polio VPg lacks methionine. This property contributed to the delay in the discovery of the protein since during the search for the methylated capping group attempts were made to label polio virion RNA with (methyl-^{3}H)Met *in vivo* (5, 21). In contrast, VPg of foot-and-mouth disease virus (FMDV) can be labeled with methionine (22).

So far the covalent attachment of a small protein to genome RNAs has been established for the following picornaviruses: poliovirus type 1 (11, 12), poliovirus type 2 (Babich, Nomoto and Wimmer, unpublished results), FMDV (22), encephalomyocarditis virus (EMC; ref. 8, 23). Preliminary data suggest that a protein is linked to the genome RNA of the rhinoviruses (Golini, Wimmer and Korant, unpublished results). Based on their migration in polyacrylamide gels, all picornavirus VPg's are smaller than polio VP4 with the exception of EMC VPg which is larger than polio VP4 (8).

It should be mentioned that the single-stranded RNA genome of the plant virus, cowpea mosaic virus has also been found to be covalently linked to a small, hydrophobic protein (G. Bruening, personal communication).

III. LINKAGE BETWEEN PROTEIN AND POLIOVIRUS GENOME

The nature of the bond between VPg and RNA has been elucidated by analyzing its stability towards acid, alkali and enzymes and by labeling VPg-RNA *in vivo* with tritiated amino acids or *in vitro* with ^{125}iodine (24, 25). These experiments have been carried out either with the nucleotidyl-protein, VPg-pUp or with proteolytic degradation products (nucleotidyl-peptides) thereof. The latter are Y(designated as (aa_m)-pUp), which is produced by pronase digestion of VPg-pUp (Lee *et al.* 1977), and Y' (designated as $(aa_n$ -pUp), which is the product of aminopeptidase treatment of Y (13). The results can be summarized as follows: (i) Y' is resistant to 0.1 M HCl at 37°C for 2 hrs and to 1 M NaOH at 100°C for 1 hr. Both, Y and Y' can be cleaved to pUp (and unidentified peptides) by snake venom 3'-exonuclease. These results exclude phosphoesters like Ser-P or phosphoramidates like nucleotidyl-(5'-N^{ω})-Arg or nucleotidyl-(5'-N^{ε})-lys as the bond between VPg and RNA. (ii) VPg linked to genome RNA can be iodinated by the Bolton and Hunter reagent (iodinated 3-(4-hydroxyphenyl) proprionic acid

N-hydroxysuccinimide ester) but not by the chloramine-T or lactoperoxidase procedures (24, 25). This observation suggested to us that VPg does not contain accessible Tyr residues. (iii) VPg-RNA can be labeled with (^{3}H)tyrosine *in vivo*. Digestion of (tyrosyl-^{3}H)VPg-pUp with Pronase yielded (tyrosyl-^{3}H)Y with almost no loss of radioactivity (24). These results are at odds with the iodination data. It is known, that O^4- derivatized tyrosine residues cannot be iodinated by the chloramine-T procedure (26). Failure to iodinate VPg by using chloramine-T or lactoperoxidase can be explained if one assumes that VPg contains only one tyrosine residue whose phenolic group is blocked. (iv) Digestion of (^{32}P) VPg-pUp with 5.6 M HCl at 110°C for 2 hr yields (^{32}P)O^4-(3' phospho-5'-uridylyl)tyrosine (Tyr-O^4-pUp) whose structure was identified by degradation with enzymes and thin-layer chromatography (24). Similarly, (^{3}H)Tyr-O^4-pUp was isolated from (tyrosyl-^{3}H) VPg-pUp (24).

These data identify the linkage between VPg and the RNA as a bond between the O^4 of tyrosine and the 5'-P atom of the terminal uridylic acid residue (24). Labeling studies of VPg with (^{3}H)tyrosine have also led Ambros and Baltimore (20) to the conclusion that the terminal pU of the genome RNA is linked to a tyrosine in VPg.

It is possible that the protein-RNA linkage in other picornavirus RNAs is also Tyr-O^4-pUp as in all cases analyzed so far (poliovirus type 1 and type 2, polio minus strands, EMC) the 5'-terminal nucleotide is pUp and the stabilities of the VPg-RNA bond towards acid, alkali or enzymes are identical.

Interestingly, the phosphodiester of nucleotidyl-O-tyrosine can be considered to be an "energy-rich" phosphate bond with a standard free energy of approximately -9.6 Kcal/mole (27).

IV. IS VPg VIRUS-CODED?

This question has been solved by two experiments:

(i) The protein can be labeled with amino acids at a time after infection when host cell protein synthesis is severely suppressed (11).

(ii) EMC-VPg isolated from virus grown in L-cells is identical to EMC-VPg from virus grown in HeLa cells, as determined by urea/NaDodSO_4/polyacrylamide gel electrophoresis. In contrast, polio-VPg isolated from virus grown in HeLa cells migrates more rapidly than EMC-VPg when analyzed in these gels, an observation indicating a lower apparent molecular weight of polio-VPg (8). These data show that the physical properties of the genome-linked proteins of two picornaviruses depend on the infecting virus rather than the host cell, suggesting that they are virus-coded. It should also be recalled that picornaviruses can replicate in the presence of actinomycin D and in enucleated cells (28, 29). Therefore, VPg cannot be a cellular protein which is induced transcriptionally after viral infection. The alternative possibility that two different VPg's are selected by EMC and poliovirus from pre-existing cytoplasmic protein pools of the same host cell seems unlikely and we therefore conclude that VPg is encoded by the viral genome.

Originally, we entertained the possibility that VPg and VP4 were related proteins (11). However, preliminary results of tryptic digests of VPg and VP4 indicate that these viral polypeptides have no overlapping amino acid sequences (19). Experiments are now in progress to map VPg within the viral genome.

V. POSSIBLE FUNCTION(S) OF VPg

The fascinating problem of the role VPg plays in picornavirus replication has not yet been solved. Based on the following evidence it seems quite certain, however, that VPg is not involved in early steps (up to and including protein synthesis) of viral replication initiated by transfection. (i) Viral mRNA, that is, viral RNA isolated from polyribosomes of the infected cell, is always 5'-terminated with pUp and never protein-linked. Thus, the protein-synthesizing machinery must select only those viral RNA molecules that are VPg-free. (ii) Viral mRNA has been reported to be infective (14). Although the specific infectivity of the mRNA was very low and details of the isolation and characterization of the mRNA were not given, these data suggest that VPg is dispensible for the penetration and translation of the transfecting RNA (14). Additional support for this conclusion comes from more convincing data by Sangar *et al.* (22) and by Notmoto *et al.* (17) who showed that virion RNA from which VPg has been cleaved proteolytically fully retains its specific infectivity. Furthermore, mengovirus RNA similarly deprived of its VPg was efficiently translated *in vitro*, and directed the synthesis of the same set of polypeptides as did the untreated mengovirus RNA (43).

Our strategy to identify the function of the genome-linked protein has been then, to analyze the step in viral replication at which VPg becomes attached to the viral RNA. For example, if VPg were to be found in replicative structures of polio RNA, its attachment to RNA (and consequently its function) might be related to replication of RNA. As polio replicative intermediate (RI) RNA is the only structure involved in synthesis of viral single-stranded progeny RNA (28) (^{32}P)RI was isolated from poliovirus infected HeLa cells (30) and analyzed for the presence of VPg. Absence of free plus strand RNA was shown by gel electrophoresis in 1% agarose (13). Digestion of denatured (^{32}P)RI with RNase T2 and separation of the products by column chromatography is shown in Figure 1C. Material eluting in the void volume was shown to be VPg-pUp (designated in Figure 1C as T2-VPg) by (i) gel electrophoresis, in which it comigrated with virion RNA-derived VPg-pUp, and (ii) digestion with Pronase and snake venom exonuclease which yielded Y and pUp, respectively. The number of nascent strands in RI preparations have been derived from the resistance of the structure to digestion with RNase (6). Based on this number and the yield of VPg-pUp we concluded that each nascent strand of the RI is protein-linked (13).

Figure 1C shows that some labeled material elutes at the position of the nucleoside diphosphates. These fractions contained all 4 pNp's and we assume that this material originates from spurious ends in the RI. Most importantly, however, no nucleoside 3'-monophosphate -5'-triphosphates (pppNp) were observed in this or in several other digests (compare ref. 5) as would be expected if initiation of RNA synthesis occurred _de novo_ with a nucleoside 5'-triphosphate. The amount of (^{32}P)RI analyzed in Figure 1C would have allowed us to detect one pppNp residue per RI molecule (13). On the basis of these considerations we have proposed that the initiation of RNA synthesis is primed and that this primer is VPg (11, 13). Flanegan _et al_. (14) and Pettersson _et al_.(31) who have also found VPg in polio RI, have entertained the same possibility.

The analysis of polio RI for VPg did not permit us to decide whether VPg was also linked to polio minus strands, since the quantitation of VPg is too inaccurate and because at 3.5 hr after infection, 90% of the RI structures produce plus strands (32). Evidence that VPg was linked to the 5'-terminal poly(U) of minus strands of RI as well as viral double-stranded RNAs (RF) was obtained by an analysis of hybrid double-stranded composed of ^{32}P-labeled minus strands and unlabeled plus strands (13). (^{32}P) poly(U) was isolated from these hybrids and had a 5'-terminal pU linked to VPg (13). This result has recently been confirmed by Pettersson _et al_. (31). It has also been extended by Wu _et al_. (15), who derivatized the VPg of polio RF with dinitrofluorobenzene and subsequently complexed it with anti-DNP IgG. Inspection of these molecules in the electron microscope showed that both ends of RF were protein-tagged. The frequency of tagging was high enough to conclude that both strands of polio RF are protein-linked.

A schematic representation of the terminal structure of various poliovirus RNAs is shown in Figure 3. All newly synthesized RNAs are VPg-linked with the exception of mRNA. RI molecules are represented in two different structures (A and B) either of which may occur _in vivo_. Structure (A) is favored by evidence obtained by Öberg and Philipson (33) and Thach _et al_. (34) and structure (B) by other more recent evidence (35).

The arrows in Figure 3 indicate the relationships between

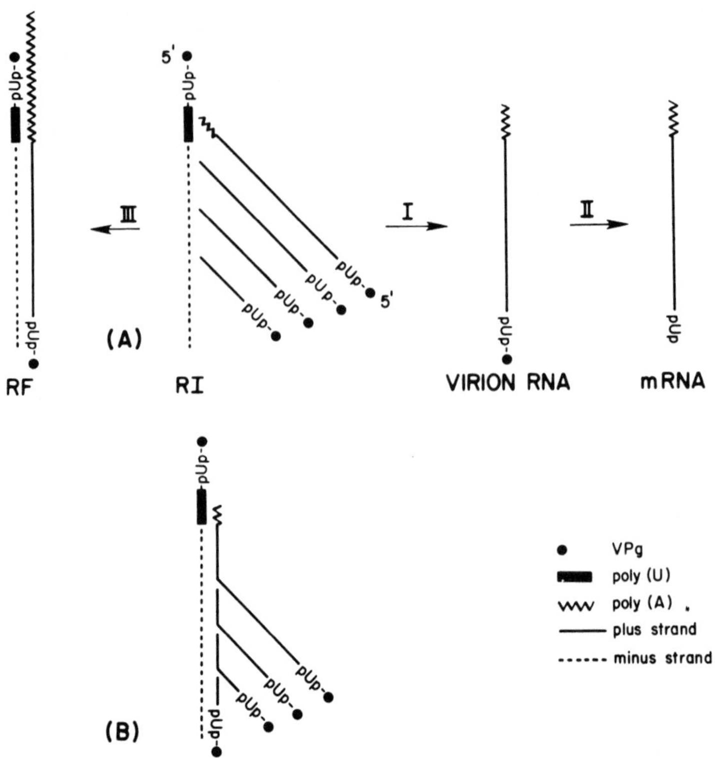

Figure 3. Schematic representation of terminal structures of poliovirus RNAs. The arrows (I, II and II) suggest the relationship between different RNA structures. Both structures (A) or (B) of polio RI may occur *in vivo*, but structure (B) certainly arises upon deproteinization of structure (A). For further discussion, see text.

different intracellular viral RNA species. Thus, VPg-linked plus strands are released from RI (Figure 3, I) and encapsidated within a few minutes (6). RNA destined to form polyribosomes may be formed by a "processing" step to yield pUp-terminated mRNA (13,17). This process (Figure 3, II) may involve cleavage of the linkage between VPg and RNA, since the 5'-terminal, RNase T1-resistant oligonucleotide of poliovirus mRNA is identical to that of virion RNA (16, 17). The mechanism leading to the formation of RF molecules (Figure 3, III)is unknown (see below). It should be stressed that the kinetics of turnover of RF and the exceptionally long poly(A) tails in plus strands of RF provide strong evidence that these molecules are by-products of replication and not artefacts of the isolation procedures (36-38).

Based on these considerations we propose the replication scheme of picornaviruses shown in Figure 4 (compare ref.13). After release of the viral genome from infecting particles the linkage between VPg and RNA may be cleaved to yield mRNA. A host cell specific protein capable of cleaving VPg from virion RNA has recently been found by Ambros _et al_. (39). This enzyme may "process" the incoming genome RNA as well as those newly synthesized RNA strands that are on their way to become mRNA (see below). Removal of the positively charged protein may be obligatory for translation _in vivo_, although this speculation rests entirely on the fact that viral polyribosomal RNA isolated 3.5 hr after infection is VPg-free.

Newly synthesized viral proteins including VPg (or a precursor of VPg - see below) may form the putative replicase, a protein complex capable of initiating RNA synthesis. In the process of initiation of RNA synthesis VPg may function as primer and be

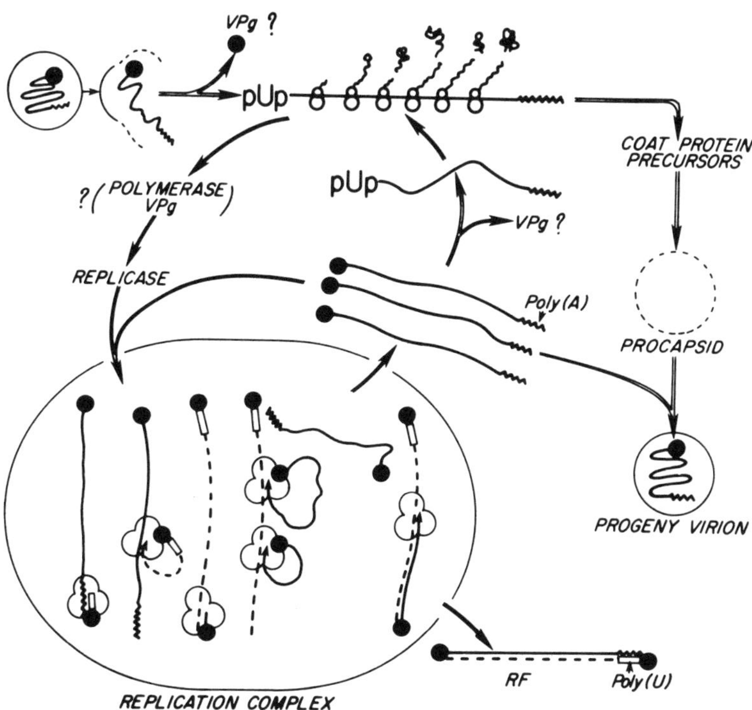

Figure 4. Replication scheme of poliovirus with emphasis on possible functions of VPg. Note that the 5' ends of the template strands in the replication complex are shown to be VPg-linked; this may not be necessary.

covalently bound to the first nucleotide (uridylic acid) of the nascent strand. Strand elongation will then proceed with VPg linked to RNA and remaining attached to the polymerase. This condition, although purely speculative, would prevent the nascent strand from annealing to the complementary template RNA. Synthesis of minus and plus strands occurs by the same mechanism, the 5'-terminal poly(U) of minus strands being transcribed from the 3'-terminal poly(A) of plus strands and vice versa (30, 40, 41).

If the RNA-linked VPg dissociates from the RNA polymerase after initiation of RNA synthesis, strand elongation could occur in the absence of the "initiator" protein. As a consequence the nascent strand may anneal to template RNA and generate double-stranded RNA (RF).

The newly synthesized plus strand RNA can serve as template or may enter the pool of RNA which is encapsidated. It should be emphasized that the model of RNA replication (Figure 4) predicts that template RNA need not be linked to the protein, so mRNA or protease-treated virion RNA could function as template at the onset of RNA replication (17).

A third function of the newly sunthesized plus strand RNA is its participation in protein synthesis. Since all mRNA molecules are VPg-free, this step must be preceeded by the cleavage of VPg from the 5' end of plus strand RNA. Based on this consideration one can predict from the number of plus strand RNA molecules synthesized (6) that a considerable number of free molecules of VPg, (at least 10^5 molecules/cell) should be found in the cytoplasm of the infected cell. This number can be increased by a factor of 10 if free VPg is produced before initiation of RNA synthesis by the processing of each polyprotein synthesized.

To our surprise, however, we have not been able to identify free VPg in extracts of infected cells (Golini and Wimmer, unpublished results). This paradox may be explained if VPg is generated by proteolytic cleavage from a larger precursor protein only at the moment of initiation of RNA synthesis, and is then covalently linked to the RNA and if cleavage of VPg from newly synthesized plus strands results in the concomitant rapid degradation of the protein. A more radical hypothesis is that VPg is synthesized as part of a precursor protein distinct from polyprotein NCVP 00, and that this precursor protein for VPg is translated at much lower frequency than polyprotein NCVP 00 (see E. Ehrenfeld, this volume, chapter 11).

A tight coupling of poliovirus RNA synthesis to viral protein synthesis exists throughout the infectious cycle even though the sites of translation and replication are physically separated (6). Our model may offer an explanation for this control mechanism.

Since there appears to exist no pool of free VPg in the infected cell the attachment of VPg to RNA requires that VPg or its precursor be continuously supplied by the translational machinery (13).

If mRNA has the same nucleotide sequence as virion RNA, why are mRNA strands not encapsidated into virions? A trivial explanation would be that viral RNA, once it has combined with ribosomes, may subsequently not exist free of them. However, free viral RNA (RNA not bound to ribosomes) that is 5'-terminated with pUp has been found in cytoplasmic extracts of infected cells (21). Thus there exists the intriguing possibility that, in addition to the role in replication, VPg linked to the 5' end of plus-strand RNA is also a prerequisite for virion formation. For example, the newly synthesized VPg-RNA may associate with one or several capsid protomers (42) prior to encapsidation. This protomer/VPg-RNA complex could then protect the VPg(Tyr)-O^4pU linkage from cleavage by the host cell enzyme. Consequently, the concentration of protomers would determine how much of the newly synthesized VPg-RNA is "processed" to mRNA, that is, early in infection most if not all progeny RNA may be cleaved at the 5' end. If VPg linked to RNA is a prerequisite for encapsidation, why do virions not contain minus strands? A simple explanation could be provided by the observation that newly synthesized minus strands are rapidly incorporated into RI and RF structures inside the membrane-bound replication complex; minus strands are not found as free RNA in the cytoplasm (5) and are thus neither cleaved at the 5' end nor encapsidated.

Most of the functions of VPg alluded to in this chapter are matter of speculation. Research during the next few years will undoubtedly produce answers to many of these questions and this, in turn, will shed light on many of the unsolved problems in picornavirus replication.

ACKNOWLEDGEMENTS

I would like to express my gratitude to my colleagues for their contribution to the work described here, particularly Alexander Babich, Barbara Morgan-Detjen, Fred Golini, Timothy J.R. Harris, Naomi Kitamura, Don van der Kolk, Yuan Fon Lee, Akiko Nomoto, Nancy Reich and Paul G. Rothberg. I thank Timothy J.R. Harris for his critical reading of the manuscript, and Sandi Donaldson for the preparation of the manuscript. This work was supported in part by Grant CA-16879 awarded by the National Cancer Institute and Grant AI-15122 awarded by the National Institute of Allergy and Infectious Diseases, Department of Health, Education and Welfare.

REFERENCES

1. SHATKIN, A.J. Cell (1976), 9, 645-653.

2. WIMMER, E. J. Mol. Biol. (1972), 68, 537-540.

3. SUMMERS, D.F. and LEVINTOW, L. Virology (1965), 27, 44-53.

4. SUMMERS, D.F., MAIZEL, J.V. and DARNELL, J.E. Virology (1967), 31, 427-435.

5. NOMOTO, A., LEE, Y.F. and WIMMER, E. Proc. Natl. Acad. Sci. U.S.A. (1976), 73, 375-380.

6. BALTIMORE, D. In The Biochemistry of Viruses, Levy, H.B. ed. (1969), pp. 101-176, Marcel Dekker, New York.

7. GROHMANN, K., AMALRIC, F., CREWS, S. and ATTARDI, G. Nucl. Acids Res. (1978), 5, 637-651.

8. GOLINI, F., NOMOTO, A. and WIMMER, E. Virology (1978), 89, 112-118.

9. HEWLETT, M.J., ROSE, J.K. and BALTIMORE, D. Proc. Natl. Acad. SCI. U.S.A. (1976), 73, 327-330.

10. FERNANDEZ-MUÑOZ, R. and DARNELL, J.E. J. Virol. (1976), 18, 719-726.

11. LEE, Y.F., NOMOTO, A., DETJEN, B.M. and WIMMER, E. Proc. Natl. Acad. Sci. U.S.A. (1977), 74, 59-63.

12. LEE, Y.F., NOMOTO, A. and WIMMER, E. Prog. Nucl. Acid. Res. and Mol. Biol. (1976), 19, 89-96.

13. NOMOTO, A., DETJEN, B., POZZATTI, R. and WIMMER, E. Nature (London) (1977), 268, 208-213.

14. FLANEGAN, J.B., PETTERSSON, R.F., AMBROS, V., HEWLETT, M.J. and BALTIMORE, D. Proc. Natl. Acad. Sci. U.S.A. (1977), 74, 961-965.

15. WU, M., DAVIDSON, N. and WIMMER, E. Nucl. Acids Res. (1978), 5, 4711-4724.

16. PETTERSSON, R.F., FLANEGAN, J.B., ROSE, J.K. and BALTIMORE, D. Nature (London) (1978), 268, 270-272.

17. NOMOTO, A., KITAMURA, N., GOLINI, F. and WIMMER, E. Proc. Natl. Acad. Sci. U.S.A. (1977), 74, 5345-5349.

18. SWANK, R. and MUNKRIES, K. Anal. Biochem. (1971), 39, 462-477.

19. DETJEN, B.M. Dissertation (1978), St. Louis University, St. Louis, Missouri.

20. AMBROS, V. and BALTIMORE, D. J. Biol. Chem. (1978), 253, 5263-5266.

21. FERNANDEZ-MUÑOZ, R. and LAVI, U. J. Virol. (1977), 21, 820-824.

22. SANGAR, D.V., ROWLANDS, D.J., HARRIS, T.J.R. and BROWN, F. Nature (London) (1977), 268, 648-650.

23. HRUBY, D.E. and ROBERTS, W.K. J. Virol. (1978), 25, 413-415.

24. ROTHBERG, P., HARRIS, T.J.R., NOMOTO, A. and WIMMER, E. Proc. Natl. Acad. Sci. U.S.A. (1978), 75, (in press) N 10.

25. HARRIS, T.J.R., DUNN, J.J. and WIMMER, E. Nucl. Acids Res. (1978), 11, 4039-4054.

26. ADLER, S.P., PURICH, D. and STADTMAN, E.R. J. Biol. Chem. (1975), 250, 6264-6272.

27. HOLZER, H. and WOHLHUETER, R. In Advances in Enzyme Regulation, Weber, G. ed. (1972), pp.121-132, Pergamon Press, Oxford.

28. REKOSH, D.M.K. In The Molecular Biology of Animal Viruses, Nayak, D.P. ed. (1977), Vol. 1, pp.63-110, Marcel Dekker, New York.

29. DETJEN, B.M., LUCAS, J.J. and WIMMER, E. J. Virol. (1978), 27, 582-586.

30. YOGO, Y. and WIMMER, E. J. Mol. Biol. (1975), 92, 467-477.

31. PETTERSSON, R.F., AMBROS, V. and BALTIMORE, D. J. Virol. (1978), 27, 357-365.

32. BISHOP, J.M., KOCH, G. and EVANS, B. J. Mol. Biol. (1969), 46, 235-249.

33. ÖBERG, B. and PHILIPSON, L. J. Mol. Biol. (1971), 58, 725.

34. THACH, S., DOBBERTIN, D., LAWRENCE, C., GOLINI, F. and THACH, R.E. Proc. Natl. Acad. Sci. U.S.A. (1974), 71, 2549-2553.

35. MEYER, J., LUNDQUIST, R.E. and MAIZEL, J.V. Jr., Virology (1978), 85, 445-455.

36. YOGO, Y. and WIMMER, E. Nature New Biol. (1973), 242, 171-174.

37. YOGO, Y., TENG, M.H. and WIMMER, E. Biochem. Biophys. Res. Comm. (1974), 61, 1101-1109.

38. SPECTOR, D.H. and BALTIMORE, D. J. Virol. (1975), 15, 1418-1431.

39. AMBROS, V., PETTERSON, R.F. and BALTIMORE, D. Cell (1978), 15, 1439-1446.

40. YOGO, Y. and WIMMER, E. Proc. Natl. Acad. Sci. U.S.A. (1972), 69, 1877-1882.

41. DORSCH-HASLER, K., YOGO, Y. and WIMMER, E. J. Virol. (1975), 16, 1512-1527.

42. RUECKERT, R.R. In Comprehensive Virology, Fraenkel-Conrat, H. and Wagner, R.R. eds. (1976), 6, pp.131-213, Plenum Press, New York.

43. PEREZ-BERCOFF, R. and GANDER, M. FEBS Lett. (1978), 96, 306-312.

SECTION IV:

ON DECODING THE VIRAL mRNA

THE MECHANISM AND CYTOPLASMIC CONTROL OF MAMMALIAN PROTEIN SYNTHESIS

RICHARD J. JACKSON

Department of Biochemistry, University of Cambridge

Cambridge, England

INTRODUCTION

My aim in this article is to summarise the state of knowledge, as I see it, concerning a number of selected topics in the field of mammalian protein biosynthesis, and to discuss the major unsolved problems in these areas. The reasons governing the choice of these topics are: (a) they are the stages at which cytoplasmic control mechanisms seem to operate, (b) they are the most controversial areas in which future research is likely to be concentrated, and (c) they are the topics of most relevance to the problems of the control of viral gene expression, the mechanism of shut-off of host cell protein synthesis, and the action of interferon. In the interests of coherence and brevity, I have made reference to recent reviews wherever possible, and I have limited the citation of published research results to those papers which have been decisively instrumental in formulating the concepts which I will discuss. The lack of reference to any particular paper should not be taken to imply that I regard it as being without merit, but merely that I do not consider it as indispensible for our current state of knowledge. For a more comprehensive survey of the literature on these topics, and for reviews on other aspects of protein synthesis omitted from this discussion, the reader should consult the relevant chapters in reference (1), and Lodish's review on translational control of protein synthesis (2).

For many years eukaryotic protein synthesis attracted much less attention than the bacterial system. This, together with the greater complexity of the eukaryotic system and the lack of available mutants, has resulted in a lag in our knowledge, but the deficit is now rapidly being made good. One eukaryotic cell-free protein synthesis system, the reticulocyte lysate, accurately reflects the intact cell in terms of the rate of protein synthesis and the controls that regulate it, and this system can be converted into an mRNA-dependent translation system of comparable activity. The solution to most of the outstanding problems therefore lies simply in unravelling and understanding events that take place as a matter of routine in our test tubes, and it is surely not beyond the wit of man to achieve this with current technology.

The greater complexity of the eukaryotic protein synthesis system (larger ribosomes containing more structural proteins of greater average size, a greater number of initiation factors, the unusual features of the termini of eukaryotic mRNA, and the fact that the mRNA is associated with proteins in mRNPs) not only impedes progress towards understanding the mechanism of mammalian protein synthesis, but also leads to speculations about its 'raison d'être' which at the moment we cannot explain. Perhaps it is only natural to think that if the bacterial system represents the basic machinery needed to synthesise proteins, the additional components of the more complex eukaryotic system are in some way involved in control mechanisms. It is certainly the case that mammalian protein synthesis is subject to regulation at the translational level, but there is so far no indication that anything approaching all the extra complexity is needed for this control.

It may be that we should be more concerned with why the bacterial machinery is relatively so simple. In rapidly growing bacterial cells a very substantial fraction of the total cellular protein is involved directly in protein biosynthesis (activating enzymes, ribosomal structural proteins, elongation factors etc). Evolutionary pressure to attain high growth rates may therefore have resulted in a streamlining of the bacterial protein synthesis mechanism to the bare minimum of components, in which case we should not look for some subtlety of regulation in every additional feature of complexity in the eukaryotic system.

I. RIBOSOME STRUCTURE

The gap in our understanding of eukaryotic as opposed to prokaryotic systems is most acute in respect of what is known about the structure of ribosomes. Nevertheless some progress has been made in the last few years, and several of the approximately 40 different proteins of the large ribosomal subunit and the 30 different proteins of the small subunit have now been purified. One can anticipate further advances along the same lines as have been exploited in the study of bacterial ribosomes: characterisation and amino acid sequencing of the purified proteins, precise determination of their stoichiometry, and investigations into the topography of the ribosomes using methods for the chemical cross-linking of proteins. There has so far been little insight into the relationship between the structure and function of the ribosomes. Progress in this direction has been blocked by our inability to reconstitute active mammalian ribosomal subunits *in vitro* from the constituent proteins and RNA. To date, the use of antibodies against ribosomal proteins is the only successful approach to the problem (3), and even this may prove not very satisfactory because mammalian ribosomal proteins sometimes have poor antigenicity.

At the present moment, the most important single question which we would wish to see answered by these studies is whether mammalian ribosomes show any structural heterogeneity which might be indicative of functional heterogeneity. The current state of the art does not allow this question to be answered definitively but the overall opinion is against the existence of fractional proteins and in favour of a simple stoichiometry of the ribosomal structural proteins (3). Moreover the rate of turnover of the individual ribosomal proteins suggests that ribosomal subunits are fixed entities which do not exchange proteins to any significant extent, with the possible exception of three proteins of the large subunit (4). These tentative conclusions drawn from studies of ribosome structure are happily in accord with investigations into their function. Cell-free translation experiments suggest that essentially all ribosomes are non-discriminatory with respect to different mRNA species. Studies of the rate of ribosome run-off after inhibition of initiation further suggest that all ribosomes have very similar rates of elongation on mRNA. Although non-functioning 80S monomeric ribosomes are present in the majority of tissues or cell lines, the evidence suggests that for the most part these are not a special class of ribosomes, fundamentally different from the polysomal ribosomes. They can virtually all be mobilised into the active state in polysomes when the rate of elongation is reduced (e.g. by low concentrations of cycloheximide) to make initiation no longer the rate limiting step (2). The proportion of non-functioning 80S ribosomes is also drastically reduced by careful treatment of the cells - nutritionally rich media, adequate serum levels, and suitable cell densities. At least the majority, if not all, of the 80S monomeric ribosomes are therefore thought to be temporarily in an inactive state solely as a consequence of insufficient initiation capacity or insufficient mRNA levels. Whilst our present level of understanding cannot exclude minor subtle effects, it seems reasonably certain that there is no gross functional or structural heterogeneity of mammalian ribosomes.

Mammalian ribosomes are subject to phosphorylation and probably also to acetylation (3). It is naturally tempting to assume that such modifications affect the activity of the ribosomes in some way. Before this idea can be evaluated it is important to be sure that the phosphorylation observed *in vitro* does in fact occur *in vivo*, and in this respect the history of the study of ribosome phosphorylation provides some cautionary lessons. The number of ribosomal proteins phosphorylated in intact cells or when native crude ribosomes are studied *in vitro*, is considerably less than can be phosphorylated using salt-washed ribosomal subunits (5, 6). Whether this increase in phosphorylation sites is due to the unmasking of sites by washing the ribosome free of associated non-ribosomal proteins, or whether it is caused by some denaturation of ribosomal structural proteins during the washing procedure, is not yet clear.

Ribosomes phosphorylated *in vitro* show no alteration in their translational activity (3, 7). Trivial explanations such as the possibility that the ribosomes are dephosphorylated in the translation assay seem to have been eliminated. It is, of course, possible that current assay methods are inadequate to detect subtle changes caused by phosphorylation, but it is equally valid to draw the straightforward conclusion that phosphorylation does not affect the activity of ribosomes. This is in accord with the fact that changes in the phosphorylation state of ribosomes *in vivo* do not appear to correlate with alterations in the efficiency of the ribosomes in protein biosynthesis. The one possible exception is the phosphorylation of one of the small ribosomal subunit proteins (S2) that occurs after infection by vaccinia virus (8). The timing of the phosphorylation of S2 seems however to be related neither to the shut-off of host cell protein synthesis nor to the switch from early viral protein synthesis to late gene expression, and it remains to be proven whether the phosphorylation has any material effect on ribosome activity or specificity.

II. INITIATION FACTORS AND THE MECHANISM OF INITIATION

Two groups, one in N.I.H. and the other under Theo Staehelin in Basel, have taken the purification and characterisation of mammalian initiation factors virtually to completion (and Hershey's group has recently achieved the same result following procedures derived from those of the Basel group). The results of others are largely in agreement, but it is only the work of these three groups that provides us with anything approaching a complete picture of how initiation takes place, and our confidence in this model is increased by the fact that the Basel and N.I.H. groups have exchanged and compared each other's factors and have established a fair measure of agreement.

It is accepted that there are seven initiation factors in rabbit reticulocytes: eIF-1,2,3,4A,4B,4C and 5 (9-13). The N.I.H. group overlooked eIF-1 because it is present in saturating amounts as a contaminant of the enzyme preparation used to support elongation in their assays, whilst the Basel group originally overlooked eIF-4C since it was present as a contaminant of their

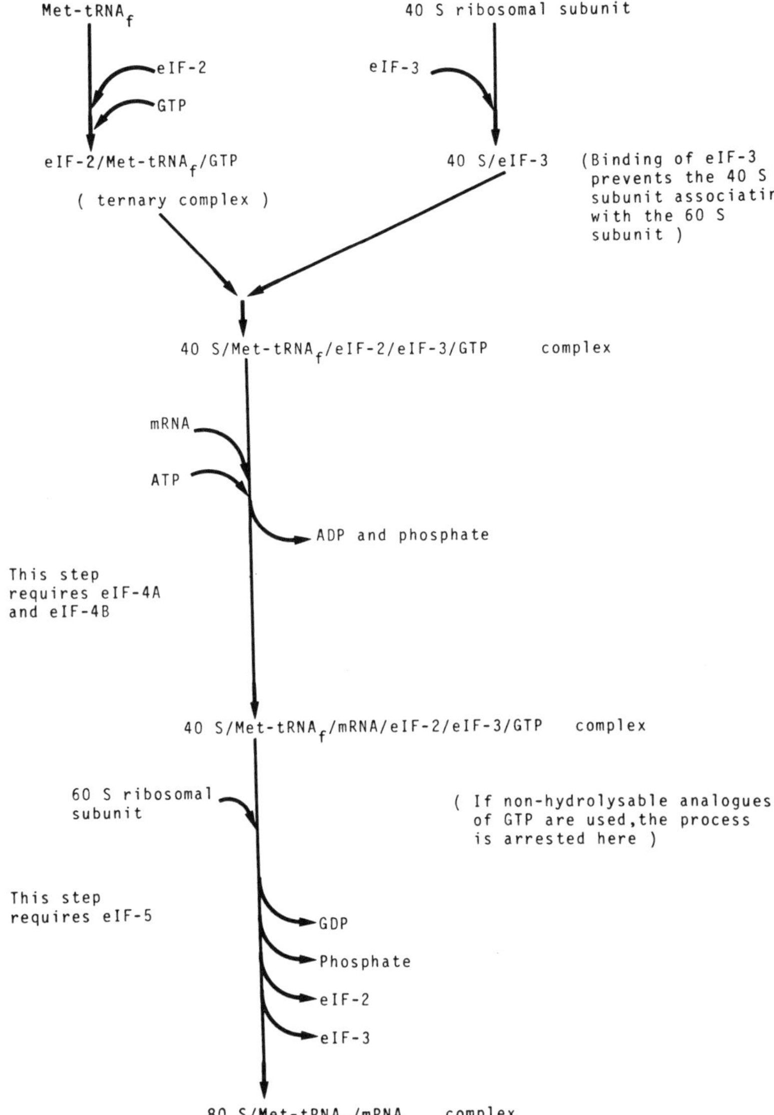

Figure 1. The sequence of events in mammalian initiation. This flow diagram shows the order of the steps, the intermediate complexes identified so far, and the different initiation factor requirements for each step (apart from eIF-1 and eIF-4C which both have rather ill-defined roles). The association of initiation factors with the intermediate complexes is known only in the cases of eIF-2 and eIF-3. The other factors do not bind sufficiently tightly for such association to be detected by the usual methods.

eIF-5 preparations (9, 11). Both of these factors, particularly eIF-1, stimulate the rate and extent of initiation complex formation, rather than being absolutely required as are the other factors. Besides these seven factors the N.I.H. group have isolated a protein known as IF-M2B$_\alpha$ (or eIF-4D, see reference (13)) which also has a stimulatory function in their hands, but it is not considered to be a bona fide initiation factor by the Basel group (9, 11) since they find that it does not stimulate in the presence of polyamines, and only shifts the Mg^{++} optimum in the absence of polyamines (the conditions used by the N.I.H. group). In addition there is the factor known as M1 which catalyses the binding of many species of aminoacyl-tRNA, or, better, N-acyl-aminoacyl-tRNA, to 40S ribosomal subunits in a codon-dependent and GTP-independent reaction (reviewed in reference 14). There is good evidence that this factor, which is present in a wide range of eukaryotic cells, is not required for natural initiation (15), and its role, if any, remains a mystery.

The way in which the seven recognised initiation factors are believed to promote initiation is shown in Figure 1. This scheme is based on experiments in which the binding of labelled ligands (eIF-2, eIF-3, Met-tRNA$_f$, mRNA) to ribosomes is studied, not only under normal conditions but also with certain individual components omitted, or with GTP and ATP replaced by non-hydrolysable analogues (10, 11, 13). The details still remain to be worked out, but there is little doubt that this scheme is correct in outline. It is fully consistent with experiments carried out using the more active crude reticulocyte lysate, where it was first observed that 40S ribosomal subunits bind initiator tRNA (Met-tRNA$_f$) before associating with the mRNA (16) - a view which was surprising and highly controversial at the time, but is now fully accepted.

Have all the proteins required for initiation now been identified? There is no doubt that the seven factors are sufficient to promote initiation, but in comparison with the crude reticulocyte lysate these systems that use purified components function at a very slow rate - of the order of 5% (11). Moreover, the amounts of initiation factors required to obtain maximal rates in the fractionated system seem very high. Although there are no firm estimates of the concentrations of initiation factors in crude lysates, few would dispute that they are probably lower than the levels used in fractionated systems, where, with one exception, the relationship between the quantities of factors added and the number of initiation events is such that each molecule of factor might function only once in a stoichiometric manner. The exception is eIF-5 which saturates at low levels and definitely acts catalytically. For the other factors, the explanation may be that there is a large fraction of denatured factor which is defective in catalysing the correct overall reaction, but can interfere with the action of the fully active factor, and thereby cause a low rate of initiation even at high levels of input. On the other hand we cannot yet exclude the possibility that there are additional factors to be discovered which have a rate-enhancing function or which act to recycle one or more of the basic seven factors so that they can operate catalytically rather than stoichiometrically (compare the role of EFT_s in the recycling of EFT_u).

Although globin mRNA has generally been used for the assay and study of initiation factors, sufficient experiments have been done with other mRNAs for it to be clear that the same set of seven initiation factors is needed for all species of mRNA (17). There is absolutely no indication that initiation on any one type of mRNA may require different factors. The only qualification is that some of the factors, particularly eIF-4A and eIF-4B, may be needed in different relative proportions and in different absolute amounts for maximal rates of initiation on different mRNAs (17-19). Thus, lower levels of eIF-4B seem to be needed for picornavirus RNA than for cellular mRNAs, but, according to the results of Staehelin's group (17), the presence of eIF-4B is not totally superfluous for initiation on the viral RNA.

With the same set of seven initiation factors, each of which has been purified to apparent homogeneity, being needed for all species of mRNA, the concept of factors specific for certain types of mRNA would seem to have little or no validity. This needs possible qualification in respect of eIF-3, a factor which acts not only as an anti-association factor (binding to 40S subunits to prevent their association with the 60S subunits), but is also needed for binding mRNA to the 40S/Met-$tRNA_f$ complex (10, 11, 13). As purified, this protein has a native molecular weight of 0.5 - 0.6 x 10^6, but on dodecylsulphate- polyacrylamide gel electrophoresis it is seen to consist of at least ten polypeptide chains which are not in equimolar proportions (as judged by staining with Coomassie Blue). All of these subunits seem to be an integral part of eIF-3; not only do they copurify, but when eIF-3 binds to 40S subunits all the constituent polypeptides bind (13). The non-equivalence of the subunit quantities may possibly be explained by partial proteolysis during the purification of the factor. The alternative is that eIF-3 may be heterogeneous, and since this factor is needed for binding mRNA to the initiation complex, the possibility of mRNA-discriminating initiation factors cannot be completely ruled out. Even if such heterogeneity exists, it is unlikely to have major significance in differentiation and development, since the fact that a cell-free system from a specialised cell such as a reticulocyte can translate mRNAs from a wide range of sources (mammalian, avian, plant, fish and protozoa) and RNAs of plant and animal viruses, all with comparably high efficiency, argues that the initiation factor complement of a cell is not narrowly and restrictively adapted to initiation on homologous mRNA. Since mRNA-discriminating factors most probably do not exist, it follows that any control mechanisms that modulate the activity of initiation factors must affect initiation on all types of mRNA, although there may be some quantitative differences resulting from competition between different mRNA species (2).

III. CONTROL OF INITIATION IN RETICULOCYTE SYSTEMS

It has been known for over 20 years that reticulocyte cells require Fe^{++} or haemin to maintain maximal rates of protein synthesis, and that the effects of Fe^{++}-deprivation are fully reversible (for reviews see reference 20, 21). Subsequent work on polysome distributions showed that it was initiation that is subject to control. The key step in working out the control mechanism was the ability to reproduce the regulatory step in the cell-free system: in the presence of haemin, reticulocyte lysates synthesise protein for up to 60 minutes at essentially the same rate as the intact cell, but in the absence of haemin there is an abrupt reduction in the rate after 5-10 minutes incubation. This shut-off can be shown to be reversible, and to be due to a reduced rate of initiation. Once the sequence of events in the initiation process (Figure 1) had been worked out, the question as to what goes wrong in the absence of haemin could be pinpointed to the formation of 40S/Met-$tRNA_f$ complexes (22).

In the meantime, this control of initiation has taken on greater significance by the finding that initiation is reversibly inhibited under a very wide variety of conditions (see references 20, 21, 23): in the absence of added haemin; in the presence of low concentrations of double-stranded RNA (dsRNA); when oxidised glutathione is added; at high temperatures (42-44°); when the lysate has been subjected to high hydrostatic pressures; when the lysate has been prepared from cells subjected to ATP depletion by incubation under anaerobic conditions or with inhibitors of oxidative phosphorylation (24, 25); when the lysate is depleted of low molecular weight compounds by gel-filtration.

In all these situations, protein synthesis proceeds at control rates for a few minutes before there is an abrupt transition to a low rate, about 2-10% of the control. This transition is immediately preceeded by the disappearance of the 40S/Met-$tRNA_f$ complexes, even though native 40S subunits, Met-$tRNA_f$ and GTP levels are normal. The inhibition can be overcome by the addition of relatively large amounts of purified eIF-2; addition of eIF-1, eIF-3, eIF-4A or eIF-5 does not overcome the inhibition (26). The inhibition can also be overcome by the addition of 5 mM 3': 5' cAMP, 2-aminopurine and several other related compounds, but not by 3': 5' cGMP.

The fact that these common characteristics are exhibited in all the inhibitory conditions listed above suggests that inhibition is

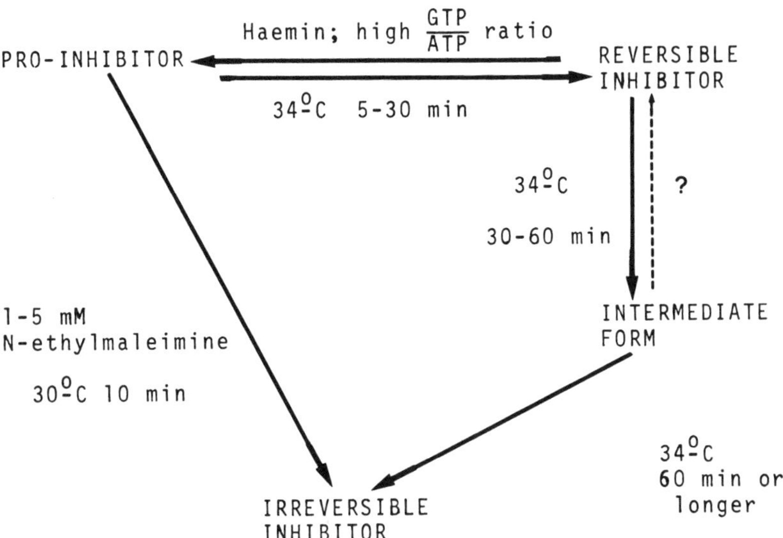

Figure 2. The interconversion of the different forms of the haem-controlled initiation inhibitor of rabbit reticulocytes.

The incubation times shown are approximate and refer to the time required to generate the particular form of the inhibitor starting from pro-inhibitor. The appearance of the irreversible form can be detected after about 60 min incubation at 34° but it requires 12-15 hours to generate the maximum levels of irreversible inhibitor. For further details see references 27-30.

due to the same proximal lesion (probably the inactivation of eIF-2 or of some factor necessary for the repeated functioning of eIF-2) in each case even though there may be differences in the detailed mechanisms. Further analysis has concentrated on the effects of haem-deficiency, which has been shown to lead to the activation of an inhibitor of initiation. This inhibitor is a protein present in the ribosome-free supernatant fraction which is activated by incubation in the absence of haemin or by brief incubation with N-ethyl maleimide (27). It exists in a bewildering variety of forms distinguishable by the methods used for activation, by whether incubation with haemin can inactivate the inhibitor, and by whether the inhibition of protein synthesis in the assay system is transient or permanent (28). Investigations into the cross-reactivity of the different forms against antibodies, the kinetics of formation of the different forms, and their purification properties suggest that they are all different forms of the same protein (29, 30), as depicted in Figure 2. The irreversible form has been extensively purified, and from the very first attempts (31) it was clear that its potency as an inhibitor was such that its mechanism must be catalytic. Subsequently it was shown to require ATP - or more specifically ATP hydrolysis - in order to inhibit the formation of 40S/Met-$tRNA_f$ complexes, and to have a protein kinase activity which phosphorylates the smallest of the three protein subunits eIF-2, but has little activity towards other protein substrates (23). The inhibitor of protein synthesis and the protein kinase activity copurify and have similar heat stabilities. Protein kinase activity is very low in extracts in which the inhibitor has not been activated, and the two properties are activated in parallel (32). High levels of cAMP and other purine derivatives (but not cGMP) are found to inhibit the protein kinase activity.

These observations lead to the hypothesis that the inhibition of initiation in lysates incubated without haemin is due to the phosphorylation of eIF-2 by the protein kinase which is activated under these conditions. The action of the purine derivatives in overcoming the inhibition is explained by their properties as inhibitors of the protein kinase activity. Attractive though this hypothesis may be, it received a set-back when it was discovered that phosphorylated eIF-2 showed absolutely no defect in the fractionated systems used to study initiation factors (33). Nevertheless the circumstantial evidence in favour of the hypothesis is sufficiently compelling that it is generally accepted as being correct, and current research is directed towards finding an explanation for the failure of the fractionated system to show any change in the activity of eIF-2 following phosphorylation. Recently two groups have reported that a defect can be observed provided an additional protein is present (34, 35). Whilst this superficially solves the problem, it would be more satisfactory if this additional protein could be shown to have a defined role in the actual mechanism of initiation.

Low concentrations of dsRNA also activate a protein kinase which phosphorylates the small subunit of eIF-2 (23). This kinase differs from the haem-controlled enzyme in its subcellular location (it is mainly associated with the ribosomes), in its somewhat broader substrate specificity, and in the nature of the agents which control its activation. Thus, whilst haemin and high GTP levels (36) each antagonise the activation of the haem-controlled kinase, high levels of dsRNA have no effect on this process, yet do specifically prevent the activation of the dsRNA-controlled kinase (23).

We know that phosphorylation of eIF-2 occurs in lysates incubated with oxidised glutathione (26, 37), and, for the reasons discussed previously, it probably also occurs in all the other inhibitory conditions. Although the kinase systems activated by haem-deficiency and by dsRNA seem to be separate entities, it seems inherently improbable that each of the other inhibitory conditions operates through a separate kinase system. It is more likely that some of the kinase systems can be activated in more than one way, and our preliminary experiments suggest that the inhibition caused by high hydrostatic pressure and by high temperature is due to an entity which so closely resembles the reversible form of the haem-controlled repressor that it must surely be the same.

A number of compounds are known that antagonise the inhibition caused by only one or two of the conditions listed above (see Table 1). It is believed that these agents exert their effect by blocking the pathway leading to activation of the inhibitory protein kinase, in which case these observations would help to define the number of independent kinase systems regulating the rate of initiation. Unfortunately the information on these antagonists is still very preliminary, and is complicated by some variability between different experiments and different laboratories. Nevertheless, the results would seem to point to the existence of at least one other system - controlled by sugar phosphates - apart from the haem-controlled and the dsRNA-activated systems. It cannot however be excluded that some of these antagonists may exert their effect by activating phosphoprotein phosphatases that dephosphorylate eIF-2. Such phosphatases are known to be present in reticulocyte lysates (23) - indeed their existence is necessary to explain the reversal of inhibition of initiation that occurs, say, when haemin is added to a lysate previously incubated without haemin - but they have not been characterised and the regulation of their activity remains an uncharted area.

Although much remains to be elucidated - the nature of the defect in phosphorylated eIF-2, the number of independent protein kinase systems, the mechanism of their activation, and the regulation of the phosphatases - the work of the past three years has shown that reticulocytes possess a complicated network of control systems which affect the rate of initiation through the reversible phosphorylation of eIF-2.

IV. REGULATION OF INITIATION IN NUCLEATED CELLS; THE MECHANISM OF INTERFERON ACTION

Although it has been generally assumed that the effects of haemin on protein synthesis might be particular to reticulocytes, recent work has shown that haemin, GTP and sugar phosphates stimulate initiation in cell-free systems from other cells in a

INHIBITORY CONDITION	Haemin (20 μM)	High levels ds RNA	GTP (1–2 mM)	$MnCl_2$ (0.5 mM)	Glucose 6-phosphate (1 mM)	Fructose 1:6-bisphosphate (2–4 mM)
	ANTAGONIST					
Haem deficiency	+++	0	+++	++	(+)	0
High hydrostatic pressure	0	0	+++	n.t.	+	0
Preincubation at 42°	0	0	+++	0	+	n.t.
Irreversible haem-controlled inhibitor	0	0	(+)	0	0	0
Low levels dsRNA	0	+++	(+)	0	0	0
Oxidised glutathione	0	0	(+)	0	+++	(+)
Lysates from ATP-depleted cells	0	n.t.	n.t.	n.t.	see below	
Gel-filtered lysates	0	0	+(+)	n.t.	see below	

Table 1. The effect of various antagonists on the inhibition of protein synthesis under different conditions.

The average effectiveness of the agents as antagonists of inhibition has been scored on a scale ranging from +++ for complete prevention of inhibition to 0 for no effect. (n.t. = not tested). This table has been compiled from the published work discussed in the text, and from unpublished work of T. Hunt, P.J. Farrell, K. Balkow and R.J. Jackson carried out in the author's laboratory.

> In extreme cases, gel-filtered lysates or lysates from ATP-depleted cells require the synergistic action of 4 mM fructose 1:6 bisphosphate, 1 mM dithiothreitol, and 5 mM cAMP for complete prevention of inhibition (24).
>
> In other cases a combination of any two of these three compounds is sufficient, and in a few experiments fructose 1:6-bisphosphate alone is sufficient to give full relief from inhibition. Glucose 6-phosphate has the same effect as a combination of fructose 1:6-bisphosphate and dithiothreitol in these experiments (24).

way which suggests that these cells possess at least some of the reticulocyte control mechanisms (38, 39). (The effect of haemin in stimulating the translation of globin mRNA injected into Xenopus oocytes probably has a different explanation, since it is only the translation of α-chain mRNA that is enhanced (40, 41), whereas the translation of all types of mRNA is controlled by haemin in the reticulocyte lysate).

The rate of initiation in muscle is stimulated by insulin, and in a variety of tissue culture cells it is inhibited in mitosis, at high temperature, in hypertonic media (e.g. with additional extracellular NaCl), and by amino acid starvation (reviewed in references 20, 42). Although the rate of initiation on some types of mRNA is affected more than on others in these various conditions, this differential effect is interpreted in terms of high and low affinity initiation sites on the different types of mRNA (2, 42). The actual regulatory process is thought to be non-discriminatory with respect to mRNA, and is therefore likely to act either before or at the mRNA binding step (see Figure 1) but apart from this we have no clue as to the mechanism of these controls.

Particularly intriguing are the effects of amino acid starvation, which have been shown to be due to a lack of aminoacyl-tRNA (or the presence of deacylated tRNA) rather than to a shortage of amino acids *per se*(43). Since starvation for any single amino acid can elicit this response, the regulatory switch must operate when no more than 5% of the total cellular complement of tRNA is in the deacylated form (assuming that tRNA is completely charged under normal conditions), regardless of which species of tRNA is concerned. These considerations would seem to rule out a simple direct mechanism in which deacylated tRNA acts as an inhibitor of some step in initiation. A more attractive idea is that the absence of an aminoacylated tRNA needed to translate the next codon in the mRNA causes some sort of 'idling' reaction on the part of the ribosomes which triggers the response, in much the same way as bacterial ribosomes synthesise the effector ppGpp when starved of the required aminoacyl-tRNA (44). Numerous efforts to detect ppGpp in mammalian cells have given negative results, and attempts to find another type of low molecular weight pleiotypic effector whose synthesis is triggered by amino acid starvation have so far failed, but the search has not yet been abandoned. A significant point is that reticulocytes do not show the same response, but merely reduce elongation rates when starved for amino acids. Do reticulocytes fail to synthesise the hypothetical effector, or do they lack the means to respond to it?

Protein synthesis in extracts of nucleated cells is not particularly sensitive to inhibition by dsRNA unless the cells have been pretreated with homologous interferon in which case the sensitivity is comparable to that of reticulocyte lysates (45). Interferon treatment seems to induce a large increase in the amount, or in the potential activity, of a dsRNA-activated protein kinase that resembles the reticulocyte counterpart in all its salient characteristics (46-48). In addition, dsRNA promotes the synthesis of ppp5'A2'p5'A2'p5'A (and higher homologues) in extracts of interferon-treated cells but not control cells (49). This unusual trinucleotide seems to activate an endonuclease which degrades both viral and cellular mRNA (50). It is the enzyme that synthesises the trinucleotide which is induced by interferon treatment, whilst the level of the latent endonuclease is relatively unchanged. As far as can be seen, the dsRNA-dependent synthesis of the trinucleotide involves an entirely distinct enzyme system from the dsRNA-activated protein kinase. Both systems are present in reticulocyte at levels comparable to interferon treated cells (51, 52). It remains a total mystery why reticulocytes should so strongly resemble the interferon-treated tissue culture cell, but this resemblance has had fortunate consequences since the work on reticulocyte control systems has had an enormous influence on understanding the changes induced by interferon. It still remains unclear which of the two dsRNA-dependent processes - the kinase system operating on initiation,and the trinucleotide-endonuclease system of mRNA degradation - is the more important in the overall anti-viral effect, nor is it by any means certain that these two effects (neither of which shows significant discrimination between viral and cellular mRNA) are the total explanation of the consequences of interferon treatment on viral mRNA translation.

V. VIRUS-INDUCED SHUT-OFF OF HOST CELL PROTEIN SYNTHESIS

In so far as such a sweeping generalisation can be accepted without qualifications, it appears that the reduction in host protein synthesis that follows virus infection of mammalian cells is not due to the destruction of the host mRNA but is caused primarily by a decrease in the rate of initiation on host mRNA (any changes in the rate of elongation being relatively minor). Host cell mRNA extracted from infected cells can usually be translated _in vitro_ with normal efficiency; if there is any reduction in translatability it is insufficient to account for the degree of shut-off of host cell protein synthesis in the cells from which the mRNA was obtained.

There are two views currently in vogue to account for the shut-off. One proposes that as the translation of viral RNA _in vitro_ is more resistant to high monovalent cation concentrations (particularly to added Na^+) than is the translation of cellular mRNA, it might be the influx of Na^+ into the infected cell (as a consequence of the membrane changes known to occur after viral infection) that suppresses host protein synthesis whilst viral protein synthesis is relatively unaffected or may even be enhanced (53, 54). It is generally the case that viral mRNA translation in infected cells is much more resistant to the inhibitory effects of additional extracellular NaCl (which causes inhibition of initiation) than is the translation of host cell mRNA either in infected or in uninfected cells (55).

The other view, which is not necessarily in contradiction,stems from the observation that in a competitive situation the translation of EMC RNA outcompetes the translation of cellular mRNA _in vitro_ under virtually all experimental conditions (56). The only exception is that the addition of extra eIF-4B tends to diminish the competitive advantage of the viral RNA (18). (According to the Basel group there is a requirement for eIF-4B for initiation on EMC RNA (17). The conclusion which is the most consistent with all these observations is therefore that the optimum ratio of eIF-4B to other factors may be lower for EMC RNA than for cellular mRNAs). The shut-off of host cell protein synthesis in this view is due to the competition by the viral RNA at the initiation step. The competitive effect might be enhanced if viral infection resulted in some inactivation or modification of eIF-4B, and some compelling evidence for this has recently been found (19), although the nature of the change in eIF-4B remains unknown.

Both of these ideas have their attractions, but they seem unable to account for all virus-host systems, particularly those in which the reduction in host cell protein synthesis occurs soon after infection. The increased permeability of the membrane to monovalent cations is generally thought to occur too late in the infection, and the out-competition by viral mRNA can only cause a significant reduction in host cell protein synthesis when viral mRNA has been produced in sufficient amounts to suppress initiation on host mRNA. As a total explanation, the competition hypothesis seems limited to those systems where the overall rate of protein synthesis remains relatively constant after infection, the rate of host protein synthesis declining progressively as the rate of viral protein synthesis increases. It cannot, for instance, explain the rapid reduction in host protein synthesis which occurs shortly after vaccinia infection (57, 58),even if u.v. inactivated virus is used or if all vaccinia mRNA synthesis is blocked. In cases like this where host protein synthesis is reduced before (and independently of) significant production of viral mRNA and virus-coded proteins, it seems necessary to postulate two processes: (1) a general reduction in the rate of initiation, probably through modification of the activity of initiation factors or ribosomes (or, less likely, through modification of the host mRNA), as well as (2) an ability of the viral mRNA to outcompete the host cell mRNA for the surviving initiation capacity. Moreover, since the first of these can be observed in the absence of viral replication or transcription, and at relatively low inputs of virus (in relation to the levels of ribosomes or initiation factors in the infected cell) we can probably reject models of a stoichiometric type (e.g. inhibition of initiation as the result of the binding of viral capsid protein to an initiation factor or two ribosomes), and must think in terms of some catalytic mechanism. Whether this has any relationship with the type of controls discovered in the reticulocyte lysate has yet to be determined.

VI. mRNPs AND UNTRANSLATED mRNA

It is probable that mRNA never exists as a free entity, but is complexed with proteins in the form of mRNPs: translated mRNAs as polysomal mRNPs, and untranslated mRNAs as free mRNPs. There is far from general agreement as to the identity of the mRNP proteins, and as there are no good criteria for judging whether any mRNP preparation represents the state of the mRNA as it occurs in the cell, or whether proteins might not have become adventitiously associated with it during the isolation procedure, it is hard to evaluate the difference found in different laboratories. This difficulty is particularly serious with respect to the free mRNPs not associated with polysomes: with the polysomal mRNPs one can at least wash the polysomes free of loosely bound proteins before releasing the mRNPs by use of chelating agents or by incubation with puromycin at high KCl concentration.

There are two predominant proteins (M.Wt. approximately 50,000 and 78,000) in polysomal mRNPs isolated from a wide variety of mammalian sources (reviewed in reference 59). Other proteins are present in minor amounts, and are far from consistent between different preparations. Those who wish to see complications at every level of mammalian protein synthesis like to hint that these minor components could be protein specific to particular mRNA species, perhaps involved in regulating the translation of the mRNA, whilst the 50,000 and 78,000 dalton proteins would be seen as invariant components common to all mRNAs. This idea seems unlikely to be true, at least as a generality, since there is no sign of the putative mRNA-specific proteins in mRNP preparations from cells which contain large amounts of a single mRNA species (e.g. reticulocytes, oviduct, etc) where the mRNA-specific proteins should be detected as major abundant mRNP proteins.

We have very little idea about the stoichiometry of the two major polysomal mRNP proteins, or their location on the mRNA, apart from the fact that the 78,000 dalton protein seems to be associated with the poly A tract at the 3' end (59). The function of the proteins is even more obscure. In cell-free translation assays the activity of polysomal mRNP is no greater than that of deproteinised mRNA (20). This result is open to the qualification that in crude cell-free systems the added deproteinised mRNA might pick up proteins from the cell-free extract and thereby be effectively converted into polysomal mRNP particles, but since the same result is obtained in highly fractionated systems (11) this reservation can probably be discounted. The major polysomal mRNP proteins are distinct from all seven recognised initiation factors (11), and initiation complex formation is found to require the same set of seven factors regardless of whether polysomal mRNP or deproteinised mRNA is used (11). In short, there is no evidence that these proteins play a role in mRNA translation.

A more intriguing problem is the nature and function of the free untranslated mRNPS. The stored maternal mRNA of amphibian oocytes (and the eggs of other higher organisms) and plant seeds is the most obvious example of such free mRNP. In addition, the mRNAs characteristic of specific differentiated cells have often been detected as free mRNPs in the stages immediately prior to differentiation, e.g. myosin mRNA is reported to be present as untranslated mRNP in pre-fusion myoblasts (60). What is often not fully appreciated is that significant levels of untranslated mRNPs are present in growing non-differentiating cells, and it is particularly this problem that will be discussed here since it has more general significance. It remains to be established whether the same basic mechanism governs the control of translation of free mRNPs in all cases - developing eggs, differentiating cells, and growing tissue culture cells - or whether each of these situations should be regarded as a distinct problem.

In cell lines such as mouse myeloma or Ehrlich ascites tumor cells some 20% of the total cytoplasmic mRNA is not associated with ribosomes but exists in an untranslated state even though deproteinised mRNA prepared from these mRNP particles is perfectly translatable _in vitro_. A more or less trivial explanation for such untranslated mRNA is that it originates from a small fraction of the cells in which the rate of initiation is very low, e.g. cells in mitosis. A comparison of the products of _in vitro_ translation of this mRNA and of the polysomal mRNA renders this explanation less than wholly satisfactory. In extreme cases, the non-translated mRNPs are found to contain an entirely different set of mRNA species from the polysomal mRNA. In other experiments there is partial overlap between the two mRNA populations. It therefore seems that in growing cells certain species of mRNA may be very poorly translated (yet are translatable *in vitro*) and exist largely, if not exclusively, as free mRNPs.

There are two types of explanation for this finding. The first is that these are mRNAs with extremely low affinity initiation sites which lose out in the competition for the initiation capacity of the cells. One can make the prediction (which to my knowledge has only been put to the test in the case of the α-globin mRNA of rabbit reticulocytes) that the presence of low concentrations of cycloheximide should result in the mobilisation of these mRNAs into small polysomes, because the rate of initiation ceases to be the rate-limiting step in protein synthesis under these conditions (2, 61). As Lodish has shown, the α-globin mRNA has a lower affinity initiation site than the β-globin mRNA of rabbits, and is translated on smaller polysomes (61). A consideration of the statistical frequency of initiation events leads to the prediction that rabbit reticulocytes should contain significant amounts of α-chain mRNA but only traces of β-chain mRNA in untranslated free mRNPs, and this is born out by the observations (62, 63). The presence of low concentrations of elongation inhibitors (cycloheximide, emeteine, etc) results in an increase in the amount of α-chain mRNA present in polysomes - an increase which can only be explained by a mobilisation of mRNA from the untranslated mRNPs into polysomes (61). Moreover, when reticulocyte ribosome-free supernatant (containing free untranslated α-chain mRNP) is added to heterologous cell-free systems, α-globin synthesis occurs (62, 63). All the evidence therefore favours the view that the free α-chain mRNPs of rabbit reticulocytes arise from competition between mRNAs at the initiation step. What is not yet clear, however, is whether this explanation can account for all the untranslated mRNPs found in nucleated cells.

The alternative type of explanation is that the translation of these mRNAs is specifically repressed by some mechanism (presumably through the binding of the mRNP proteins) and that their translation requires some sort of cytoplasmic induction. In this case, mobilisation of the mRNA into polysomes would not necessarily follow treatment of the cells with low levels of cycloheximide, but would require exposure of the cells to the appropriate inducer. One possible candidate for an inducer may be Fe^{++} which causes a doubling in the rate of ferritin synthesis in tissue culture cells and in liver (64, 65). In the control situation only 50% of the cytoplasmic ferritin mRNA in liver is associated with polysomes, and the rest occurs as free mRNPs. Administration of Fe^{++} results in the rapid assimilation of all the ferritin mRNA into polysomes (65). If further investigation confirms these results, and shows that Fe^{++} selectively controls ferritin mRNA translation and has little effect on other mRNAs, this will be the first proven case of 'cytoplasmic induction'.

If the free mRNPs are a specifically repressed form of mRNA one might expect that they would prove to be unstranslatable *in vitro* when added as mRNPs, although the corresponding deproteinised mRNA is active. Several attempts to test this prediction have been made, but the results are highly controversial and the innocent bystander has very little idea of the true picture. Scherrer and his colleagues find that duck reticulocyte free mRNPs are not only untranslatable *in vitro*, but actually inhibit the translation of deproteinised mRNA (66). They attribute this to specific repression of the activity of the added deproteinised mRNA, but this idea raises some conceptual problems, if, as one imagines, the inhibition is caused by the transfer of proteins from the free mRNPs to the deproteinised mRNA. Assuming that the free mRNPs are a repressed form of the mRNA with presumably a defined complement of 'repressor' proteins bound to the mRNA, it is difficult to envisage why this transfer does not liberate the free mRNP from its repressed state. One would like to see a detailed analysis of the nature of the inhibitory action of free mRNPs, with some demonstration that it was caused solely by the sequestration of the added deproteinised mRNA and not by inhibition of some other step in protein synthesis. In addition, the protein component of the mRNP responsible for the inhibition should be identified, and some demonstration would be desirable that this is a true stoichiometric component of the mRNP particle and not some protein which bound adventitiously to the mRNP during isolation.

In contrast to Scherrer's results, the free α-globin mRNP of rabbit reticulocytes seems efficiently translated *in vitro*, as mentioned above. Experiments in our laboratory on the mRNA of myeloma cells exposed to hypertonic media also serve to rule out that free mRNPs are necessarily untranslatable *in vitro*. Under these conditions of inhibited initiation the mRNA is released from polysomes as free mRNPs which we find have a template activity comparable to that of deproteinised polysomal mRNA isolated from cells grown under normal conditions (B.Mechler, in preparation). It is, of course, legitimate to argue that these mRNPs, and the free α-chain mRNP of rabbit reticulocytes, represent a special class of free mRNP different from those in duck reticulocytes, but the results should at least caution against the extreme view that all mRNPs are untranslatable in cell-free protein synthesis systems.

If the free mRNPs represent mRNAs in a state of specific repression, one might expect that the proteins would not only differ from the polysomal mRNP proteins, but would also differ according to the mRNA species present in the mRNP. Virtually all free mRNPs, including the α-chain mRNP of rabbit reticulocytes which is almost certainly not a repressed form of α-globin mRNA, lack the 78,000 dalton polypeptide characteristic of polysomal mRNPs and associated with the poly A tract (67). A more interesting observation is that two different species of mRNA of duck reticulocytes have different proteins in their respective free mRNPs (68). If this proves to be general it means that there are cytoplasmic mRNA-binding proteins that are discriminatory with respect to the species of mRNA, from which it is but a short step to the concept of specific cytoplasmic repression of mRNA translation.

The problem of free untranslated mRNPs is thwarted by inadequate technology for their isolation and characterisation. The field is sufficiently confused that it is tempting to dismiss free mRNPs as an artefact or as an uninteresting oddity. This is a mistaken attitude, in my opinion. The presence in free mRNPs of specific mRNA types, representing a significant proportion of the cellular mRNA of tissue culture cells, should convince us that a real problem exists that is worth investigating.

VII. INITIATION SITE SELECTION ON mRNA; THE ROLE OF "CAPS"

What features of the mRNA structure are important to the selection of initiation sites, and how is this recognition achieved? As a first step towards answering these questions, the nucleotide sequences of several eukaryotic mRNA initiation sites have been determined through the traditional approach of isolating the fragment of mRNA which is protected in the initiation complex against nuclease digestion. All these experiments are based on the'shift' system described by our group in 1973 (16): this is essentially a blocked translation system using crude cell-free extracts and high concentrations of elongation inhibitors to lock the ribosomes at the initiation sites of the added mRNA. The fragment of mRNA protected in the 40S/Met-$tRNA_f$/ mRNA complex turns out to be invariably longer than the fragment from the 80S initiation complex (69, 70, 71). The explanation for this result - which is at first sight rather surprising - almost certainly lies in the large size of eIF-3 (9, 13); complexes of 40S subunits with eIF-3 look quite as large, if not larger than 80S ribosomes in electronmicrographs. The fragments of a given species of mRNA protected in the two types of initiation complex can be more or less aligned at their 3' ends, and the sequences which are unique to the larger 40S complex fragment are at the 5' end (69, 70, 71).

These experiments have been dissapointing in the sense that the only features which seem to be universal are the AUG initiation codon and the 5' cap structure (71, 72), and even the latter is missing from some viral RNAs, notably picornavirus RNAs (both the virion RNA (73) and the viral mRNA (74)), Cowpea Mosaic Virus (CPMV) RNA (75), and Satellite Tobacco Necrosis (STNV) RNA (76). The role of the cap has been studied by examining the properties of mRNA from which the cap has been chemically removed, and of vaccinia and reovirus mRNAs transcribed from viral cores in the presence of S-adenosyl homocysteine, which prevents synthesis of the m^7GpppX.... cap structure, and causes synthesis of mRNA with ppX... and GpppX... end groups (reviewed in reference 77). The absence of the methylated cap structure reduces the rate of initiation to some 5-30% of the rate obtained with the capped form of the same mRNA (77). The first such experiments were done with systems that showed such a marked reduction that caps were thought to be absolutely necessary for initiation. Considerable controversy arose when more modest effects were found in other systems, but it is now accepted that there are real differences according to the particular species of mRNA, the cell-free system and the incubation conditions (although the reasons for these differences are not understood). Caps are considered to enhance the efficiency of initiation, but they are not absolutely required, nor do they affect the precision of initiation site selection (71). The ribosomes bind to the same sites and synthesise the same proteins regardless of whether they are presented with the mRNA in the capped or uncapped form (78).

Another discovery which highlights the function of caps is the inhibition of cell-free translation of capped mRNAs by cap analogues such as m^7GMP, m^7GTP or m^7GpppX (76). Inhibition is seen only if the G residue in the analogue is methylated, and it is initiation that is the sensitive step. Moreover, the analogues inhibit initiation only when capped mRNAs are tested: with uncapped mRNAs they either have no effect, or may even stimulate (76). The cap analogues are therefore believed to inhibit the binding of the native 40S ribosomal subunits to the capped end of the mRNA by a competitive or pseudo-competitive mechanism. This argues that the 5' capped end is a feature which is positively recognised by the initiating ribosome, and does not play a mere passive or negative role such as preventing the region of the mRNA around the initiation site from adopting a conformation unfavourable for initiation.

In the first mRNAs to be studied the AUG initiation codon was situated not more than some 50 nucleotide residues from the 5' capped terminus (reviewed in reference 72), so that in all cases the fragment of mRNA protected in the 40S initiation complex contained the 5' cap (although in some cases the cap was missing from the shorter fragment from 80S initiation complexes). Thus it was possible to imagine that when the native 40S subunit bound to the mRNA, both the 5' cap and the AUG initiation codon were recognised simultaneously. Now that a wider range of mRNAs has been examined, it turns out that the separation of the AUG codon from the cap is much greater in some cases - at least 80 residues in Rous Sarcoma Virus RNA (79), and probably of the order of some 200 residues in some late adenovirus mRNAs. In such cases it seems impossible that the initiating 40S subunit could recognise and bind to both the 5' end and the initiation codon simultaneously, yet by all the usual criteria initiation on these mRNAs is as cap-dependent as on any other messages. If we assume that this means that the 40S subunit does bind to the cap at some stage during initiation, it seems that there must be at least two steps in initiation site selection: (1) binding to the 5' capped terminus, and (2) movement of the 40S subunit from the cap to the AUG initiation codon. It is, of course, conceivable that the conformation of the mRNA causes the 5' end to lie back close to the initiation codon so that the distance involved in the second step would be quite short, but there is no direct evidence for this and we have to admit the possibility of quite substantial movement. The mRNA binding step requires ATP hydrolysis, and one is tempted to speculate that this might be the driving force for the movement of the ribosome from the cap to the AUG codon (particularly as there are no other obvious reasons, such as the release of bound initiation factors, for this requirement for ATP hydrolysis). It would be interesting to know the stoichiometry of ATP hydrolysis per 40S/Met-$tRNA_f$/ mRNA complex formed, and whether this stoichiometry differs from one type of mRNA to another in a way that correlates with the distance separating the 5' end from the initiation codon.

Movement of the ribosome has also been invoked to explain the action of edeine, an inhibitor of initiation that leads to the formation of aberrant 40S/Met-$tRNA_f$/mRNA complexes which are unable to bind a 60S subunit (80). Edeine can promote the binding of several 40S subunits to a single mRNA molecule, and even if one of these subunits may be located at the true initiation site, the others surely have random locations. The binding of all of these 40S subunits is sensitive to inhibition by cap analogues and seems dependent on their 'entry' at the 5' capped terminus (81). In the model in which the 40S subunit binds at the 5' end and then moves to the correct initiation site, edeine is envisaged as interfering with the arrest of this movement at the correct site, thereby allowing several 40S subunits to be accomodated at random positions on the mRNA (81).

What signals would cause the 40S subunit to arrest at the correct initiation site in normal circumstances? In the vast majority of mRNAs examined the initiation site is the 5' proximal AUG codon (72), but there are hints that this is not so absolutely universal that it could be the sole criterion for initiation site selection. For example, in Rous Sarcoma Virus RNA it is the second AUG codon from the 5' end that is thought to be the site of initiation of translation (79). If this is substantiated the necessary structure of an initiation site must include other features besides the AUG codon. Nevertheless, studies of a variety of viral mRNAs show that we need to retain the concept of the overriding importance of the first correct initiation site from the 5' end. Many virion RNAs (Tobacco Mosaic Virus RNA, Rous Sarcoma Virus RNA, and others) and viral mRNAs (particularly SV40 late mRNA, and late adenovirus mRNA) are polycistronic and contain information coding for more than one protein, yet it is only the 5' proximal cistron that is translated *in vitro*. Recent work on the SV40 mRNAs (reviewed in reference 82) and preliminary studies on the adenovirus mRNAs point to the existence of families of mRNA species bearing the following relationship to each other:

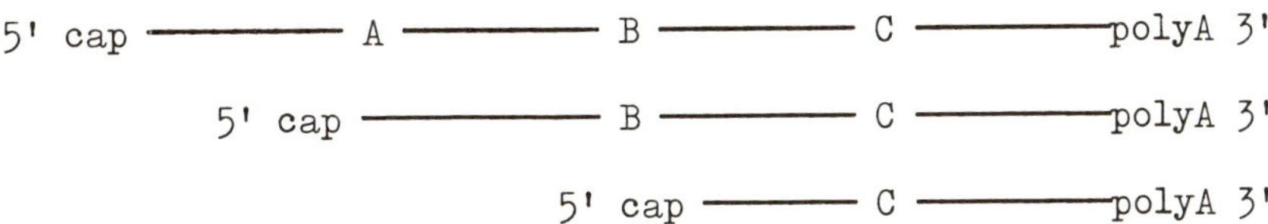

where A, B and C are the coding segments for three different proteins (83). In cell-free systems the first of these mRNAs is translated to yield apparently only protein A (and not B or C), the second gives only protein B (and not C), whilst C is translated from the third type of mRNA. These results would be explained if, after binding to the 5' cap, the 40S subunits moved systematically towards the 3' end searching for a correct initiation site and stopping at the first such site that was encountered. The same result might also be given by a random search if one postulates that the further the 40S subunit moves from the 5' capped terminus the higher is the probability of it dissociating from the mRNA end of the search being therefore abandoned.

How is the correct AUG initiation codon selected in the second stage of initiation, in moving from the cap to the initiation site? We have seen that this site may not be unambiguously defined as the AUG codon that is nearest to the 5' end. Is there any evidence for additional recognition features in the neighbouring nucleotide sequence? In bacterial systems there is good evidence that the correct initiation site is defined, at least in part, by a sequence of bases which is situated on the 5' side of the initiation codon, and which can base-pair with a -CCUCC- sequence very near the 3' end of the 16S ribosomal RNA of the *E.coli* small ribosomal subunit. There is, moreover, good evidence that this base-pairing does occur when initiation complexes are formed (84). The nucleotide sequences of the 3' ends of the small ribosomal subunit RNAs from *E.coli* and a wide range of eukaryotic organisms have recently been compared (72). They show a remarkable degree of homology, but the mammalian RNAs differ from *E.coli* 16S rRNA in one crucial respect: the -CCUCC-sequence is deleted. Thus, if similar base-pairing plays a role in eukaryotic initiation site selection, it must be a completely different segment of the 18S rRNA and a different sequence that is involved. There is a purine-rich sequence near the end of the 18S rRNA which has been proposed as a possible candidate (72), but the potential complementarity that can be found between this sequence and the known eukaryotic initiation sites seems considerably less compelling an argument for base-pairing as a means of initiation site selection than is the case with bacterial systems. Thus if there are what might be termed 'nucleotide sequence recognition elements' in eukaryotic initiation sites their exact nature remains to be established.

One further problem requires consideration: initiation on the naturally occuring uncapped RNAs, such as picornavirus RNAs, CPMV RNA and STNV RNA. Unlike the initiation on uncapped derivatives of mRNAs that normally have methylated caps, initiation on picornavirus RNA is efficient and the rate of initiation _in vitro_ is high. It appears that the same set of seven initiation factors is needed for the translation of picornavirus RNA as for capped mRNAs (17), even though the viral RNA may require lower levels of eIF-4B for maximal rates of initiation (18, 19). Since there are no indications of 'cap-specific' initiation factors whose presence is totally redundant for initiation on picornavirus RNAs, it is difficult to envisage completely different mechanisms for initiation on the two classes of RNA, capped and uncapped. One can only guess that the 'nucleotide sequence recognition elements' of picornavirus RNAs are of such high affinity that an efficient initiation site selection can operate directly, in contrast to capped mRNAs where these elements may be of lower affinity so that an initial focussing of the 40S subunit to the 5' cap is needed for efficient initiation at the nearest initiation site.

When a 5' cap is added _in vitro_ to STNV RNA using the enzyme system from vaccinia cores, the rate of initiation complex formation is increased by a factor of about 2.5 (85). The sensitivity of initiation complex formation to inhibition by cap analogues is now as high as is the case with TMV RNA, suggesting that the presence of a 5' cap somehow suppresses the potential for intiation in a cap-independent mechanism, as occurs when the RNA is in its normal (uncapped) form.

The phasing of the ribosome to the AUG initiation codon - the final stage in initiation site selection - gives rise to fewer conceptual problems, because the initiating ribosome binds Met-$tRNA_f$ before associating with the mRNA. As we pointed out when this was first discovered (16), this means that the anticodon of Met-$tRNA_f$ can assist in the phasing of the ribosome precisely to the initiation codon, through normal codon-anticodon interaction.

VIII. THE EXPRESSION OF VIRAL POLYCISTRONIC GENOMES

The inability of mammalian ribosomes to translate efficiently all but the 5'-proximal cistrons of a polycistronic mRNA raises problems in the expression of viral genomes. With many viruses these problems are solved at the stages of transcription, processing and splicing of the mRNA. This is obviously applicable to viruses that are transcribed in the nucleus, but it is also exploited by some viruses which replicate in the cytoplasm. For example, VSV seems to consist of a single transcription unit but the RNA product is processed to give monocistronic mRNAs which are capped and polyadenylated by virion enzymes (77). Vaccinia, on the other hand, seems to synthesise capped and polyadenylated monocistronic mRNAs directly, without a polycistronic precursor, according to experiments on the sensitivity of protein synthesis in a coupled cell-free transcription-translation system to u.v. inactivation (86).

A somewhat different type of mechanism is employed by Tobacco Mosaic Virus for the expression of the coat protein gene which is situated towards the 3' end of the virion RNA and is not expressed when intact virion RNA is translated *in vitro*. In infected plants, a low molecular weight RNA coding for the coat protein is produced, either by selective transcription or by selective processing of the viral genome (87).

RNA viruses have also evolved mechanisms of obtaining more than one gene product from a single RNA genome without going through a stage of selective transcription or processing. TMV RNA is translated *in vitro* to yield two products of 110,000 and 160,000 daltons, whose synthesis appears to be initiated at the same site (88). The larger is thought to arise by read-through of a 'leaky' UAG terminator codon at the end of the cistron coding for the smaller(110,000 dalton) protein, since the addition of yeast amber suppressor tRNA to the translation system increases the yield of the larger protein at the expense of the smaller (88).

Picornaviruses seem to have a major high efficiency initiation site specifying a large translation unit, yet produce many different small protein products (89, 90), not only in the infected cell but also in cell-free translation systems. Processing of the protein product in cell-free systems seems to follow a very similar (if not identical) pattern as occurs in the infected cell, even if the cell-free system is from reticulocytes (91), a cell which seems unlikely to encounter picornavirus infection. Some evidence from studies of infected cells, and our own recent work on cell-free translation of EMC RNA suggests that at least some of the processing steps are carried out by an activity encoded in the viral genome itself, and located some two-thirds of the way along the translation unit (91). Whereas the incorporation of amino acid analogues into proteins usually renders them more susceptible to intracellular proteolysis, the translation product of picornavirus RNA synthesised in the presence of analogues is more resistant to proteolytic processing (89, 90). It seems possible that this somewhat unusual consequence of amino acid analogue incorporation may be due to the synthesis of non-functional virus-coded processing protease (90). Recent work by Hugh Pelham in our laboratory suggests that a similar type of auto-processing occurs when Cowpea Mosaic Virus RNAs are translated *in vitro* (92).

Besides these various devices exploited by viruses in the expression of their polycistronic genomes, there is in principle the additional possibility of splicing of the RNA to delete termination codons, or to delete 5' -proximal cistrons and thereby construct an RNA in which the 5' capped end is located near another cistron. Whether this possibility deserves serious consideration in respect of viruses which replicate in the cytoplasm depends on whether RNA splicing occurs uniquely in the nucleus, or whether the appropriate enzyme activities are also present in the cytoplasm. This question has not yet been satisfactorily answered, but seems likely to be resolved in the near future.

IX. CONCLUDING REMARKS

In the eight years since mammalian mRNA was first shown to have template activity *in vitro* and the first progress towards understanding the mechanism of mammalian initiation was made, there have been considerable advances on many fronts. There is every reason to expect that most of the questions raised in this review will be solved in the next eight years using more detailed exploitation of existing technology. Some technical problems do, however, persist. To my mind, there are three important experiments that one would wish to be able to do, but which are currently impossible or unsatisfactory: (1) to reconstitute active mammalian ribosomal subunits from purified rRNA and ribosomal structural proteins, (2) to obtain cell-free systems from nucleated cells which, like the reticulocyte lysate accurately reflect the intact cell in terms of the rate of protein synthesis and its sensitivity to control processes, and (3) to isolate mRNPs in the same state as they exist in the cell, and to characterise them. It is also my belief that we would learn a great deal about eukaryotic protein synthesis if we had good yeast cell-free protein synthesis systems with which to investigate and analyse the tantalisingly large number of different yeast temperature-sensitive protein synthesis mutants available.

ACKNOWLEDGEMENTS

I wish to thank Tim Hunt and High Pelham for numerous helpful discussions, Cathy Land and Liz Campbell for technical assistance in the unpublished work discussed here, which was supported by grants from the Medical Research Council and the Cancer Research Campaign.

REFERENCES

1. WEISBACH, H. and PESTKA, S. eds., Molecular mechanisms of protein biosynthesis (1977), Academic Press, New York.

2. LODISH, H.F. Ann. Rev. Biochem. (1976), 45, 39-72.

3. WOOL, I.G. and STOFFLER, G. In Ribosomes. Nomura, M., Tissieres, A. and Lengyel, P. eds., pp.417-460, Cold Spring Harbor Laboratory, Cold Spring Harbor, New York.

4. LASTICK, S.M. and McCONKEY, E.H. J. Biol. Chem. (1976), 251, 2867-2875.

5. GRESSNER, A.M. and WOOL, I.G. J. Biol. Chem. (1974), 249, 6917-6925.

6. TRAUGH, J.A. and PORTER, G.G. Biochemistry (1976), 15, 610-616.

7. EIL, C. and WOOL, I.G. J. Biol. Chem. (1973), 248, 5130-5136.

8. KAERLEIN, M. and HORAK, I. Nature (London) (1976), 259, 150-152.

9. SCHREIER, M.H., ERNI, B. and STAEHELIN, T. J. Mol. Biol. (1977), 116, 727-753.

10. TRACHSEL, H., ERNI, B., SCHREIER, M.H. and STAEHELIN, T. J. Mol. Biol. (1977), 116, 755-767.

11. ERNI, B. Ph.D. Thesis (1976), ETH, Zurich.

12. SAFER, B., ADAMS, S.L., KEMPER, W.M., BERRY, K.W., LLOYD, M. and MERRICK, W.C. Proc. Nat. Acad. Sci. U.S.A. (1976), 73, 2584-2588.

13. BENNE, R. and HERSHEY, J.W.B. J. Biol. Chem. (1978), 253, 3078-3087.

14. WEISSBACH, H. and OCHOA, S. Ann. Rev. Biochem. (1976), 45, 191-216.

15. FILIPOWICZ, W., SIERRA, J.M., NOMBELA, C., OCHOA, S., MERRICK, W.C. and ANDERSON, W.F. Proc. Nat. Acad. Sci. U.S.A. (1976), 73, 44-48.

16. DARNBROUGH, C.H., LEGON, S., HUNT, T. and JACKSON, R.J. J. Mol. Biol. (1973), 76, 379-403.

17. STAEHELIN, T., TRACHSEL, H., ERNI, B., BOSCHETTI, A. and

SCHREIER, M.H. Proceedings 10th FEBS Meeting (1975), 10, 309-323.

18. GOLINI, F., THACH, S.S., BIRGE, C.H., SAFER, B., MERRICK, W.C. and THACH, R.E. Proc. Nat. Acad. Sci. U.S.A. (1976), 73, 3040-3044.

19. ROSE, J.K., TRACHSEL, H., LEONG, K. and BALTIMORE, D. Proc. Nat. Acad. Sci. U.S.A. (1978), 75, 2732-2736.

20. JACKSON, R.J. MTP International Review of Science: Biochemistry, Series One (1975), 7, 89-135.

21. HUNT, T. Brit. Med. Bull. (1976), 32, 257-261.

22. LEGON, S., JACKSON, R.J. and HUNT, T. Nature New Biol. (1973), 241, 150-152.

23. FARRELL, P.J., BALKOW, K., HUNT, T., JACKSON, R.J. and TRACHSEL, H. Cell (1977), 11, 187-200.

24. GILOH, H. and MAGER, J. Biochem. Biophys. Acta (1975), 414, 293-308.

25. GILOH, H., SCHOCHOT, L. and MAGER, J. Biochem. Biophys. Acta (1975), 414, 309-323.

26. FARRELL, P.J. Ph. D. Thesis (1977), University of Cambridge.

27. GROSS, M. and RABINOVITZ, M. Biochem. Biophys. Acta (1972), 287, 340-352.

28. GROSS, M. Biochem. Biophys. Acta (1974), 366, 319-332.

29. GROSS, M. Biochem. Biophys. Acta (1974), 340, 484-497.

30. GROSS, M. Biochem. Biophys. Res. Commun, (1974), 57, 611-619.

31. GROSS, M. and RABINOVITZ, M. Biochem. Biophys.Res. Commun. (1973), 50, 830-835.

32. GROSS, M. and MENDELEWSKI, J. Biochem. Biophys.Res. Commun. (1977), 74, 559-569.

33. TRACHSEL, H. and STAEHELIN, T. Proc. Nat. Acad. Sci. U.S.A. (1978), 75, 204-208.

34. DE HARO, C., DATTA, A. and OCHOA, S. Proc. Nat. Acad. Sci. U.S.A. (1978), 75, 243-247.

35. RANU, R.S., LONDON, I.M., DAS, A., DASGUPTA, A., MAJUMDAR, A., RALSTON, R., ROY, R. and GUPTA, N.K. Proc. Nat. Acad. Sci. U.S.A. (1978), 75, 745-749.

36. BALKOW, K., HUNT, T. and JACKSON, R.J. Biochem. Biophys. Res. Commun. (1975), 67, 366-375.

37. FARRELL, P.J., HUNT, T. and JACKSON, R.J. Eur. J. Biochem. (1978), 89, 517-521.

38. WEBER, L., FEMAN, E. and BAGLIONI, C. Biochemistry (1975), 14, 5315-5321.

39. LENZ, J.R., CHATTERJEE, G.E., MARONEY, P.A. and BAGLIONI, C. Biochemistry (1978), 17, 80-87.

40. GIGLIONI, B., GIANNI, A.M., COMI, P., OTTOLENGHI, S. and RUNGGER, D. Nature New Biology (1973), 246, 99-102.

41. LANE, C.D, GURDON, J.B. and WOODLAND, H.R. Nature (London) (1974), 251, 436-437.

42. REVEL, M. In Molecular mechanisms of protein biosynthesis. Weissbach, H. and Pestka, S. eds., (1977), pp.246-321, Academic Press, New York.

43. VAUGHAN, M.H. and HANSEN, B.B. J. Biol. Chem. (1973), 248, 7087-7096.

44. BLOCK, R. and HASELTINE, W.A. In Ribosomes. Nomura, M., Tissieres, A. and Lengyel, P. eds., (1974), pp.747-761, Cold Spring Harbor Laboratory, Cold Spring Harbor, New York.

45. KERR, I.M., BROWN, R.E. and BALL, .L.A. Nature (London) (1974), 250, 57-59.

46. ROBERTS, W.K., HOVANESSIAN, A., BROWN, R.E., CLEMENS, M.J. and KERR, I.M. Nature (London) (1976), 264, 477-480.

47. LEBLEU, B., SEN, G.C., SHAILA, S., CABRER, B. and LENGYEL, P. Proc. Nat. Acad. Sci. U.S.A. (1976), 73, 3107-3111.

48. ZILBERSTEIN, A., FEDERMAN, P., SHULMAN, L. and REVEL, M. FEBS Lett. (1976), 68, 119-124.

49. KERR, I.M. and BROWN, R.E. Proc. Nat. Acad. Sci. U.S.A. (1978), 75, 256-260.

50. RATNER, L., SEN, G.C., BROWN, G.E., LEBLEU, B., KAWAKITA, M., CABRER, B., SLATTERY, E. and LENGYEL, P. Eur. J. Biochem.

(1977), 79, 565-577.

51. HOVANESSIAN, A.G. and KERR, I.M. Eur. J. Biochem. (1978), 84, 149-159.

52. CLEMENS, M.J. and VAQUERO, C. Biochem. Biophys. Res. Commun. (1978), 83, 59-68.

53. CARRASCO, L. and SMITH, A.E. Nature (London) (1976), 264, 807-809.

54. CARRASCO, L. FEBS Lett. (1977), 76, 11-15.

55. NUSS, D.L., OPPERMANN, H. and KOCH, G. Proc. Nat. Acad. Sci. U.S.A. (1975), 72, 1258-1263.

56. LAWRENCE, C. and THACH, R.E. J. Virol. (1974), 14, 598-610.

57. MOSS, B. J. Virol. (1968), 2, 1028-1037.

58. PERSON, A. and BEAUD, G. J. Virol. (1978), 25, 11-18.

59. SHAFRITZ, D.A. In Molecular mechanisms of protein biosynthesis. Weissbach, H. and Pestka, S. eds., (1977), pp.555-601. Academic Press, New York.

60. BUCKINGHAM, M.E., CAPUT, D., COHEN, A., WHALEN, R.G. and GROS, F. Proc. Nat. Acad. Sci. U.S.A. (1974), 71, 1466-1470.

61. LODISH, H.F. J. Biol. Chem. (1971), 246, 7131-7138.

62. BONANOU-TZEDAKI, S.A., PRAGNELL, I.B. and ARNSTEIN, H.R.V. FEBS Lett. (1972), 26, 77-82.

63. JACOBS-LORENA, M. and BAGLIONI, C. Eur. J. Biochem. (1973), 35, 559-565.

64. CHU, L.L.H. and FINEBERG, R.A. J. Biol. Chem. (1969), 244, 3847-3854.

65. ZAHRINGER, J., BALIGA, B.S. and MUNRO, H.N. Proc. Nat. Acad. Sci. U.S.A. (1976), 73, 857-861.

66. CIVELLI, O., VINCENT, A., BURI, J.F. and SCHERRER, K. FEBS Lett. (1976), 72, 71-76.

67. VAN VENROOIJ, W.J., VAN EEKELEN, C.A.G., JANSEN, R.T.P. and PRINCEN, J.M.G. Nature (London) (1977), 270, 189-191.

68. VINCENT, A., CIVELLI, O., BURI, J.F. and SCHERRER, K.

FEBS Lett. (1977), 77, 281-286.

69. LEGON, S. J. Mol. Biol. (1976), 106, 37-53.

70. KOZAK, M. and SHATKIN, A.J. J. Biol. Chem. (1976), 251, 4259-4266.

71. KOZAK, M. and SHATKIN, A.J. Cell (1978), 13, 201-212.

72. HAGENBUCHLE, O., SANTER, M., STEITZ, J.A. and MANS, R.J. Cell (1978), 13, 551-563.

73. FLANEGAN, J.B., PETTERSSON, R.F., AMBROS, V., HEWLETT, M.J. and BALTIMORE, D. Proc. Nat. Acad. Sci. U.S.A. (1977), 74, 961-965.

74. HEWLETT, M.J., ROSE, J.K. and BALTIMORE, D. Proc. Nat. Acad. Sci. U.S.A. (1976), 73, 327-330.

75. KLOOTWIJK, J., KLEIN, I., ZABEL, P. and VAN KAMMEN, A. Cell (1977), 11, 73-83.

76. HICKEY, E.D., WEBER, L.A. and BAGLIONI, C. Proc. Nat. Acad. Sci. U.S.A. (1976), 73, 19-23.

77. SHATKIN, A.J. Cell (1976), 9, 645-653.

78. PELHAM, H.R.B., SYKES, J.M.M. and HUNT, T. Eur. J. Biochem. (1978), 82, 199-209.

79. HASELTINE, W.A., MAXAM, A.M. and GILBERT, W. Proc. Nat. Acad. Sci. U.S.A. (1977), 74, 989-993.

80. HUNT, T. Ann. N.Y. Acad. Sci. (1974), 241, 223-231.

81. KOZAK, M. and SHATKIN, A.J. J. Biol.Chem. (1978), 253, 6568.

82. FIERS, W., CONTRERAS, R., HAEGMAN, G., ROGIERS, R., VAN DE VOORDE, A., VAN HEUVERSWYN, H., VAN HERREWEGHE, J., VOLCKAERT, G. and YSEBAERT, M. Nature (London) (1978), 273, 113-120.

83. NEVINS, J.R., and DARNELL, J.E. J. Virol. (1978), 25, 811-823.

84. STEITZ, J.A. and JAKES, K. Proc. Nat. Acad. Sci. U.S.A. (1975), 72, 4734-4738.

85. BROOKER, J. and MARCUS, A. FEBS Lett. (1978), 83, 118-124.

86. PELHAM, H.R.B. Nature (London) (1977), 269, 532-534.

87. HUNTER, T.R., HUNT, T., KNOWLAND, J. and ZIMMERN, D. Nature (London)(1976), 260, 759-764.

88. PELHAM, H.R.B. Nature (London)(1978), 272, 469-471.

89. JACOBSON, M.F. and BALTIMORE, D. Proc. Nat. Acad. Sci. U.S.A. (1968), 61, 77-84.

90. KIEHN, E.D. and HOLLAND, J.J. J. Virol. (1970), 5, 358-367.

91. PELHAM, H.R.B. Eur. J. Biochem. (1978), 85, 457-462.

92. PELHAM, H.R.B. Ph. D. Thesis (1978), University of Cambridge.

IN VITRO TRANSLATION OF PICORNAVIRUS RNA

ELVERA EHRENFELD

Department of Biochemistry and Microbiology
The University of Utah
Salt Lake City, Utah 84132, U.S.A.

I. TRANSLATION OF PICORNAVIRUS RNA IN CELL-FREE EXTRACTS

A. EMC Virus

Although studies of picornavirus proteins and their biosynthesis *in vivo* have been conducted with approximately equal intensity for both poliovirus and encephalomyocarditis (EMC) virus (1), the majority of *in vitro* translation studies have been performed using the mouse virus RNA. Beginning in the early 1970's, several different laboratories reported the synthesis of virus-specific proteins in preincubated cell-free systems derived from Krebs II, Ehrlich or mouse plasmacytoma ascites cells, or from cultured L cells (2-13). In general, a variety of different sized polypeptides, including those with molecular weights over 100,000 daltons, were produced, but, unlike the various viral proteins synthesized *in vivo*, the *in vitro* products appear to arise from premature termination of translation rather than from proteolytic cleavage of a complete translate. Complete translation of the entire viral RNA molecule may occur in rare instances (5), but most of the polypeptides synthesized *in vitro* contain amino acid sequences derived from a common amino terminus, and the majority of the products contain the tryptic (or cyanogen bromide) peptide fragments which are present in the virus capsid proteins. These proteins are known to be encoded by nucleotide sequences within the 5' half of the viral RNA (1). A diagram of the overlap in amino sequences of the major translation products (I-VI) of EMC virus RNA by a Krebs or Ehrlich ascites cell-free system is shown below. The data are from Hunt (13).

EMC RNA

5' — — — — — — — — — — — — — 3'

H_2N ——— EMC POLYPROTEIN ——— COOH

δ β γ α F C

——— I

——— II In vitro translation

——— III products

——— IV

——— V

——— VI

The reason for such extensive premature termination in these systems is unclear. Possibly, codons are encountered for which the concentration of the corresponding tRNA in the extracts or the reaction mix is extremely low. Alternatively, there may be restrictive conformations or possibly proteins bound at particular sites on the RNA which cause a blockage of ribosome movement. A third possibility is that endogenous nucleolytic activities cause breakage of the RNA at limited numbers of preferentially sensitive sites. Several investigators have presented arguments favoring the last mechanism (14, 4), but direct evidence is lacking. RNA re-extracted after translation in a protein-synthesizing system is often extensively degraded, but it is likely that only a small fraction of the input RNAs in these systems is actually being translated, and it has not been possible to distinguish these from the "inactive" molecules. Similar results have been obtained for the translation of mengovirus RNA in a fractionated cell-free system prepared from uninfected or infected Ehrlich ascites tumor cells (15). Extracts were reconstituted from preparations of ribosomal subunits, crude initiation factors, pH 5 enzyme fractions, and mRNA.

The existence of many prematurely terminated translation products has made an analysis of protein processing difficult. However, pulse-chase experiments _in vitro_, or the addition to extracts of inhibitors of proteolytic enzymes, have failed to provide any indication that cleavage or correct processing of large polypeptides occurred. However, one consistent finding is that uninfected cell extracts translate EMC virus RNA to produce a major polypeptide approximately 20,000 daltons larger than is produced by translation of the same RNA in extracts from infected cells (5, 12). The smaller product from infected cells is the same size as the capsid precursor protein made *in vivo*. A comparison of product sizes and peptide content with those of authentic capsid proteins suggests that the larger product synthesized by uninfected cell extracts contains an extra amino-terminal sequence which precedes the translation of capsid protein sequences (13), and which could possibly correspond to labile, amino-terminal peptides detected in small amounts in EMC virus-infected cells (3, 7). The loss of this "extra" sequence from the product synthesized by infected cell extracts suggested that these extracts contained a proteolytic activity necessary for processing the larger precursor. This activity could be demonstrated by the addition of infected cell extract to viral polypeptides previously synthesized by uninfected cell extracts, and upon partial purification it appeared to behave similarly to capsid protein γ (16).

More recently, EMC virus RNA was translated in a rabbit reticulocyte lysate made mRNA-dependent by prior treatment with micrococcal nuclease (17). This system has been shown to maintain a high elongation rate, to be relatively free of ribonuclease activity, and to avoid premature termination on other mRNAs (18). In contrast to the results obtained previously with other translation systems, EMC virus RNA appeared to be translated efficiently and completely. Additional mouse liver tRNA was added to the treated reticulocyte lysate. The absence of products arising from premature termination indicate that most, if not all, ribosomes completed translation of the RNA. In addition, cleavage of the nascent polypeptide occurred to products which corresponded well with those synthesized *in vitro*. Significantly, correct processing appeared to be dependent upon a virus-specific product synthesized during the translation reaction.

One unsolved question arising from this study concerns the order of removal of the amino terminal "extra" sequence referred to above in the processing reaction. When the amino terminus of the translation products were specifically labeled with formyl [^{35}S]met $tRNA_f^{met}$, both the large and the smaller capsid protein precursor retained the radiolabel, suggesting that the putative lead-in sequence which is made *in vitro* but which is not present in the final capsid proteins is removed at a later stage of maturation, and is not responsible for the difference in size between the larger and smaller capsid protein precursors.

In summary, the complete translation of EMC virus RNA and correct processing of the polypeptide products appear to occur in mRNA-dependent reticulocyte lysates supplemented with tRNA. Additional data presented at this meeting show that the RNA of another picornavirus, foot-and-mouth disease virus (FAMD), is also completely translated in the same reticulocyte system with excellent fidelity of both translation and processing (D. Black and D. Sangar, unpublished observations). Although the absence of difficulties previously encountered with other mammalian cell-free systems is most likely due to the choice of the reticulocyte lysate, it is unfortunate that no direct comparison of the translation abilities of several different systems with the same (presumably completely intact) RNA preparations has been made under conditions where the reticulocyte lysate functions with good fidelity. This point is emphasized by the statement in a recent report (19) that conditions have been found for the translation of EMC virus RNA in extracts from Krebs II ascites cells under which almost all major virus-specific proteins are synthesized and normally processed _in vitro_.

B. Poliovirus

The translation of poliovirus RNA in cell-free systems has been less well studied. An early report describing translation of EMC virus RNA in Ehrlich ascites cell and L-cell extracts claimed that poliovirus RNA was not translated in these systems (11), and other unpublished but less than thorough efforts to translate poliovirus RNA in other systems suggested that this RNA was a relatively poor messenger compared with the mouse virus RNA. This observation remains unsatisfying since poliovirus grows to high titers _in vivo_, and the replication cycle is thought to be normal in mouse cells when the adsorption block is by-passed by infection with purified RNA (20). It is noteworthy that extracts prepared from wheat germ (21), which are extremely efficient at translating other mRNAs, translate poliovirus RNA very poorly if at all (22). No published reports of efforts to translate EMC virus RNA in wheat germ extracts are available. It has been suggested that the wheat germ system exhibits a high requirement for a capped 5' terminus on RNAs which are translated (23), and the lack of the usual cap group as well as the inability of the extract to provide the cap for poliovirus RNA, which carries a protein at its 5' end (24, 25), may be responsible for its poor translation in this system.

At this time, the only detailed analyses of the _in vitro_ translation products of poliovirus RNA have been performed following translation in extracts from either uninfected or infected HeLa cells (26, 27). The earlier of these two reports (26) shows that addition of poliovirus RNA to a preincubated HeLa cell extract causes the formation of a spectrum of polypeptides very similar to those made in infected cells in which cleavage was inhibited by amino acid analogs, and it is suggested that the entire viral RNA was translated with good efficiency. In these studies, the salt (KCl) concentration was increased after the first 15 minutes of incubation from 90mM (the optimum for initiation) to 155 mM (the optimum for elongation), since this was found to stimulate the synthesis of larger polypeptides, including the largest one detected *in vivo* after treatment of infected cells with amino acid analogues. Extensive premature termination appeared not to occur, and it was suggested that uninfected HeLa cell extracts were capable of catalyzing the initial cleavages of nascent polyprotein. The later report (27), from the same laboratory, utilized additional digestion of the extracts with micrococcal nuclease following the preincubation step to further reduce endogenous mRNA translation. No shift in salt concentration was performed: KOAc was maintained at 80 mM. The translation products from these reactions did not co-migrate on SDS-polyacrylamide gels with those synthesized in infected cells, and contained predominantly capsid precursor peptides derived from translation of the 5' terminal half of the RNA. These results are reminiscent of the premature termination events found earlier with EMC virus RNA.

Preliminary experiments have been performed using the mRNA-dependent reticulocyte lysate to translate poliovirus RNA (Brown and Ehrenfeld, unpublished). A spectrum of polypeptides of varying molecular weights is produced, but these have not yet been characterized, as was done for the products of EMC virus RNA in the reticulocyte system. The results are promising, however, that faithful and perhaps complete translation can occur in this system for poliovirus RNA as well as for EMC RNA. Similar experiments have been performed in another laboratory which clearly demonstrates that complete translation and processing can occur (R.R. Rueckert, in press). Earlier reports of poliovirus RNA translation in reticulocyte lysates showed stimulation of amino acid incorporation, but the products were not examined (28).

The viral RNAs used as mRNAs in all of these studies were extracted from purified virions. It is known that these RNAs carry a covalently-linked small protein at their 5' termini which is absent from intracellular viral mRNAs extracted from polysomes in infected cells. The latter molecules terminate in 5' pUp (29, 30, 31), and thus, in this respect, they are different from the virion RNA molecules used for *in vitro* translation. It is claimed, however, that the 5' terminal protein is rapidly removed in the HeLa cell extracts (27) as well as in the reticulocyte lysate (V. Ambros and D. Baltimore, personal communication), to generate an apparently authentic mRNA.

Other investigators have studied translation of poliovirus RNA in non-preincubated cell-free systems prepared from infected HeLA cells, in which ribosomes initiate and elongate on endogenous viral mRNAs which were in the cell at the time of extract preparation (32, 33). These extracts synthesize polypeptides which show excellent correlation with those made in infected cells (32, 34, Ehrenfeld, unpublished observations). Neither detectable premature termination nor cleavage deficiencies are apparent.

Although translation of picornavirus RNAs in a variety of cell-free systems derived from mammalian cells has presented more difficulties than have been encountered with other viral mRNAs, it appears that complete translation of the viral genome can occur *in vitro*, accompanied by correct post-translational processing. The factors required and the mechanism responsible for the later event have not yet been elucidated. A major area of undetermined importance in these studies is the role and function of cellular initiation factors. Earlier studies did not involve tests of initiation factor effects; more recent work has yielded conflicting reports in which addition of crude or partially purified initiation factor preparations either stimulated translation slightly or had no effect (26, Ehrenfeld, unpublished observations). However, a number of preliminary studies suggest that some modification of a component(s) in preparations of factors may occur in infected cells (27, 33, 34, 35), and the effects of these changes on the specificity of virus translation are just beginning to be examined. Specific experimentation on the initiation of virus translation are discussed in a separate section, below.

II. INITIATION

A. The Number of Initiation Sites

Analysis of virus-specific proteins synthesized in the infected cell originally revealed the presence of a complex set of polypeptides, whose summed molecular weights exceeded the theoretical coding capacity of the viral genome by a factor of at least four (1). This puzzling observation was rapidly clarified by analysis of pulse-labeled proteins and their metabolic fate following a chase. Although few and predominantly high molecular weight polypeptides are synthesized during a very short labeling period, these proteins rapidly disappear from the infected cell during a chase period, and several smaller molecular weight polypeptides appear. The precursor-product relationship of the various viral proteins was confirmed by tryptic peptide maps which conclusively showed that the amino acid sequences present in accumulated viral proteins were also present in transient larger proteins (1,20). In addition, treatment of infected cells with amino acid analogues or with inhibitors of specific proteases could cause the accumulation of new polypeptides, larger than any normally detected in the infected cell. The largest of these was a polypeptide whose mobility in SDS-polyacrylamide gels indicated a molecular weight of approximately 210,000 daltons. This size is sufficiently close to the maximum coding capacity of an RNA containing 7600 nucleotides that it was immediately suggested that the synthesis of poliovirus proteins occurred by a mechanism in which a single initiation site was utilized to generate a single polycistronic protein which subsequently underwent cleavage to produce intermediates and eventually functional viral proteins. On the assumption that only one initiation site was active, the various viral proteins have been mapped with respect to the relative position of their coding sequences on the viral genome.

Efforts to confirm this model by translation of viral RNA *in vitro* were made by Öberg and Shatkin (36) and by Smith (37). Preincubated protein-synthesizing extracts from cells were incubated with EMC viral RNA in the presence of [^{35}S]-formyl-methionyl-$tRNA_f^{met}$. The charged initiator tRNA thus served to donate radiolabel to the amino terminus of newly-initiated polypeptide chains; the methionine was enzymatically formylated by extracts from *E.coli* in order to insure that only amino termini would carry the radiolabel and to increase the stability of the amino terminus of the newly-synthesized proteins. Translation products were digested with trypsin or with chymotrypsin and the fragments representing the amino terminal sequences were analyzed by anion-exchange chromatography or by electrophoresis. Although predominantly one initiated tryptic or chymotrypic peptide was identified in both studies, a small proportion (about 10%) of a second initiated chymotryptic peptide is evident in the published data of Öberg and Shatkin (36). More recently, similar experiments were performed on the translation products of poliovirus RNA in a HeLa cell extract with a single fmet-labeled tryptic peptide detected by high voltage electrophoresis at pH 3.5 (26). However, other investigators working with the same strain of poliovirus reported the utilization of two different initiation sites during *in vitro* translation by HeLa cell extracts prepared from infected cells, which resulted in the production of two different fmet-labeled trypsin digestion products (38). The two amino-terminal tryptic peptides were readily resolved by high voltage electrophoresis at pH 1.9 or 3.5 by paper chromatography. Furthermore, the penultimate amino acids adjacent to the N-terminal fmet are different in the two tryptic peptides which directly demonstrates that they result from initiation at two sites. Both sites were utilized when purified exogenous viral RNA was translated in a preincubated uninfected cell extract, as well as when endogenous RNA was translated in an infected cell extract.

Analysis of a second picornavirus was deemed important in light of the conflicting data from two laboratories, to demonstrate that the utilization of two initiation sites for *in vitro* translation was not unique to a particular laboratory virus stock. Consequently, experiments similar to those described above were performed with protein-synthesizing extracts from LSc poliovirus-infected cells. The LSc strain is a multi-step, temperature-sensitive mutant derived from the Mahoney strain of poliovirus type 1. It was selected for these studies because previous work had suggested that initiation of translation of LSc virus proteins *in vivo* showed a greater resistance to hypertonic salt treatment than did translation of Mahoney virus proteins (39), and other workers have interpreted this finding as a reflection of a difference in the efficiencies of the initiation process between the two RNAs (40). The results showed that translation of LSc virus RNA *in vitro* also occurs at two different initiation sites, and the initiated tryptic peptides are identical with those produced by translation of Mahoney virus RNA (41).

A striking finding in these studies was that the relative rates of initiation at the two sites varied markedly as a function of the Mg^{++} concentration in the reaction mixture (38). Although the optimum Mg^{++} concentration for total fmet incorporation is approximately 2.0 mM, at Mg^{++} concentrations between 1.6 mM and 2.5 mM, initiation occurs predominantly at the site which generates tryptic peptide II; higher Mg^{++} concentrations up to about 4.0 mM shift the proportion in favor of trypic peptide I. Thus the relative rates of initiation at each of the two sites can be experimentally manipulated. It is interesting that translation of LSc virus RNA always results in an increased relative proportion of tryptic peptide II, at any Mg^{++} concentration, when compared with the proportion synthesized by Mahoney virus RNA (41).

B. The Tryptic Peptides

Characterization of the two fmet-labeled trypic peptides which are derived from the products of initiation on poliovirus RNA shows several differences in their properties. Gel filtration on BioGel P-4 or P-6 columns reveals two peaks, each corresponding to an individual tryptic peptide. Estimated molecular weights of 1550 daltons (tryptic peptide I) and 1300 daltons (tryptic peptide II) were obtained from a P-4 column calibrated with oligosaccharides, peptide and peptidyloligosaccharide markers (41). Thus, both tryptic peptides are small, and probably contain about 13 and 15 amino acids, respectively. As expected, each contains a basic amino acid arginine or lysine, at the carboxyl terminus (38). Electrophoresis at pH 1.9 or 3.5 shows that tryptic peptide II has a greater mobility than tryptic peptide I, and it has a slower chromatographic R_f in a butanol/acetic acid/water solvent.

C. The Initiated Polypeptides

When the fmet-labeled polypeptides synthesized *in vitro* are analyzed directly by SDS-PAGE, two major bands are resolved (Knauert and Ehrenfeld, submitted). According to their mobilities in the gel, one has an apparent molecular weight of 115,000 daltons, whereas the second is a small polypeptide of approximately 6-9,000 daltons. Resolution of small polypeptides on these gels is poor, and the correlation between mobility and molecular weight in this range often breaks down. In addition, several minor bands are often detected with intermediate mobilities. Analysis of the products of reactions incubated at different Mg^{++} concentrations shows that at lower Mg^{++} concentrations, when the initiation site that generates tryptic peptide II is utilized most efficiently, the high molecular weight fmet-labeled polypeptide predominates, whereas higher Mg^{++} concentrations produce almost exclusively the small fmet-labeled polypeptide. These results suggested a correlation between the two tryptic peptides which defined the two initiation sites and the two polypeptides representing the protein products of initiation. The intensities of the minor bands co-varied with that of the 115,000 dalton polypeptide.

Verification of the relationship between the polypeptides and the tryptic peptides was achieved by eluting the individual fmet-labeled proteins from a polyacrylamide gel and analyzing their trypsin digestion products. The high molecular weight polypeptide contains only tryptic peptide II, whereas the small polypeptide contains tryptic peptide I. Three minor bands were individually analyzed in this way and all contained only tryptic peptide II. Thus, the minor bands appear to result from either premature termination or aberrant cleavage, but all represent initiation at the same site that generates the 115,000 dalton polypeptide.

The isolation of fmet-labeled polypeptides which represent the products of initiation of translation at two unique sites on the viral genome provided the material for studies to identify these two proteins. Comparison of the mobility of the 115,000 dalton polypeptide in SDS-polyacrylamide gels with those of viral proteins in cytoplasmic extracts of infected cells showed that this polypeptide co-migrated with the previously identified precursor to capsid proteins, NCVPla. However, an additional series of experiments were undertaken to establish that the high molecular weight fmet-labeled polypeptide was indeed NCVPla. In 1971, Cole _et al_. (42) reported on the isolation of defective interfering (DI) particles from cells which had been infected with serial high multiplicity passage stocks of poliovirus. They subsequently characterized these DI particles as being internal deletion mutants which were missing a segment of RNA in the region which encoded the capsid proteins (43). Analysis of the proteins synthesized by cells infected with purified DI particles show no NCVPla, but instead a new protein is detected of approximately 68,000 daltons, which apparently represents the coat protein precursor containing the deletion. In our laboratory, DI-infected cells were used to prepare protein-synthesizing extracts and the products were labeled with fmet and analyzed by SDS-PAGE. Again, two major labeled products were resolved, but the high molecular weight polypeptide from the DI translation reaction exhibited a mobility corresponding to a molecular weight of 68,000 daltons, and this fmet labeled polypeptide co-migrated with the coat protein precursor containing the deletion seen in DI-infected cells. The mobility of the smaller fmet-labeled polypeptide appeared to be unchanged. Thus, the larger polypeptide which results from initiation of translation at the site which generates tryptic peptide II is NCVPla, the precursor to capsid proteins. Coding sequences for this protein have been previously mapped near the 5' end of the viral RNA (1, 20).

The identity of the small molecular weight fmet-labeled polypeptide which is independently initiated in vitro has not yet been identified. Its small size suggested the possibility that it was VP4, the smallest of the capsid proteins, whose sequences are derived from the amino terminal portion of NCVPla. Several considerations argue against this possibility, however. First, cleavage of the intermediate VPO to generate VP4 and VP2 appears not to occur except during the final stages of morphogenesis (44), and has never been seen to occur in vitro. Analyses of the total translation products in these extracts (labeled with internal amino acids) show no production of VP4. Second, the DI particles do not produce VP4; nevertheless, translation of DI particle RNA in vitro does result in normal synthesis of the fmet-labeled small molecular weight polypeptides. Third, if VP4 is derived from the extreme 5' terminal end of NCVPla, it would be expected to contain the same initiated tryptic peptide, whereas the small fmet-labeled polypeptide contains a different terminal tryptic peptide from the fmet-labeled NCVPla.

A second small viral protein has recently been described which becomes covalently attached to the 5' end of viral RNA (24, 25). This protein, VPg, also co-migrates with the small fmet-labeled polypeptide, but further identification has not been made. The map position of the gene sequence for VPg is not known.

D. Preferential Utilization of One Initiation Site Stimulated by Initiation Factors from Infected Cells

In order to stimulate fmet incorporation by cell-free extracts, crude preparations of initiation factors prepared from the 0.5 M KCl wash of HeLa cell ribosomes were added to the protein-synthesizing reactions. The ribosomal salt wash from uninfected cells stimulated total fmet incorporation about two-fold at the concentrations utilized, and stimulated initiation at both sites equally. Ribosomal salt wash prepared from infected cells also stimulated total fmet incorporation to about the same extent as the uninfected cell preparation; however, SDS-PAGE analysis of the fmet-labeled products revealed that only synthesis of the smaller polypeptide was increased. No stimulation of NCVPla synthesis occurred (Knauert and Ehrenfeld, submitted).

Alteration of at least one initiation factor following infection with poliovirus has been implicated by experiments designed to study the inhibition of host cell protein synthesis induced by poliovirus (27, 33, 34, 35). Whether the factor in infected cells responsible for the failure to stimulate initiation of translation of host cell messenger RNA is the same as that responsible for the preferential utilization of only one initiation site on viral RNA is not known at this time.

E. Location of the Two Initiation Sites on the Viral Genome

Of equal importance to identifying the two polypeptides that are initiated *in vitro* is the determination of the location of the two initiation sites on the viral RNA. If the two sites are located near one another, at or near the 5' end of the mRNA, then having two ribosome binding sites might serve merely to ensure efficient translation of an mRNA initially present in small amounts in the infected cell. Subsequent cleavage and trimming of the amino terminus could yield identical protein products from initiation at both sites. It is also formally possible that two initiation sites close together could result in out-of-phase reading to produce a unique, non-overlapping protein from an overlapping RNA sequence. Alternatively, the two initiation sites could be located at some distance from one another. If this were the case, the second site could be used to initiate translation independently of the first site to produce a unique translate.

From the available data collected from studies of a variety of plant and animal viral RNAs, it appears that initiation is usually limited to a region of the mRNA near the 5' terminus of the chain, although the maximum allowable distance has not been defined. Thus, if an internal initiation site does exist on poliovirus RNA, it would appear to be an exception to the general preference for sites located near the 5' end. If the proximity of the initiation site to the 5' end is usually determined by a requirement to be near the "cap" group, however, this restriction may not apply to an "uncapped" mRNA. At this time, there is no information regarding the location of the two initiation sites utilized for translation of poliovirus RNA *in vitro*.

F. Significance of Two Initiation Sites

The biological significance of the existence of two initiation sites for translation of poliovirus RNA *in vitro* depends upon whether or not these sites are utilized for translation *in vivo*. Although all of the identified translational products do appear to be made in equal molar amounts, when corrected for alternate cleavage pathways and protein degradation (45-47, and see chapter 7 of this book), the relative synthesis of VPg (which has not been detected free of RNA in the cytoplasmic extracts of infected cells), has not been measured. Although no evidence has ever been presented for specific early-late proteins, it is also true that a careful examination of viral proteins synthesized early during infection has not been made since it is difficult to detect viral proteins over the large background of host cell protein synthesis which persists for the first 1 or 1.1/2 hours post-infection. Thus a model which proposes that input, parental viral RNA is translated differently from progeny RNA molecules cannot be conclusively ruled out. In this regard, it is worth noting several unexplained observations concerning the timing of poliovirus-specific functions. First, the inhibition of host cell protein synthesis which appears to result from some expression of the viral genome, occurs early after infection, even when synthesis of all detectable viral RNA and protein synthesis is prevented by the presence of low concentrations of guanidine in the medium (48). Second, Lundquist _et al_.have reported that NCVP4 is a viral component of the RNA replicase activity in infected cells (49); however, although synthesis of NCVP4 parallels total protein synthesis throughout the infection cycle, the NCVP4 molecules that appear in the replication complex at any time are all synthesized between 90 and 135 minutes post-infection, a time period which precedes the bulk of viral RNA and protein synthesis (50). Lastly, Cole and Baltimore have reported that replication of DI particle RNA exhibits a cold-sensitive step only at early times post-infection (51). This suggests that some essential viral function occurs early after infection which either does not occur or is not required at later times. These observations, taken together with the demonstration of two initiation sites recognized for translation _in vitro_, raise some question about the mechanism of overall regulation of viral protein synthesis.

REFERENCES

1. RUECKERT, R.R. On the structure and morphogenesis of picornaviruses. In Comprehensive Virology. Fraenkel-Conrat, H. and Wagner, R.R. eds., (1976), Vol.6, pp.131-213, Plenum Press, New York.

2. KERR, I.M. and MARTIN, E.M. Virus protein synthesis in animal cell-free systems: Nature of the products synthesized in response to RNA of EMC virus. J. Virol. (1971), _7_, 438-447.

3. DOBOS, P., KERR, I.M. and MARTIN, E.M. Synthesis of capsid and non-capsid viral proteins in response to EMC virus RNA in animal cell-free systems. J. Virol. (1971), _8_, 491-499.

4. KERR, I.M., BROWN, R.M. and TOVELL, D.R. Characterization of the polypeptides formed in response to EMC virus RNA in a

cell-free system from mouse ascites tumor cells. J. Virol. (1972), 10, 73-81.

5. ESTEBAN, M. and KERR, I.M. The synthesis of EMC virus polypeptides in infected L-cells and cell-free systems. Eur. J. Biochem. (1974), 45, 567-576.

6. MATHEWS, M.B. and KORNER, A. Mammalian cell-free protein synthesis directed by viral RNA. Eur. J. Biochem. (1970), 17, 328-338.

7. SMITH, A.E., MARCKER, K.A. and MATHEWS, M.B. Translation of RNA from EMC virus in a mammalian cell-free system. Nature (London) (1970), 225, 184-187.

8. MATHEWS, M.B. and OSBORN, M. The rate of polypeptide chain elongation in a cell-free system from Krebs II ascites cells. Biochem. Biophys. Acta. (1974), 340, 147-152.

9. BOIME, I., AVIV, H. and LEDER, P. Protein synthesis directed by EMC virus RNA. II. The in vitro synthesis of high molecular weight proteins and elements of the viral capsid. Biochem. Biophys. Res. Comm. (1971), 45, 788-795.

10. BOIME, I. and LEDER, P. Protein synthesis directed by EMC virus RNA. III.Discrete polypeptides translated from a monocistronic messenger in vitro. Arch. Biochem. Biophys. (1972), 153, 706-713.

11. EGGEN, K.L. and SHATKIN, A.J. In vitro translation of cardiovirus ribonucleic acid by mammalian cell-free extracts. J. Virol. (1972), 9, 636-645.

12. LAWRENCE, C. and THACH, R.E. Encephalomyocarditis virus infection of mouse plasmacytoma cells. I. Inhibition of cellular protein synthesis. J. Virol. (1974), 14, 598-610.

13. HUNT, L.A. In vitro translation of EMC viral RNA: Synthesis of capsid precursor-like polypeptides. Virology (1976), 70, 484-492.

14. HUNT, L.A. Mechanism of premature polypeptide termination in a mouse ascites cell-free system programmed by EMC RNA. J. Virol. (1976), 18, 788-792.

15. HACKETT, P.B., EGBERTS, E. and TRAUB, P. Translation of ascites and mengovirus RNA in fractionated cell-free systems from uninfected and mengovirus-infected Ehrlich ascites tumor cells. Eur. J. Biochem. (1978), 83, 341-352.

16. LAWRENCE, C. and THACH, R.E. Identification of a viral protein involved in post-translational maturation of the EMC virus capsid precursor. J. Virol. (1975), 15, 918-928.

17. PELHAM, H.R.B. Translation of EMC virus RNA in vitro yields an active proteolytic processing enzyme. Eur. J. Biochem. (1978), 85, 457-462.

18. PELHAM, H.R.B. and JACKSON, R.J. An efficient mRNA-dependent translation system from reticulocyte lysates. Eur. J. Biochem. (1978), 85, 457-462.

19. SVITKIN, Y.V., GINEVSKAYA, V.A., UGAROVA, T.Y. and AGOL, V.I. A cell-free model of the EMC virus-induced inhibition of host cell protein synthesis. Virology (1978), 87, 199-203.

20. LEVINTOW, L. The reproduction of picornaviruses. In Comprehensive Virology. Fraenkel-Conrat, H. and Wagner, R.R. eds. (1974), Vol.2, pp.109-169, Plenum Press, New York.

21. ROBERTS, B.E. and PATERSON, B.M. Efficient translation of tobacco mosaic virus RNA and rabbit globin 9S in a cell-free system from commercial wheat germ. Proc. Nat. Acad. Sci. U.S.A. (1973), 70, 2330-2334.

22. EHRENFELD, E. and LUND, H. Untranslated vesicular stomatitis virus messenger RNA after poliovirus infection. Virology (1977), 80, 279-308.

23. LODISH, H.F. and ROSE, J.K. Relative importance of 7-methyl guanosine in ribosome binding and translation of VSV mRNA in wheat germ and reticulocyte cell-free systems. J. Biol. Chem. (1977), 252, 1181-1188.

24. LEE, Y.F., NOMOTO, A., DETJEN, B.M. and WIMMER, E. A protein covalently linked to poliovirus genome RNA. Proc. Nat. Acad. Sci. U.S.A. (1977), 74, 59-63.

25. FLANEGAN, J.B., PETTERSSON, R.F., AMBROS, V., HEWLETT, M. and BALTIMORE, D. Covalent linkage of a protein to a defined nucleotide sequence at the 5' terminus of virion and replicative intermediate RNA of poliovirus. Proc. Nat. Acad. Sci. U.S.A. (1978), 74, 961-965.

26. VILLA-KOMAROFF, L., GUTTMAN, N., BALTIMORE, D. and LODISH, H.F. Complete translation of poliovirus RNA in a eukaryotic cell-free system. Proc. Nat. Acad. Sci. U.S.A. (1975), 72, 4157-4161.

27. ROSE, J.K., TRASCHEL, H., LEONG, K. and BALTIMORE, D. Inhibition of translation by poliovirus: Inactivation of a

specific initiation factor. Proc. Nat. Acad. Sci. U.S.A. (1978), 75, 2732-2736.

28. SHAFRITZ, D.A., PADILLA, M., CANAANI, D., GRONER, Y., WEINSTEIN, J.A., BAR-JOSEPH, M. and MERRICK, W.C. Initiation factor eIF-4B-dependent recognition and translation of capped versus uncapped eukaryotic mRNA. J. Biol. Chem. (1978), in press.

29. FERNANDEZ-MUÑOZ, R. and DARNELL, J.E. Structural difference between the 5' terminus of viral and cellular mRNA in poliovirus-infected cells: Possible basis for the inhibition of host protein synthesis. J. Virol. (1976), 18, 719-726.

30. HEWLETT, M.J., ROSE, J.K. and BALTIMORE, D. 5' terminal structure of poliovirus polyribosomal RNA is pUp. Proc. Nat. Acad. Sci. U.S.A. (1976), 73, 327-330.

31. NOMOTO, A., LEE, Y.F. and WIMMER, E. The 5' end of poliovirus mRNA is not capped with $m^7G(5')ppp(5')Np$. Proc. Nat. Acad. Sci. U.S.A. (1976), 73, 375-380.

32. CELMA, M.L. and EHRENFELD, E. Effect of poliovirus double-stranded RNA on viral and host cell protein synthesis. Proc. Nat. Acad. Sci. U.S.A. (1974), 71, 2440-2444.

33. KAUFMAN, Y., GOLDSTEIN, E. and PENMAN, S. Poliovirus-induced inhibition of polypeptide initiation *in vitro* on native polyribosomes. Proc. Nat. Acad. Sci. U.S.A. (1976), 73, 1834-1838.

34. HELENTJARIS, T. and EHRENFELD, E. Control of protein synthesis in extracts from poliovirus-infected cells. J. Virol. (1978), 26, 510-521.

35. HACKETT, P.B., EGBERTS, E. and TRAUB, P. Selective translation of mengovirus RNA over host mRNA in homologous fractionated, cell-free translation systems from Ehrlich ascites tumor cells. Eur. J. of Biochem. (1978), 83, 353-361.

36. ÖBERG, B.R. and SHATKIN, A.J. Initiation of picornavirus protein synthesis in ascites cell extracts. Proc. Nat. Acad. Sci. U.S.A. (1972), 69, 3589-3593.

37. SMITH, A.E. The initiation of protein synthesis directed by the RNA from encephalomyocarditis virus. Eur. J. Biochem. (1973), 33, 301-313.

38. CELMA, M.L. and EHRENFELD, E. Translation of poliovirus RNA *in vitro*: Detection of two different initiation sites. J. Mol. Biol. (1975), 98, 761-780.

39. TERSHACK, D.R. Effect of hypertonic medium on protein synthesis of Mahoney and LSc poliovirus. J. Virol. (1976), 20, 597-603.

40. NUSS, D.L., OPPERMAN, H. and KOCH, G. Selective blockage of initiation of host protein synthesis in RNA virus-infected cells. Proc. Nat. Acad. Sci. U.S.A. (1975), 72, 1258-1262.

41. JENSE, H., KNAUERT, F. and EHRENFELD, E. Two initiation sites for translation of poliovirus RNA *in vitro*: Comparison of LSc and Mahoney

42. COLE, C.N., SMOLER, D., WIMMER, E. and BALTIMORE, D. Defective interfering particles of poliovirus. I. Isolation and physical properties. J. Virol. (1971), 7, 478-485.

43. COLE, C.N. and BALTIMORE, D. Defective interfering particles of poliovirus. II. Nature of the defect. J. Mol. Biol. (1973), 76, 325-343.

44. JACOBSON, M.F. and BALTIMORE, D. Morphogenesis of poliovirus. I. Association of viral RNA with coat protein. J. Mol. Biol. (1968), 33, 369-378.

45. BUTTERWORTH, B.E., HALL, L., STOLTZFUS, C.M. and RUECKERT, R.R. Virus-specific proteins synthesized in encephalomyocarditis virus-infected HeLa cells. Proc. Nat. Acad. Sci. U.S.A. (1971), 68, 3083-3087.

46. LUCAS-LENARD, J. Cleavage of mengovirus polyproteins *in vivo*. J. Virol. (1974), 14, 261-269.

47. PAUCHA, E. and COLTER, J.S. Evidence for control of translation of the viral genome during replication of mengovirus and poliovirus. Virology (1975), 67, 300-305.

48. HELENTJARIS, T. and EHRENFELD, E. Inhibition of host cell protein synthesis by UV-inactivated poliovirus. J. Virol. (1977), 21, 259-267.

49. LUNDQUIST, R., EHRENFELD, E. and MAIZEL, J.V. Isolation of a viral polypeptide associated with poliovirus RNA polymerase. Proc. Nat. Acad. Sci. U.S.A. (1974), 71, 4773-4777.

50. LUNDQUIST, R. and MAIZEL, J.V. *In vivo* regulation of the poliovirus RNA polymerase. Virology (1978), 89, 484-493.

51. COLE, C.N. and BALTIMORE, D. Defective interfering particles of poliovirus. IV. Mechanisms of enrichment, J. Virol. (1973), 12, 1414-1426.

SECTION V:

TRANSLATIONAL REGULATION: MECHANISM OF ACTION OF INTERFERON

THE INTERFERON SYSTEM: STUDIES ON THE MOLECULAR MECHANISM OF INTERFERON ACTION

M. REVEL, A. KIMCHI, A. SCHMIDT, L. SHULMAN and
A. ZILBERSTEIN

Department of Virology, Weizmann Institute of Science
Rehovot, Israel

INTRODUCTION

Exposure of sensitive cells to purified interferon, from a compatible species, induces a series of biological effects, the best known among these being the antiviral state and the decrease in the rate of cell proliferation. At least one of interferon's antiviral effects is to inhibit viral protein synthesis (1-3). This inhibition is not immediate but requires several hours of active cellular RNA and protein synthesis before it fully develops. In the past year, interferon was shown to induce, in the treated cells, several new enzymatic activities which regulate protein biosynthesis and produce the translational inhibition observed. These biochemical mechanisms induced by interferon are reviewed here, and tentatively shown as three "pathways" in Figure 1.

I. THE PROTEIN KINASE (PK-i) PHOSPHORYLATING eIF-2

A. Isolation of PK-i

Crude lysates (S10) or ribosomal extracts from L cell cultures treated for 24 h by 200 U/ml mouse interferon, when incubated with (^{32}P)-γ-ATP and low concentrations of dsRNA (0.04 - 0.4 μg/ml poly rI:rC), demonstrate an increased phosphorylation of several endogenous proteins (4-9). Among these, proteins 67 and 35 (67,000 and 35,000 MW in SDS) are most prominent. Protein 35 comigrates by polyacrylamide gel electrophoresis with the small subunit of initiation factor eIF-2 (Figure 2). Interferon-treated cell extracts contain a dsRNA-dependent protein kinase phosphorylating this subunit of eIF-2, this activity being more

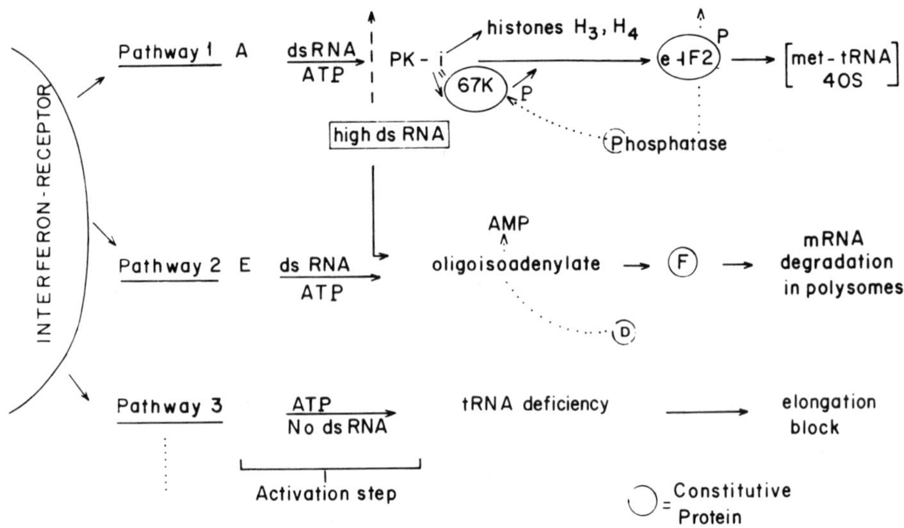

Figure 1

than 5-fold higher than in fractions from untreated cells (10) (Table 1). Two third of this eIF-2 kinase (PK-i) is in the cell sap and one third in ribosomal extracts, the latter containing most of protein 67 and 35 substrates (4). PK-i is not retained on DEAE-cellulose, elutes from phosphocellulose (pH6.7) at 330mM KCl, and from hydroxylapatite at 200mM phosphate; overall purification was about 500-fold. The endogenous protein 67 phosphorylation activity was still present in the purified PK-i (Figure 2) as well as a dsRNA-dependent histone kinase selective for the arginine-rich H_3, H_4 (3, 10).

DsRNA stimulates the rate, but not the extent of protein 67 phosphorylation, and is required for eIF-2 phosphorylation. This requirement can be partly explained by an activation of the eIF-2 kinase.

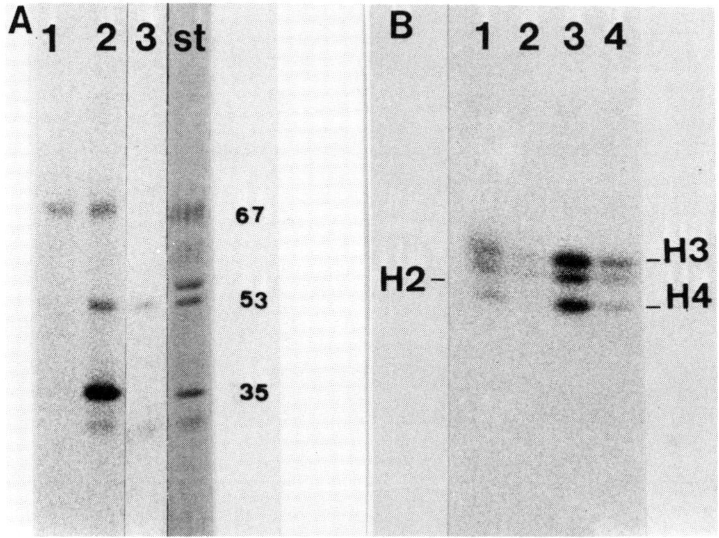

Figure 2. Interferon-induced protein kinase PK-i (10). A: Incubation with ^{32}P-γ-ATP and 0.4 μg/ml poly I:C of PK-i (slot 1), PK-i+eIF-2 (slot 2), eIF-2 (slot 3) and autoradiography; slot st: Coomassie Blue stain of slot 2. Incub. 30 min 30°C, followed by SDS-acrylamide gel. B: Calf thymus histones phosphorylated by PK-i pre-activated with 1mM ATP and 0.4 μg poly I:C alone (slot 1) or with factor A (slot 3). Slots 2 and 4 as 1 and 3 but with 20 μg/ml poly I:C.H1 was not phosphorylated.

B. Activation of eIF-2-kinase: dsRNA, ATP and Factor A

The eIF-2-kinase of interferon-treated cells (PK-i) is not sensitive to cyclic nucleotides or to calcium, but is activated by pre-incubation with ATP and low dsRNA. Study of the activation process is facilitated by the fact that high concentrations of dsRNA (20 μg/ml) inhibit the eIF-2-kinase unless it is pre-incubated with both ATP and low dsRNA (11). The activated protein kinase can therefore be defined as high-dsRNA resistant. PK-i actively phosphorylates histones only if pre-incubated with ATP and low dsRNA. Histones and eIF-2 compete for the activated PK-i suggesting that the same enzyme phosphorylates both substrates. The dsRNA-ATP activation of eIF-2 and histone H_3, H_4 phosphorylation is not seen in fractions from cells not treated by interferon (3,10) (Table 1).

Table 1. Interferon induced protein kinase PK-i

Addition	eIF-2 35K phosphoryl. area	Globin mRNA translation Retic.lys. cpm	L cell S10 cpm
None	0	705,930	167,950
PK-i (interferon):			
-dsRNA-ATP activated	8.4	175,750	79,500
-not activated	1.4	589,320	n.d
-ATP activated (no ds)	1.4	833,750	n.d
Control preparation:			
-dsRNA-ATP activated	0.5	646,750	143,120

2.5 μg PK-i, purified through DEAE and phosphocellulse (10) or similar fraction from control cell was used. Assay in high dsRNA. Conditions as Figure 2 or Table 2.

How is PK-i activated by dsRNA-ATP? For activation, ATP cannot be replaced by the non-hydrolyzable analog AMPPNP. DsRNA seems to interact with PK-i, protein 67 and eIF-2 as shown by their binding to poly I:C Sepharose (3, 6). Recent work indicates that with purified preparations of PK-i, activation is dependent on the presence of an additional factor A, which elutes from DEAE-cellulose at 300mM KCl (Figure 3). Factor A appears to be an acidic protein and is not retained on phosphocellulose (pH6.7; 25mM KCl). Figure 2B shows the strong stimulation of PK-i's H_3, H_4 kinase activity by A; this activation by A requires low dsRNA and ATP and is inhibited by high dsRNA. Identical results were obtained with eIF-2 as substrate. Factor A could act either as a protein kinase-kinase or by a yet more complex mechanism.

C. Translation Inhibition by PK-i

Purified preparations of PK-i inhibit translation of exogenous mRNA in L cell extracts and endogenous globin synthesis in reticulocyte lysates (10). Globin mRNA translation appears slightly more sensitive to PK-i than mengo RNA. Protein synthesis inhibition and eIF-2 phosphorylation are lost after heating PK-i 15 min at 47°C.

Table 2. Requirement of F for oligo-isoadenylate inhibition of protein synthesis

Addition	Mengo RNA translation in L cell S10 cpm	S10+iso-A cpm
None	95,660	76,530 (0.80)
F (DE-INT, 3 μg)	149,600	41,685 (0.28)
F (DE-CONT, 4.5 μg)	134,360	54,480 (0.40)

Factor F purified to DEAE-cell. step (10) (Figure 3) from interferon-treated or control L cells. Isoadenylate trimers (Figure 4) used at 6.5×10^{-9}M. ^{35}S-methionine incorporation measured (4, 10) after 60 min 30°C.

Analysis of the cell-free translation system by sucrose gradient centrifugation, shows that PK-i causes a marked decrease in the amount of (35S-met-tRNA 40S ribosome) complexes (12). Exogenous eIF-2 overcomes part of the dsRNA-dependent translational inhibition of mengo or globin mRNA in extracts from interferon treated cells (13).

It is, therefore, likely, that phosphorylation of eIF-2 impairs one of its function required for met-tRNA binding to 40S ribosomes, as shown recently in reticulocyte lysate for the haemin-regulated eIF-2 phosphorylation (14, 15). A role for the other ribosome-associated proteins phosphorylated by the dsRNA-dependent protein kinase, e.g. protein 67 and 17 (4), in the translational inhibition is, however, not excluded. Factor M1 may also be involved (3).

D. A Protein Phosphatase Dephosphorylating Protein 67 and eIF-2

Cell sap contains a specific phosphoprotein phosphatase (P) which dephosphorylates protein 67, eIF-2 and several other dsRNA-dependent phosphorylated proteins (3). The protein phosphatase P is separated from PK-i on DEAE-cellulose and elutes at 250mM KCl. As shown in Figure 3, P stimulates mengo RNA translation in interferon-treated cell extracts (S10-int) supplemented with dsRNA, but not in control cell extracts (S10-cont). This suggests a regulatory mechanism which controls the level of PK-i dependent inhibition of protein synthesis.

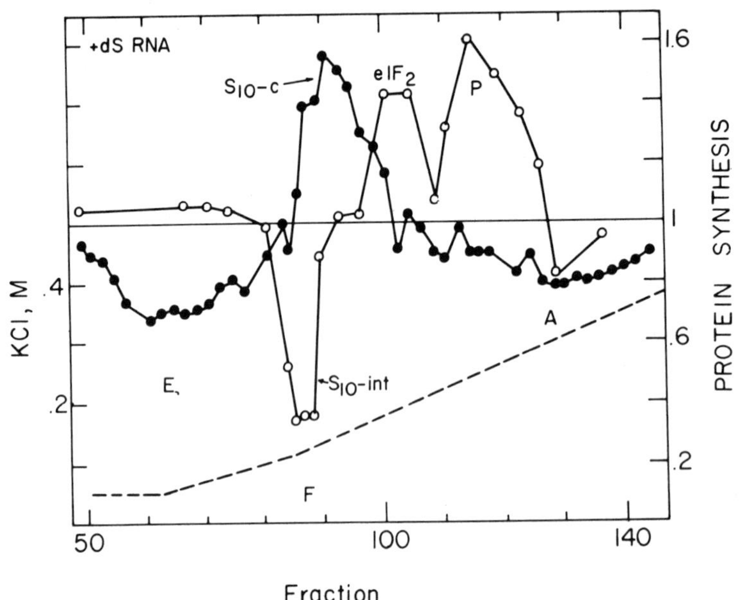

Figure 3. Separation of various translation regulatory activities on DEAE cellulose. Cell sap from interferon-treated cultures was fractionated and assayed on mengo RNA translation (as in Table 2) in S10 from control (c) and interferon-treated cells (int) with 0.04 μg/ml dsRNA. Activity without fraction = 1. Oligo-isoadenylate synthetase E (10), factor F (18). Protein phosphatase (P) and protein kinase activator A, are discussed in the text.

E. What Causes the Increase in PK-i after Interferon?

The mechanism by which interferon treatment increases the enzymatic phosphorylation of protein 67 and eIF-2 was investigated. The increase is seen at 3-6 hours after addition of interferon to mouse L cells and is blocked by actinomycin D. Since protein 67 is found in purified preparations of PK-i we wondered whether this protein is induced by interferon. Not having protein 67 substrate free of PK-i, we made use of the fact that in human cells the interferon- and dsRNA-induced equivalent of protein 67 migrates more slowly on SDS polyacrylamide gels (protein 69) (16). Addition

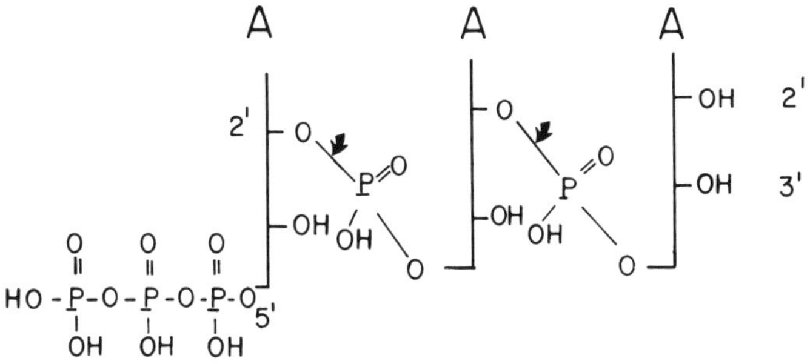

OLIGO-ISOADENYLATE,(TRIMER)

Figure 4. Arrow shows bond split by nuclease (see text).

of the mouse PK-i to human cell extracts not treated by interferon, caused phosphorylation of protein 69. The reaction required dsRNA. This strongly suggests that the 67-69K substrate is present before interferon-treatment but not phosphorylated. This would also indicate that PK-i is not simply protein 67. Protein 67 phosphorylation is probably not directly involved in PK-i activation since phosphorylation of protein 67 is not inhibited by high levels of dsRNA and could not be stimulated by factor A. Protein 67 appears rather as a ribosome-bound protein closely associated with PK-i and serving as substrate for this enzyme.

We also examined whether factor A could be alone responsible for the increased PK-i activity after interferon treatment. Factor A stimulated slightly eIF-2 and histone phosphorylation in control cell extracts but not nearly as much as in interferon-treated cell extracts. Factor A was found in control cell preparations. Some component of the protein kinase PK-i itself seems, therefore, to change after interferon treatment. Finally, protein phosphatase P is found both in control cell sap and, in lower amounts, in interferon-treated cell sap.

II. OLIGO-ISOADENYLATE SYNTHESIS AND ACTIVATION OF NUCLEASE F

A. Oligo-isoadenylate-Synthetase E

During purification of the protein synthesis inhibitors from interferon-treated cells, the dsRNA-ATP activated protein kinase PK-i, was separated from another dsRNA-ATP dependent activity. This second enzyme, E, produces a heat-stable, small molecular weight translation inhibitor (6, 3, 10) recently identified (17) as (2'-5')pppApApA (Figure 4). Enzyme E, which binds to dsRNA, slowly polymerizes ATP into a series of oligo-nucleotides with the general structure ppp5' (A2'p)n5'A, called oligo-isoadenylate (iso-A) (18). After 1 hour at 30°C, the products are 85% dimers; after 8-16 hours, trimers represent 22-34%, tetramers 3-6% and longer forms 1.5-3%. The oligo-isoadenylate synthetase E is not retained on DEAE-cellulose at 25mM KCl, and elutes from phosphocellulose, after PK-i, at 450mM KCl. The enzyme itself is destroyed by heating to 60°C, 10 min. There are only very small amounts of E in control L cell preparations (10). In contrast to PK-i, the iso-A synthetase E is very active in the presence of 20 μg/ml dsRNA.

B. Oligo-isoadenylate Effect on Translation: Requirement for Factor F

Factor F was first detected as an activity which inhibits translation only when added to extracts from interferon-treated cells in the presence of dsRNA, but not in control cell extracts or without dsRNA. Figure 3 shows that F elutes from DEAE-cellulose at 120mM KCl. It is not well separated at this stage from a stimulator of protein synthesis, but could be purified on phosphocellulose pH 7.9, from which F eluted at 200-300mM KCl. Overall purification was 120-fold (18).

In extracts from control cells, F can be made to inhibit translation if purified iso-A trimers are added. The nucleotide by itself inhibits translation only to a small extent unless factor F is added. With F, concentrations of $1\text{-}10 \times 10^{-9}$M trimers become efficient inhibitors (Table 2). Iso-A dimers had 10-20 times less activity, but tetramers were as active as trimers (17, 18).

F is present in untreated cell extracts, but its activity may be higher by 1.5-2 fold after interferon treatment (Table 2). This change is small as compared to the more than 10-fold increase in iso-A synthetase E activity after interferon (10). F is present

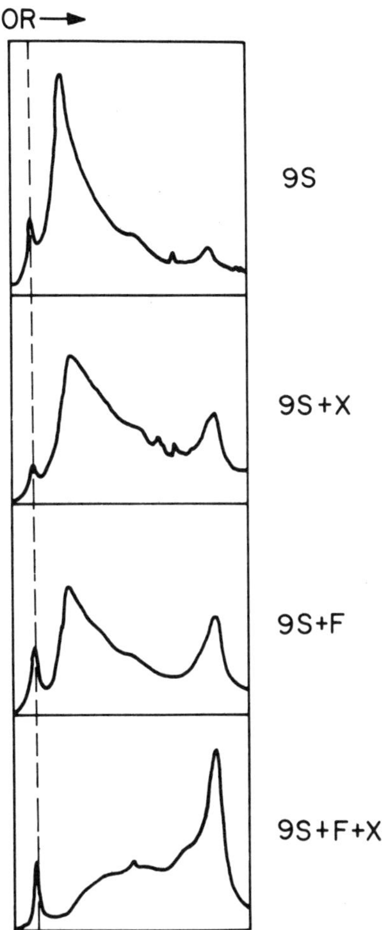

Figure 5. Oligo-isoadenylate activated nuclease activity of factor F. ^{125}I-globin mRNA (0.5 μg) incubated (18) 30 min, 30°C with 0.3 μg F (PC-purified) and 0.16 pmoles, iso-A trimers (X) was analyzed by polyacrylamide gel electrophoresis, 7M urea. Scanning of autoradiography is shown. Arrow shows direction of migration.

in limiting amounts in crude cell extracts: we observed that many cell extracts were not sensitive to iso-A or to dsRNA, unless F was added.

C. Oligo-isoadenylate Dependent Nuclease Activity of Factor F

The mechanisms by which F+iso-A trimers inhibit translation was investigated by pre-incubating the different components of the cell-free system in the presence of F+iso-A before measuring protein synthesis. These experiments clearly showed (18) that the sensitive component was the mRNA template rather than the ribosomal and soluble components of the extract. In the presence of F+iso-A, (^{35}S-met-tRNA 40S ribosome) complexes are formed normally, but no 80S ribosomal initiation complexes or polysomes developed, resulting even in an accumulation of (met-tRNA 40S ribosome mRNA) intermediates (12). The mRNA in these complexes is probably fragmented since upon incubation of mengo or globin mRNA with F+iso-A, a significant amount of the RNA was degraded. Figure 5 shows that globin mRNA is fragmented to half size molecules and smaller pieces when incubated with a purified preparation of F from control cells and nanomolar concentrations of iso-A trimers, but not with F, or iso-A, alone. Iso-A dimers, and ApApA (2'-5') without the 5' triphosphate end, had very little effect on the F nuclease (18). These data are in line with the reported effects of (2'-5')- pppApApA in reticulocyte lysates (19) and with the dsRNA-ATP dependent nuclease reported by Lengyel (20, 21).

The rate of degradation of different RNAs by nuclease F was compared: TMV RNA and mengo RNA were found most sensitive; Car MV RNA and 9S globin mRNA were sensitive to a lesser degree; 18S and 28S rRNA had a low sensitivity but were degraded; ØX DNA was resistant. These results establish that F is a ribonuclease, but do not allow to conclude whether *in vivo* it could discriminate between different mRNA templates. It was, however, constantly observed that when F+iso-A were added to the complete protein synthesis system, mengo RNA was more inhibited than globin mRNA (18). It is likely that during translation, mRNA is protected by the ribosomes and less susceptible to the action of the iso-A-dependent nuclease.

Iso-A could stimulate the nucleolytic activity of F by a direct effect on the protein, or by an indirect effect, for example, on the mRNA. Iso-A does not bind to mRNA, but a complex with F was seen by gel filtration. The nucleotide was, however, rapidly degraded to 5' AMP and ATP, indicating that the 2' phosphate bond is cleaved (Figure 4). The degradation of iso-A was accompanied by a loss of the translational inhibition: the activation of nuclease F appears therefore transient, a constant resynthesis of iso-A by E being needed to keep the nuclease F active. This may insure an autoregulation of F activity (Figure 1) (18).

The nuclease which degrades oligo-isoadenylate copurified with the iso-A activated ribonuclease F over several steps, but was separated on hydroxylapatite. The iso-A degrading enzyme (nuclease D of Figure 1) splits (2'-5')ApA to yield 5'-AMP, and more slowly (3'-5')ApA. It clearly differs from known ribonucleases that are inhibited by 2'-5' dinucleotides (22).

Studies of the relation between endonuclease F and exonuclease D should throw light on the function of oligo-isoadenylate in interferon-treated cells.

III. ELONGATION BLOCK REVERSED BY tRNA

Extracts from cells treated with interferon show in many cases a reduced ability to translate exogenous mRNAs (23), even in the absence of dsRNA. Analysis of the products shows that incomplete polypeptide chains are formed (24) and the defect can be corrected by addition of tRNA (23-26).

In mouse L cells, addition of one or two minor species of mammalian leucyl-tRNA were shown to be sufficient to restore translation (27). A similar conclusion was reached with yeast tRNA (28), even with poly(U,C) as template. The type of tRNA to be added appears to depend on the mRNA used (25, 27) and also on the extract: Friend cells were reported to require lysyl-tRNA (29). These tRNA species were, however, not absent from interferon-treated cell extracts (24, 27, 30), but had to be added in excess over what is normally needed. A decreased charging of leu-tRNA or an increased deacylation were reported (31, 32). Prolonged pre-incubation for 1-2 h of the extracts can increase the interferon-induced elongation block reversed by tRNA (33). We have recently observed that ATP is required to activate the translational inhibition seen in interferon-treated cell extracts without dsRNA. This could be shown by comparing extracts freed of endogenous mRNA activity either by micrococcal nuclease (34) or by pre-incubation with ATP. Table 3 shows that tRNA reversible-inhibition developed when ATP was present. It is not clear yet whether the protein kinase, or E and F may play a role in this inhibition without dsRNA. We also observed an inhibitor of leu-tRNA charging in ribosomal extracts of interferon-treated cells (Schmidt, A. and Revel, M., unpublished).

IV. THE "MULTIPHASE ANTIVIRAL STATE" HYPOTHESIS

The multiple effects by which interferon affects protein

Table 3. Translational inhibition reversed by tRNA

L cell extract	Mengo RNA translation Control S10 cpm	Interferon S10 cpm
Micrococcal nuclease-treated	45,210	46,315 (1.0)
ATP-preincubated	51,695	18,865 (0.32)
" " +tRNA	80,000	117,128 (1.46)

Extract prepared by the micrococcal nuclease method (34) or by 30 min pre-incubation with an energy source (23). Rabbit liver tRNA used at 8 μg/ml. ^{35}S-methionine incorporation measured 60 min after mRNA.

synthesis could each have its function in the cells. The tRNA sensitive inhibition,seen in the absence of dsRNA, may play its role mainly in the non-infected cell, where dsRNA is very low or absent. After infection, in the case of many viruses, dsRNA is formed and accumulates (6, 35). The protein kinase pathway is activated at low dsRNA concentrations but at higher dsRNA levels, it is switched off and the iso-adenylate-nuclease system may act. This "multiphase antiviral state", could insure that the interferon induced control of gene expression is appropriate to the state of the cell. It is striking that the three pathways of translation control share a common requirement for ATP and are regulated by antagonists (Figure 1). What is not understood is how these regulations discriminate between host and viral functions. Work has to be continued, in intact cells and in cell-free systems, to determine how the pathways already found or additional effects of interferon control viral replication and cell proliferation.

ACKNOWLEDGEMENTS

We thank Dr. D. Shafritz for gifts of eIF-2. The excellent assistance of Ms. P. Federman and H. Berissi is gratefully acknowledged. Work supported by NCRD (Israel) and GSF (München, Germany).

REFERENCES

1. METZ, D.H. Cell (1975), 6, 429-439.

2. YAKOBSON, E., PRIVES, C., HARTMAN, J.R., WINOCOUR, E. and REVEL, M. Cell (1977), 12, 73-81.

3. REVEL, M., GILBOA, E., KIMCHI, A., SCHMIDT, A., SHULMAN, L., YAKOBSON, E. and ZILBERSTEIN, A. Proceedings FEBS 11th Meeting, Clark, B.F.C. *et al* eds, (1978), Vol.43, pp. 47-58, Pergamon Press, Oxford.

4. ZILBERSTEIN, A., FEDERMAN, P., SHULMAN L. and REVEL, M. FEBS Letts. (1976), 68, 119-124.

5. LEBLEU, B., SEN, G.C., SHAILA, S., CABRER, B. and LENGYEL, P. Proc. Natl. Acad. Sci. U.S.A. (1976), 73, 3107-3111.

6. ROBERTS, W.K., HOVANESSIAN, A., BROWN, R.E., CLEMENS, M.J. and KERR, I.M. Nature (London) (1976), 264, 477-480.

7. COOPER, J. and FARRELL, P.J. Biochem. Biophys. Res. Comm. (1977), 77, 124-131.

8. SAMUEL, C.E., FARRIS, D.A. and EPPSTEIN, D.A. Virology (1977), 83, 56-71.

9. LEWIS, J.A., FALCOFF, E. and FALCOFF, R. Eur. J. Biochem. (1978) in press.

10. ZILBERSTEIN, A., KIMCHI, A., SCHMIDT, A. and REVEL, M. Proc. Natl. Acad. Sci. U.S.A. (1978) in press.

11. FARRELL, P.J., BALKOW, K., HUNT, T. and JACKSON, R.J. Cell (1977), 11, 187-200.

12. CHERNAJOVSKY, Y., KIMCHI, A., SCHMIDT, A., ZILBERSTEIN, A. and REVEL, M. In preparation.

13. KAEMPFER, R., ISRAELI, R., ROSEN, H., KNOLLER, S., ZILBERSTEIN, A., SCHMIDT, A. and REVEL, M. Nature (London), submitted.

14. RENU, R.S., LONDON, I.M., DAS, A., DASGUPTA, A., MAJUMDAR, A., RALSTON, R., ROY, R. and GUPTA, N.K. Proc. Natl. Acad. Sci. U.S.A. (1978), 75, 745-749.

15. DEHARO, C., DATTA, A. and OCHOA, S. Proc. Natl. Acad. Sci. U.S.A. (1978), 75, 243-247.

16. SLATE, D., LAWRENCE, D., SHULMAN, L., REVEL, M. and RUDDLE, F.H. J. Virol. (1978), 25, 319-325.

17. KERR, I. and BROWN, R.E. Proc. Natl. Acad. Sci. U.S.A. (1978),

75, 256-260.

18. SCHMIDT, A., ZILBERSTEIN, A., SHULMAN, L., FEDERMAN, P., BERISSI, H. and REVEL, M. FEBS Letts. in press.

19. CLEMENS, M.J. and WILLIAMS, B.R.G. Cell (1978), 13, 565-572.

20. SEN, G.C., LEBLEU, B., BROWN, G.E., KAWAKITA, M., SLATTERY, E. and LENGYEL, P. Nature (London) (1976), 264, 370-373.

21. RATNER, L., WIEGAND, R.C., FARRELL, P.J., SEN, G.C., CABRER, B. and LENGYEL, P. Biochem. Biophys. Res. Comm. (1978), 81, 947-954.

22. WHITE, M.D., BAUER, S. and LAPIDOT, Y. Nuc. Acid. Res. (1978), 4, 3029-3038.

23. CONTENT, J., LEBLEU, B., NUDEL, U., ZILBERSTEIN, A., BERISSI, H. and REVEL, M. Eur. J. Biochem. (1975), 54, 1-10.

24. REVEL, M., CONTENT, J., ZILBERSTEIN, A., NUDEL, U., BERISSI, H. and DUDOCK, B. Colloques INSERM: In vitro transcription and translation of viral genomes, Haenni, A. and Beaud, G. eds., (1975), Vol. 47 pp. 397-406, INSERM Paris.

25. CONTENT, J., LEBLEU, B., ZILBERSTEIN, A., BERISSI, H. and REVEL, M. FEBS Letts. (1974), 41, 125-130.

26. GUPTA, S.L., SOPORI, M.L. and LENGYEL, P. Biochem. Biophys. Res. Comm. (1974), 57, 763-770.

27. ZILBERSTEIN, A., DUDOCK, B., BERISSI, H. and REVEL, M. J. Mol. Biol. (1976), 108, 43-54.

28. WEISSENBACH, J., DIRHEIMER, G., FALCOFF, R., SANCEAU, J. and FALCOFF, E. FEBS Letts. (1977), 82, 71-76.

29. MAYR, U., BERMAYER, H.P., WEIDINGER, G., JUNGWIRTH, C., GROSS, H.J. and BODO, G. Eur. J. Biochem. (1976), 76, 541-551.

30. COLBY, C., PENHOET, E. and SAMUEL, C. Virology (1976), 74, 262-264.

31. SEN, G.C., GUPTA, S.L., BROWN, G.E., LEBLEU, B., REBELLO, M.A. and LENGYEL, P. J. Virol. (1976), 17, 191-203.

32. SELA, I., GROSSBERG, S.E., SEDMARK, J.J. and MEHLER, A.H. Science (1976), 194, 527-529.

33. FALCOFF, R., FALCOFF, E., SANCEAU, J. and LEWIS, J.A.

Virology (1978), 86, 507-515.

34. PELHAM, H.R.B. and JACKSON, R.J. Eur. J. Biochem. (1976), 67, 242-256.

35. CARTER, W.A. and DE CLERCQ, E. Science (1974), 186, 1172-1178.

ON THE ACTION OF INTERFERON IN UNINFECTED AND MENGOVIRUS-INFECTED CELLS

R. FALCOFF, O. AUJEAN, J.A. LEWIS,[+] and E. FALCOFF

Fondation Curie-Institut du Radium,

26, rue d'Ulm, 75005 Paris, France

INTRODUCTION

Treatment of cells with interferon leads to the development of an antiviral state. In a number of cases this inhibition of virus production appears to be due to an inhibition of viral protein synthesis but the exact nature of the defect is unknown.

It appeared, therefore, germane to the purpose of this book briefly to review our present understanding on the mechanism of action of interferon at the molecular level, for there is little doubt left it might play a crucial role in the regulation of the expression of the viral genome.

Some years ago, several groups demonstrated that *in vitro* translation of viral mRNAs was severely impaired in cell-free, protein-synthesizing systems prepared from interferon-treated cells (4-6, 29). The cell free system proved soon to be an unvaluable tool, because it was possible to analyze the alterations induced in the protein synthesizing machinery by the treatment with interferon.

Although at the present time the general picture is far from complete, several well established features seem to characterize the action of interferon.

i) tRNA: The protein synthetic activity of cell-free systems

+ Present address: Dept. Anatomy & Cell Biology, Downstate Med. Center, S.U.N.Y., Brooklyn, New York 11203, U.S.A.

prepared from interferon-treated cells depends critically upon the preparation of the extracts: inhibition due to interferon treatment is manifested only after a certain time of incubation, a step normally used to reduce translation of endogenous mRNA. During such a pre-incubation certain tRNA species are inactivated much more rapidly in extracts from interferon-treated than control cells (1, 2), the consequence being an impairment of translation of exogenous natural (3-8) and synthetic mRNA (9). This inhibition is reversed by the addition of eukaryotic tRNA (9-11) and the restoring tRNA species can be charged with leucine (9, 12, 13).

ii) dsRNA and iso (A): Kerr *et al*. have shown that interferon treatment leads to enhanced sensitivity to inhibition of protein synthesis by double stranded (ds) RNA (14). This group has also reported that when cell sap extracts from interferon-treated cells are incubated with dsRNA and ATP a small oligonucleotide is formed which can inhibit protein synthesis in cell-free extracts from L cells or rabbit reticulocytes (15, 16).

iii) dsRNA and phosphorylation: Several groups have shown that dsRNA stimulates the phosphorylation of two proteins (M_r 67,000 and 35,000) in cell-free extracts of interferon-treated cells (17-19). This situation is somewhat similar to that in rabbit reticulocyte lysates where dsRNA, also an inhibitor of protein synthesis in this system, stimulates phosphorylation of a protein of M_r 67,000 and also that of a component of initiation factor eIF-2 (M_r 35,000 (20). In the reticulocyte system dsRNA-induced inhibition of protein synthesis appears to be due to a block in the formation of methionyl-tRNA 40-S subunit complexes (21) and can be reversed by addition of purified protein synthesis initiation factor, eIF-2 (22).

iv) dsRNA and nuclease: Another reported effect of dsRNA in extracts from interferon-treated cells is the activation of a nuclease activity (23, 24). This activation requires the presence of ATP but its relationship to the phosphorylation of proteins is unknown. Thus the action of dsRNA in extracts of interferon-treated cells might be due to a direct effect upon the protein synthetic machinery, as in the reticulocyte system, or to degradation of mRNA.

In this chapter we will shortly review our current knowledge concerning: a) those modifications which are independent of added dsRNA; b) then those appearing in the presence of dsRNA, and c) some observations done in extracts prepared from interferon-treated, virus-infected cells.

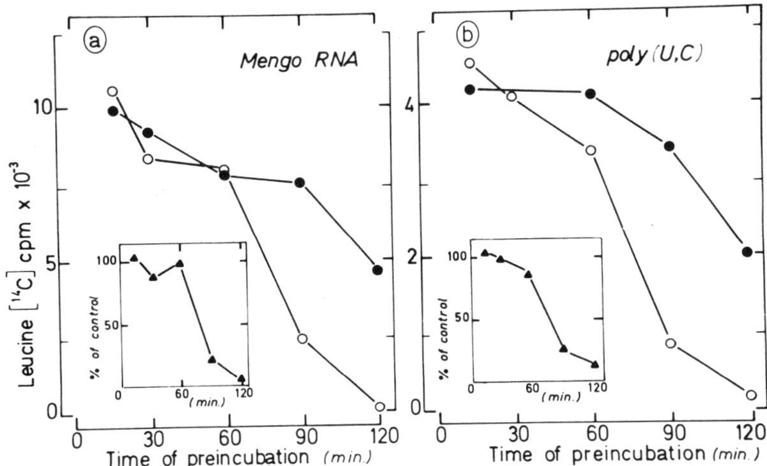

Figure 1. Polypeptide synthesis directed by (a) Mengo virus RNA and (b) poly (U,C) in Cont. S-10 and Int. S-10 pre-incubated for different times. Results are given as the counts per minute of (^{14}C) leucine incorporation for 20 μl of assay. •, Cont. S-10; o, Int. S-10.

I. THE ROLE OF PRE-INCUBATION IN THE INTERFERON-INDUCED TRANSLATIONAL DEFECT

A. Length of the Pre-Incubation and Protein Synthetic Activity

Most of the cell-free extracts used in these studies had been pre-incubated to reduce the translation of endogenous mRNA. Since different preparations of lysates exhibited varying degrees of interferon-induced inhibition, we studied the influence of the length of time of pre-incubation on the interferon-induced inhibition of in vitro protein synthesis. We could establish that the ability of cell-free extracts to translate mengo virus RNA was a function of the time of pre-incubation. The ability of an extract from interferon-treated (Int. S-10) cells to translate mengo virus RNA was not impaired when compared with an extract from control cells until between 60 and 90 min of pre-incubation. At this point, although the activity of the control extract declined somewhat, the activity of the Int. S-10 declined much more rapidly.

Similar results were obtained when translation of the synthetic mRNA poly (U,C) was studied (Figure 1b).

B. Effect of Added tRNA

The impairment of the ability of Int.S-10 to translate exogenous mRNA may be overcome if tRNA is added to the lysates

Preincubation Time - (min)	Mengo RNA directed ^{14}C leucine incorporation*					
	- tRNA		+ Total tRNA		+ $tRNA^{Leu}_{(UAG)}$	
	Control	Interferon	Control	Interferon	Control	Interferon
15	8,478	11,341 (133)	10,605	14,890 (140)	8,508	8,577 (100)
30	7,654	7,382 (96)	8,799	10,130 (115)	6,803	5,815 (85)
60	7,947	7,742 (97)	9,067	10,082 (110)	6,071	6,650 (110)
90	6,696	1,895 (28)	9,068	10,357 (110)	5,692	6,102 (103)
120	4,259	307 (7)	10,434	10,288 (99)	5,378	4,722 (90)

* cpm/20 ul/60 min.

Table 1. Effect of total calf liver tRNA on the translation of mengo virus RNA by extracts from control and interferon-treated cells preincubated for different times.

Protein synthesis directed by mengo virus RNA (60 µg/ml) was assayed in the absence and presence of added total calf liver tRNA (60 µg/ml) and $tRNA^{Leu}_{(UAG)}$ (30 µg/ml). Results are shown as the incorporation of radioactivity into hot trichloroacetic acid-insoluble material. Figures in parentheses represent incorporation by interferon extracts as a percentage of the appropriate control extract.

(9-10). Thus, pre-incubation might cause inactivation of endogenous tRNA. Since addition of total calf liver tRNA to extracts from control cells eliminated the loss of activity observed when pre-incubation exceeded 60 min, it was suggested that pre-incubation of cell-free extracts causes a progressive impairment in the ability to utilize endogenous tRNA and that pre-treatment of the cells with interferon enhances this effect.

Total calf liver tRNA (60 µg/ml) completely restored the protein synthetic capacity of control extracts which had been pre-incubated for 90 to 120 min, whereas - 30 µg/ml of yeast $tRNA^{Leu}$ (higher concentrations were inhibitory) had only a slight effect upon protein synthesis in extracts from control cells, suggesting that species other than $tRNAs^{Leu}$ were affected (9, 11, 13).

Therefore, while the pre-incubation of control extracts resulted in inactivation of more than one tRNA species, the pre-treatment of cells with interferon appears to affect specifically $tRNA^{Leu}$ species. Although $tRNA^{Leu}$ species appear to be affected in control extracts, the interferon treatment seems to enhance considerably the inactivation of some or all $tRNA^{Leu}$ (Table 2).

Translation of other synthetic polynucleotides with leucine codons, e.g. poly (A,U) and poly (U,G) (13) is also inhibited in pre-incubated Int. S-10, demonstrating a strong impairment of leucine tRNA utilisation. However for the time being, minor effects upon other tRNA species, cannot be ruled out.

Preincubation Time (min)	Poly (U,C) directed ^{14}C leucine incorporation*			
	- $tRNA^{Leu}_{(UAG)}$		+ $tRNA^{Leu}_{(UAG)}$	
	Control	Interferon	Control	Interferon
15	4,114	4,500	4,459	5,136
60	4,080	3,323	3,948	4,771
90	2,263	268	3,827	4,152

* cpm/20 ul/60 min.

Table 2. Effect of $tRNA^{Leu}_{(UAG)}$ on the translation of poly (U,C) by extracts from control and interferon-treated cells pre-incubated for different times. Protein synthesis directed by poly (U,C) was assayed in the absence or presence (30 µg/ml) of $tRNA^{Leu}_{(UAG)}$.

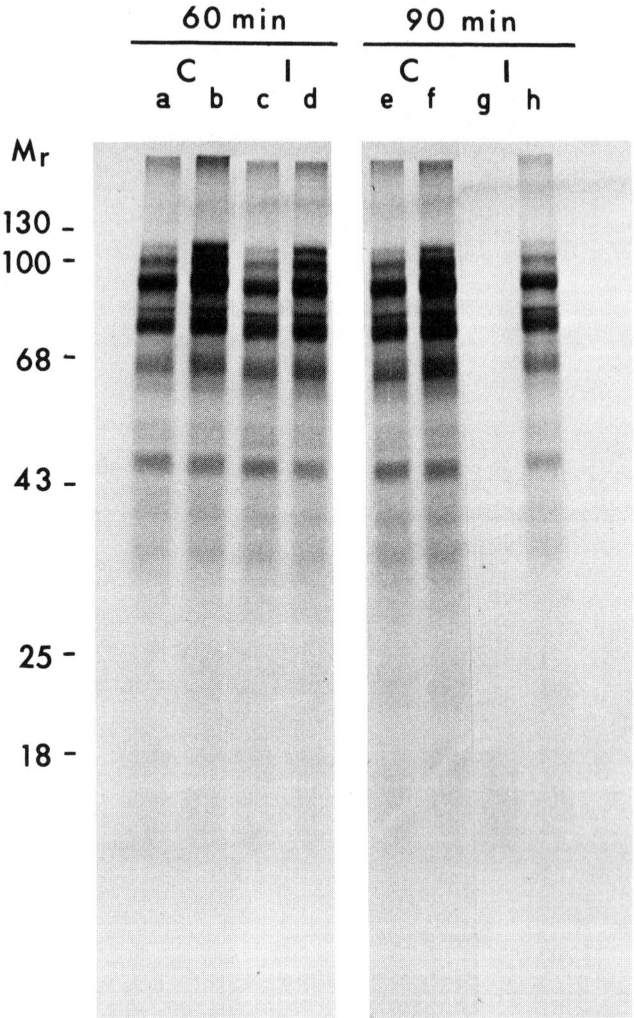

Figure 2. SDS-polyacrylamide gel analysis of proteins synthesized in response to mengo virus RNA by Cont.S-10 and Int.S-10 pre-incubated for different times. (^{35}S) methionine is the label. Extracts incubated for 60 min: (a) Cont. S-10; (b) Cont.S-10 plus 80 μg/ml total calf liver tRNA; (c) Int.S-10; (d) Int.S-10 plus tRNA. Extracts pre-incubated for 90 min: (e) Cont.S-10; (f) Cont.S-10 plus tRNA; (g) Int. S-10; (h) Int.S-10 plus tRNA.

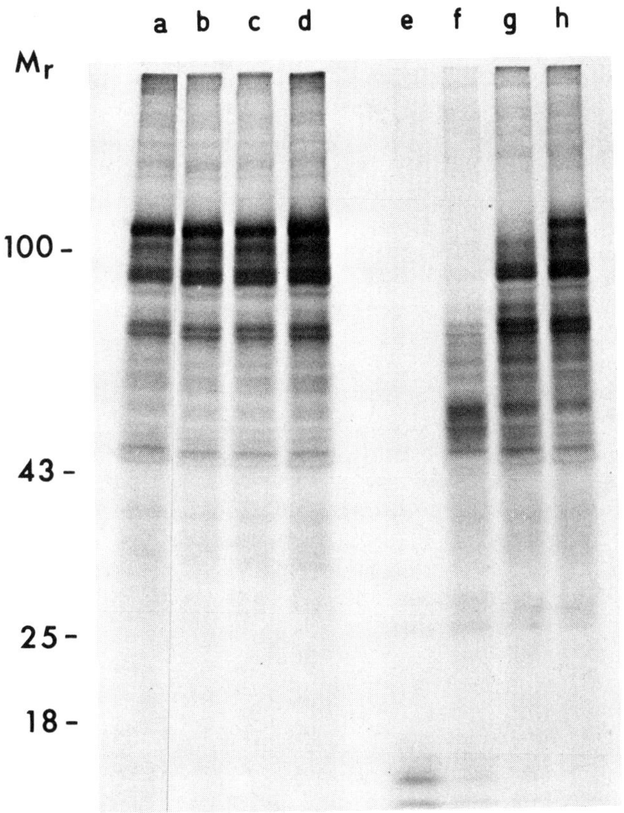

Figure 3. Influence of added tRNA on the length of mengo virus RNA directed polypeptides synthesized in Cont.S-10 and Int. S-10, pre-incubated for 90 min. (a) Cont.S-10; Cont. S-10 plus total tRNA: (b) 8 μg/ml; (c) 40 μg/ml; (d) 80 μg/ml; (e) Int.S-10; Int.S-10 plus tRNA: (f) 8 μg/ml; (g) 40 μg/ml; (h) 80 μg/ml.

C. Analysis of the Polypeptides Synthesized with Mengo Virus RNA as Messenger

The inactivation of tRNA species in Int. S-10 might lead to the synthesis of smaller amounts of the same polypeptides observed in control cell extracts or to the synthesis of shorter molecules. Polyacrylamide-SDS gel electrophoresis of the peptides synthesized *in vitro* by extracts of controls and interferon-treated cells showed that the pre-incubation step had no *qualitative* effect on the products of the control extracts, but resulted in a substantial modification of the peptides produced by the lysates of interferon-treated cells: the latter were able to synthesize only small amounts of low molecular weight peptides (Figure 2 and ref. 12).

Addition of tRNA restored the synthesis of large polypeptides such that the pattern of proteins synthesized was identical to that of control cell extracts (track h). The restoration of protein synthesis by tRNA is dose dependent (Figure 3), and increasing amounts of tRNA alter not only the quantity but also the nature of the polypeptides synthesized.

Premature termination may result from a reduction in the overall rate of protein synthesis, due to the lack of particular tRNA species (e.g. $tRNA^{Leu}$), thus increasing the probability of a ribosome falling off the mRNA. Alternatively, a dramatic limitation of leucine-tRNAs might cause ribosomes to stop at the corresponding mRNA codons.

II. EFFECT OF ADDED dsRNA

A. Inhibition of Protein Synthesis by dsRNA in Extracts Pre-incubated for Various Times

Addition of dsRNA to extracts prepared from interferon-treated cells caused 80-90% inhibition of mengo virus RNA translation, regardless of the time of pre-incubation (Table 3). The same dose of dsRNA caused an inhibition of only 8-15% in control extracts. The inhibitory effect of dsRNA is unaffected by addition of tRNA (Table 3).

The nature of the polypeptides synthesized by control extracts was unaffected by addition of dsRNA even though there was frequently some reduction in their amount. In extracts from interferon-treated cells dsRNA caused an extensive reduction in the amount of polypeptides even in extracts pre-incubated for 15-60 min where the products synthesized in the absence of dsRNA were identical to those of control extracts.

Nonetheless, small amounts of peptides of up to M_r 50,000

TIME OF PREINCUBATION (MIN)	^{14}C- LEUCINE INCORPORATION BY INTERFERON EXTRACT WITH :		
	NO ADDITION	+ POLY I : POLY C	+ POLY I : POLY C + tRNA
	% of Control Extract		
15	109	26	25
30	90	33	22
60	102	25	24
90	23	--	22
120	4	8	20

Table 3. Effect of added dsRNA and tRNA on mengo virus RNA translation by extracts of control and interferon-treated cells pre-incubated for different times.

Extracts were prepared and pre-incubated for the times indicated. Protein synthesis directed by mengo RNA (40 μg/ml) was assayed in the absence or presence of poly(I), poly(C) (40 ng/ml) and total calf liver tRNA (60 μg/ml). Results for extracts of interferon-treated cells are given as a percentage of the control extract containing identical additions. Synthesis by control extracts was inhibited by less than 15% in all cases by added poly(I) poly(C).

could just be detected. Extracts of interferon-treated cells, which had been pre-incubated for 90 min, synthesized small amounts of very short polypeptides but in the presence of dsRNA these could no longer be detected.

In marked contrast with mengo virus RNA, poly (U,C) translation was essentially unaffected by the addition of dsRNA to cell-free extracts from interferon-treated cells and in the presence of dsRNA, tRNA was still able to reverse the interferon-induced inhibition apparent in extracts pre-incubated for 75 min (Table 4). Thus, with two doses of poly(I) poly(C), which gave 80% inhibition of mengo RNA translation in interferon-treated cell extracts pre-incubated for 15 min, only a slight effect (12%) on poly (U,C) translation was observed. After pre-incubation for 75 min, translation of poly (U,C) by the interferon-treated cell extract was

ADDITIONS	^{14}C-LEUCINE INCORPORATION BY EXTRACTS PREINCUBATED FOR					
	15 MIN			75 MIN		
	CONTROL	INTERFERON		CONTROL	INTERFERON	
	cpm	cpm	%	cpm	cpm	%
NO ADDITION	5506	5685	(103)	4084	1131	(28)
+ tRNA	5062	5576	(110)	4513	3606	(80)
+ POLY I : POLY C (40 ng/ml)	4707	4601	(98)	3336	780	(23)
+ POLY I : POLY C (40 ng/ml) + tRNA	5070	4730	(93)	4833	3236	(67)
+ POLY I : POLY C (100 ng/ml)	4064	3588	(88)	2299	725	(32)
+ POLY I : POLY C (100 ng/ml) + tRNA	4615	5093	(110)	4345	3038	(70)

Table 4. Effect of Added dsRNA and tRNA on poly(U,C) translation by extracts of control and interferon-treated cells pre-incubated for different times.

Protein synthesis was assayed with poly(U,C) (0.8 mg/ml) as messenger in the absence or presence of poly(I) poly(C) (40 or 100 ng/ml) and total calf liver tRNA (40 μg/ml). Results are expressed as percentage of the value for the control extract pre-incubated for the same period of time.

strongly inhibited (72%) compared with the control but addition of poly(I) poly(C) did not further reduce this level of synthesis. Upon addition of tRNA to such an extract, poly(U,C) translation was restored to 70-80% of that in the control extract even in the presence of poly(I) poly(C).

Apparently, the inhibition of protein synthesis by dsRNA in extracts prepared from interferon-treated cells was quite distinct from the inhibition observed in the absence of dsRNA in those extracts which had been pre-incubated for times permitting an inactivation of certain tRNA species.

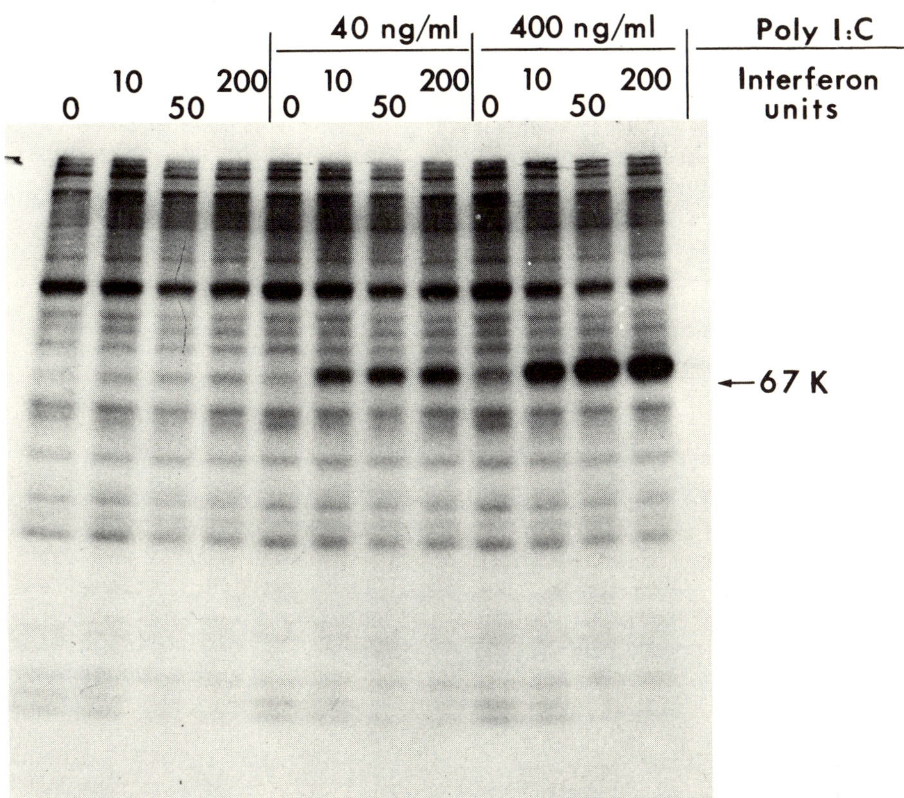

Figure 4. *In vitro* phosphorylation of proteins in extracts of cells treated with 10, 50 or 200 interferon units/ml, or untreated, in the presence (40 or 400 ng/ml) or absence of poly(I) poly(C).

B. Stimulation of the Phosphorylation of Proteins by dsRNA in Interferon-Treated Cell Extracts

In extracts of interferon-reated cells dsRNA stimulates the phosphorylation of a protein of $M_r \sim 67,000$ (referred to as P67 (17-19). Phosphorylation of P67 was shown to be independent of the presence of mRNA, and addition of tRNA had no effect on it. The extent of phosphorylation of P67 correlated with the doses of interferon and dsRNA employed.

C. Effects of dsRNA on the Initiation of Protein Synthesis

Since poly(U,C), a synthetic mRNA, does not require a normal initiation step, a likely explanation of the difference between the effect of dsRNA on mengo and poly(U,C) translation was that dsRNA inhibited initiation.

When the extent of initiation was directly measured as the amount of N-terminal methionine present in peptides synthesized with mengo virus RNA as messenger (Table 5), it was shown that addition of dsRNA to extracts from interferon-treated cells reduced the amount of N-terminal methionine by 50%. Incorporation into the residual peptides or into total protein was similarly reduced by approximately 50%. The initiation of protein synthesis was considerably inhibited by dsRNA under these conditions.

To locate the cause of this apparent inhibition of initiation more precisely, the formation of a complex of methionyl-tRNA with the 40-S ribosomal subunit in the absence of added mRNA was assayed. Addition of poly(I) poly(C) to control extracts had little effect on the amount of material sedimenting with the 40-S peak (30% reduction). However, the amount of radioactivity associated with the 40-S peak in interferon-treated cell extracts was reduced by 80% when poly(I) poly(C) was added.

The association of (^{35}S) methionyl-tRNA with the 40-S subunit was strongly inhibited by poly(I) poly(C) in extracts of interferon-treated cells. In extracts of control cells poly(I) poly(C) had no significant effect upon the formation of methionyl-tRNA 40-S subunit complexes. Addition of crude rabbit reticulocyte initiation factors greatly reduced the poly(I) poly(C)-induced inhibition of (^{35}S) methionyl-tRNA 40-S subunit complex formation. The effect was most marked when the initiation factors were added during the incubation with dsRNA, strongly suggesting that in interferon-treated cell extracts dsRNA acts at the level of a protein synthesis initiation factor which binds methionyl-tRNA to native 40-S ribosomal subunits, a situation analogous to that of the rabbit reticulocyte lysate (20-22 and chapter 10 of this volume).

D. Induction by dsRNA of a Nuclease Activity in Extracts of Interferon-Treated Cells

Although the observations described above showed that poly(I) poly(C) was capable of inhibiting steps in the initiation of protein synthesis other effects were not precluded. The appearance of a dsRNA-dependent nuclease activity in extracts of interferon-

TIME OF INCUBATION (MIN)	^{35}S -cpm INCORPORATED/40 ul BY : CONTROL	CONTROL + POLY IC		INTERFERON	INTERFERON + POLY IC	
a) N-TERMINAL EXTRACT						
10	764	875	(115)	921	540	(59)
15	1026	1003	(98)	1276	563	(44)
b) RESIDUAL POLYPEPTIDES						
10	60,684	49,571	(82)	61,691	28,177	(46)
15	112,074	100,584	(90)	111,950	43,039	(38)

Table 5. Double-stranded-RNA-induced inhibition of incorporation of (^{35}S) methionine into the N-terminal position of peptides synthesized in extracts of interferon-treated cells.

Extracts of control and interferon-treated cells pre-incubated for 60 min at 37°C were further incubated for 30 min at 32°C in the presence or absence of poly(I) poly(C) (100 ng/ml) with the usual components of the protein synthesis assay and with (^{35}S) methionine (168 μCi/ml; 34 Ci/mmol) but without added mRNA. To each incubation mixture (96 μl), 4 μl mengo RNA (1.5 mg/ml) was then added and incubation continued for either 10 or 15 min. Aliquots (40 μl) were pipetted onto Whatman 3 MM discs and radioactivity in the N-terminal position and residual peptides was analysed. Figures in parentheses represent percentage of values obtained in the absence of poly(I) poly(C).0.1 pmol methionine is equivalent to 5300 counts/min. The values given are the means of two determinations.

treated Ehrlich ascites tumour cells and L cells was reported (23, 24).

Addition of poly(I) poly(C) to the control extract did not modify the rate of degradation of (^{3}H) mengo RNA, but in the interferon extract hydrolysis was 2-fold higher after only 5 min (Figure 6). By 30 min virtually no material remained in the cold acid precipitable fraction. Sucrose density gradient analysis showed that addition of poly(I) poly(C) to the interferon-treated cell extract led to the complete disappearance of the 34-S peak and the appearance of most of the radioactivity at the top of the gradient. Almost identical results were obtained when the nuclease

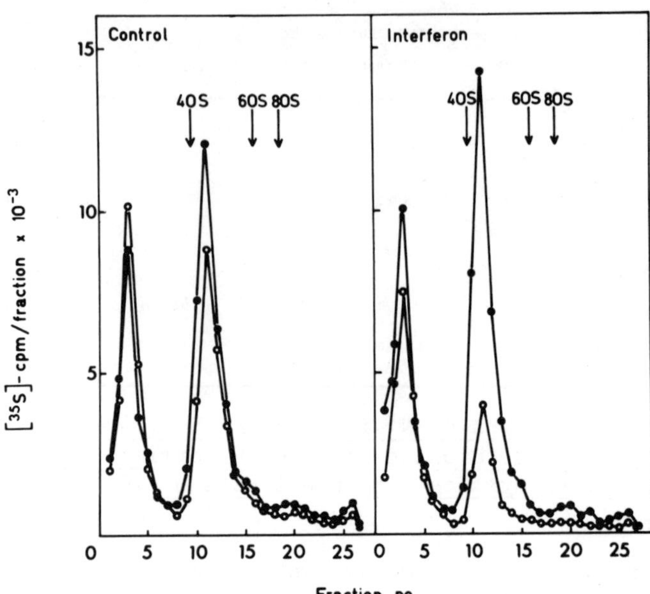

Figure 5. Effect of dsRNA on the formation of methionyl-tRNA. 40-S subunit complexes by (A) control and (B) interferon-treated cell extracts.

Extracts of control and interferon-treated cells were pre-incubated at 37°C for 60 min and subjected to Sephadex G-25 gel filtration. The extracts were incubated for 30 min at 32°C in the absence (● —— ●) or presence (O —— O) of dsRNA (100 ng/ml of poly (I) poly(C)) under the conditions used for protein synthesis but in the absence of mRNA and methionine. Then (^{35}S) methionine was added and incubation continued for 15 min. Formation of methionyl-tRNA 40-S subunit complexes was assayed by sucrose density gradient centrifugation (direction of centrifugation is from left to right) followed by precipitation of aminoacyl-tRNA in each fraction with cetyltrimethylammonium bromide. The position of ribosomal subunits is shown by the arrows.

assay was performed under conditions in which protein synthesis occurred, suggesting that the mRNA was rapidly degraded in the extracts of interferon-treated cells containing dsRNA. This destruction, however, was not a consequence of the inhibition of protein synthesis.

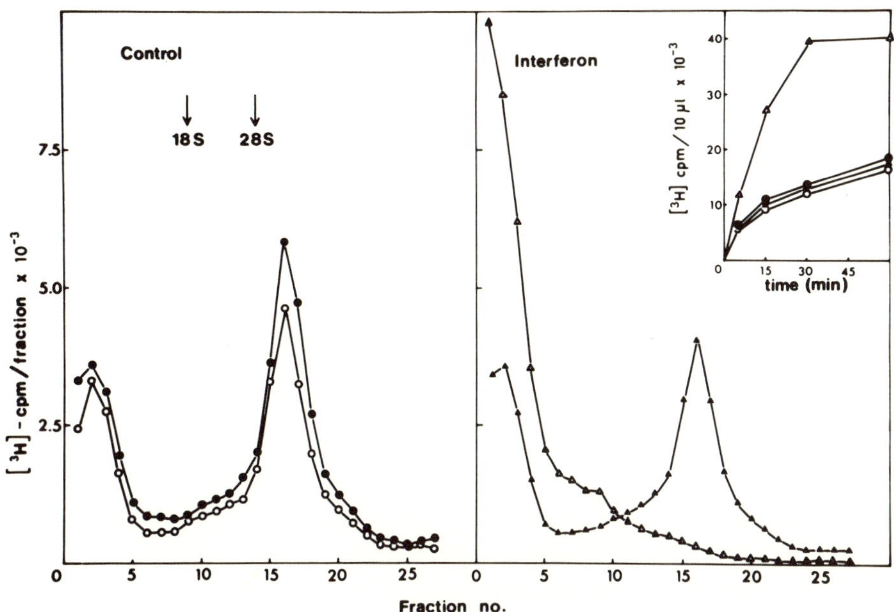

Figure 6. Effect of dsRNA on the degradation of mengo virus RNA in (A) control and (B) interferon-treated cell extracts.

Cell-free extracts were pre-incubated for 60 min at 37°C and filtered over Sephadex G-25. The extracts were incubated at 32°C for 30 min in the absence (●, ▲) or presence (O, △) of dsRNA (30 ng/ml) under the conditions used for protein synthesis but without methionine and with 5 μM pactamycin. At the end of this period mengo (^{3}H) RNA (90452 counts min^{-1} μg^{-1} ; 60300 counts/min for 10 μl final concentration) was added and the incubation continued. After 15 min incubation an aliquot (25 μl) was removed for analysis by sucrose density gradient centrifugation (centrifugation is from left to right). The position of marker ribosomal RNA is shown by the arrows. Inset: The kinetics of release of trichloroacetic-acid-soluble material is shown for control (O, ●) and interferon-treated (△, ▲) cell extracts. At various times an aliquot (10 μl) of the incubation mixture was removed and added to trichloroacetic acid containing serum albumin as carrier. After centrifugation the radioactivity remaining in the supernatant was determined (total radioactivity in the aliquot used for the determination was 45606 counts/min).

Cells[a] Infected with:	Labelling of P 67 (arbitrary units)
Control cells	
None	20
Mengo virus (3 h)[b]	50
Interferon pretreated[c] cells	
None	71
Mengo virus (1 h)	260
Mengo virus (3 h)	370
Mengo virus (6 h)	450

a) Mouse L-929 cell monolayers

b) 50 PFU/cell 3 h before harvesting

c) 18 h pretreatment with 500 units/ml

Table 6. Phosphorylation of P67 in interferon-treated mengo virus infected cells during the viral multiplication cycle.

To summarize, it seems that dsRNA inhibits protein synthesis in extracts of interferon-treated cells by preventing initiation and by inducing the destruction of mRNA. The mechanism by which dsRNA produces these effects remains to be demonstrated, although it seems likely that activation of a specific protein kinase is involved in the initiation defect. The formation of a small oligonucleotide which inhibits protein synthesis has been reported in interferon-treated cell extracts incubated with dsRNA (16). Such a molecule could act by activating a nuclease (see chapter 14 of this volume).

III. THE ROLE OF VIRUS-INDUCED dsRNA IN INTERFERON-TREATED MENGOVIRUS-INFECTED CELLS

Under proper conditions (500 U/ml of interferon, 50 PFU/ml), the yield of mengovirus after a single replication cycle was reduced by 99%. Thus, the multiplication of the virus was not completely abolished; however, the amount of virus-induced dsRNA structures was too low to permit an accurate determination.

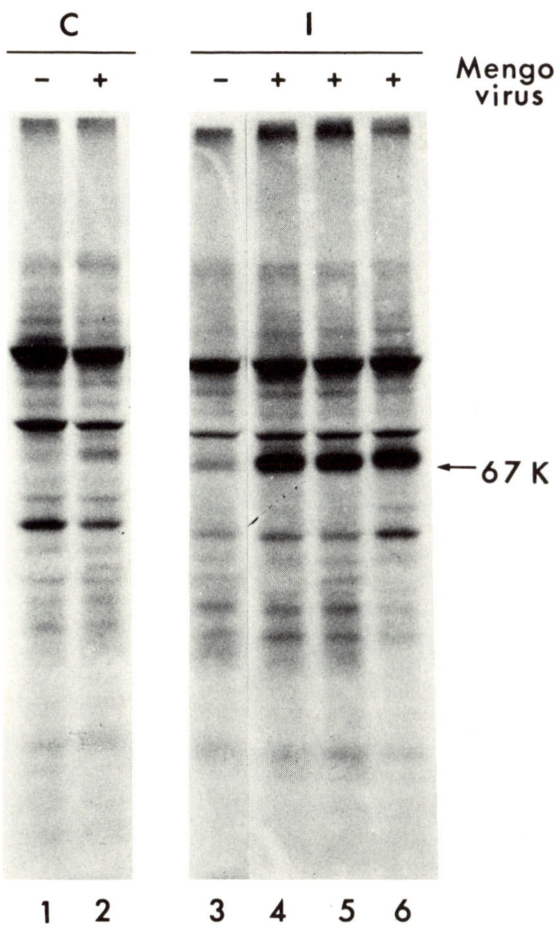

Figure 7. In vitro phosphorylation of proteins in extracts of control and interferon-treated cells, infected or not with mengo virus. Cell extracts were incubated with (γ- ^{32}P) ATP at 32° for 60 min and analysed by SDS-polyacrylamide gel electrophoresis followed by autoradiography. Control cell extracts are shown in tracks 1 and 2; interferon-treated cell extracts are in tracks 3 to 6. Incubation time after infection was 1 h (track 4) 3 h (tracks 2 and 5) and 6 h (track 6). Non infected cells tracks 1 and 3. The arrow indicates the position of a phosphorylated band of apparent molecular weight of 67, 000.

dsRNA from infected cells [a]	total cpm	µg	Specific Activity (cpm/µg)
interferon untreated	1×10^7	71	1.4×10^5
interferon treated	3.5×10^5	2.5 [b]	----

a) 1×10^9 untreated or interferon treated cells

b) calculated

Table 7. Quantification of dsRNA in mengo virus infected cells.

Assuming that the synthesis of viral RNA is reduced by the interferon pre-treatment of the cell but the pattern of uridine incorporation per RNA molecule remained unchanged, the amount of virus-induced dsRNA in 1×10^9 interferon pre-treated,mengo virus-infected L cells was estimated to be 2.5 µg (Table 7). This amount is sufficient to explain the enhanced phosphorylation of P67 reported in the preceeding paragraphs.

Viral RNA synthesis in the picornavirus group is associated with membrane structures (25, 26) and can be isolated in the form of a replication complex by centrifugation of the cytoplasmic extracts at 20,000 g (27). The phosphorylase activity was shown to be associated with the replication complexes (Table 7), a result expected, as dsRNA structures were in the P-20 fraction.

The low response to the addition of poly(I) poly(C) to the S-20 fraction of interferon-treated, mengovirus-infected cells, suggested that the location (or relative concentration) of some of the components of the phosphorylating system (protein kinase, substrate P67, phosphatase) was modified by the infection. In interferon-treated non-infected cells extracts, P67 was found in S-20 as well as in S-100.

A. Protein Phosphorylation in Extracts of Interferon-Treated Cells After Infection with Mengovirus

Infection of L cells with mengovirus results in the enhancement of a phosphorylating activity, whose major product appears in polyacrylamide gel electrophoresis as a band with a M_r of 67,000 (Figure 7). Interferon treatment *per se* enhances such an activity to a comparable extent and mengovirus infection of interferon-treated cells further increases such an effect (Table 6).

The enhancement of phosphorylation after longer periods of infection suggested that some early steps of viral multiplication were required. This finding was confirmed using UV irradiated virus: with irradiated virus no significant enhancement of phosphorylation was detected while with the same non-irradiated virus P67 was phosphorylated 5-fold more than in the interferon-treated non-infected cells as measured by scanning the autoradiograph.

B. Location of a Specific Ribonuclease Activity

It has been reported that a ribonuclease is induced when extracts of interferon-treated non-infected cell extracts are incubated with dsRNA and ATP (23). Since dsRNA is present in interferon-treated mengo virus-infected cells the induction of a dsRNA dependent ribonuclease might occur. Kinetic assays for ribonuclease activity showed a marked increase in ribonuclease activity of the P-20 from interferon-treatment, mengo virus-infected cells. Poly C, Poly U and poly A were also degraded by the enzyme, whereas (^{3}H) replicative form of mengo virus was not degraded. Extracts from interferon-treated non-infected cells did not show enhanced ribonuclease activity compared with non-infected control cells.

It seems, therefore, that infection of interferon-treated cells with mengo virus results in modifications similar to some of those reported as dsRNA dependent effects in cell-free extracts: a) an enhanced phosphorylation of a protein of M_r 67,000 and

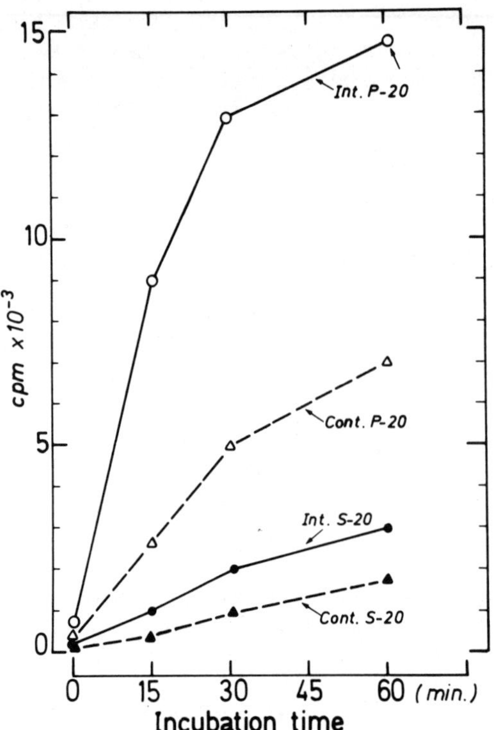

Figure 8. Degradation of mengo virus (^{3}H) RNA by fractions from mengo virus-infected cells, treated (Int.P-20 and Int. S-20) and untreated (Cont.P-20 and Cont.P-20) with interferon.

b) an increased ribonuclease activity.

The ribonuclease activity was found in a sedimentable fraction known to contain the replication complex where synthesis of virus-coded RNA takes place. Therefore, the antiviral action of interferon may be explained in this particular system by the ability of such a nuclease to degrade viral mRNA. Although the nuclease is not only active on mengo virus RNA its location within the cell would limit its action to the virus-coded single stranded RNA, the final consequence being some discrimination between viral and host cell messengers. This finding does not exclude the possibility that the inhibition of protein synthesis also occurs by other mechanisms such as impaired initiation of translation.

C. Do Other Viruses Induce Similar Response?

The question was examined in two different virus/cell systems: VSV and vaccinia virus infected, interferon-treated cells.

i) VSV infected cells: while the characteristic phosphorylation of P67 was enhanced, it was difficult to correlate the phosphorylating activity with the multiplicity if infection or with some particular stage of virus development. Furthermore, results varied from one experiment to another. The addition of poly(I) poly(C) to interferon-treated, VSV-infected cell extracts, enhanced the phosphorylation of P67, indicating that the kinase system remained functional.

The level of nucleolytic activity did not change in any subcellular compartment of the interferon-treated VSV-infected cells.

ii) The effect of infection with vaccinia virus was quite different: there was no increase in the phosphorylating activity and the nuclease was somehow enhanced if infection was carried out with at least 6 PFU/cell.

Taken together, these results provide additional evidence of the pleiotropic effect of interferon. Although interferon-treated cells exhibited the antiviral state when infected with each of the viruses used throughout this study, the cell responded in different ways to the infection, at least at the level of ribonuclease activity and of the extent of phosphorylation of P67. In the case of mengo virus the enhanced ribonuclease activity and its location, triggered by the infection may explain the interferon action. The relationship between the expression of the antiviral state and the protein kinase system remains to be demonstrated.

ACKNOWLEDGEMENTS

Work reviewed in this paper was supported by grants from INSERM (CRL 77.4.064.1 and ATP 77-82) and DGRST (77.7.1743). J.A.L. was in receipt of an Embo long term Fellowship. O.A. is in receipt of an DGRST long term Fellowship.

REFERENCES

1. FALCOFF, R., FALCOFF, E., SANCEAU, J. and LEWIS, J.A. Virology (1978), 86, 507-515.

2. SEN, G.C., GUPTA, S.L., BROWN, G.E., LEBLEU, B., REBELLO, M.A. and LENGYEL, P. J. Virol.(1976), 17, 191-203.

3. KERR, I.M. J. Virol. (1971), 7, 448-459.

4. FRIEDMAN, R.M., METZ, D.H., ESTEBAN, R.M., TOVELL, D.R., BALL, L.A. and KERR, I.M. J. Virol. (1972), 10, 1184-1198.

5. FALCOFF, E., FALCOFF, R., LEBLEU, B. and REVEL, M. J. Virol. (1973), 12, 421-430.

6. GUPTA, S., SOPORI, M.L. and LENGYEL, P. Biochem. Biophys. Res. Commun. (1973), 54, 777-783.

7. SAMUEL, C.E. and JOKLIK, W.K. Virology (1974), 58, 476-491.

8. HILLER, G., WINKLER, I., VIEHHAUSER, G., JUNGWIRTH, C., BODO, G., DUBE, S. and OSTERTAG, W. Virology (1976), 69, 360-363.

9. FALCOFF, R., LEBLEU, B., SANCEAU, J., WEISSENBACH, J., DIRHEIMER, G., EBEL, J.P. and FALCOFF, E. Biochem. Biophys. Res. Commun. (1976), 68, 1323-1331.

10. CONTENT, J., LEBLEU, B., ZILBERSTEIN, A., BERISSI, H. and REVEL, M. FEBS Lett. (1974), 41, 125-130.

11. GUPTA, S.L., SOPORI, M.L. and LENGYEL, P. Biochem. Biophys. Res. Comm. (1974), 57, 763-770.

12. ZILBERSTEIN, A., DUDOCK, B., BERISSI, H. and REVEL, M. J. Mol. Biol. (1976), 108, 43-54.

13. WEISSENBACH, J., DIRHEIMER, G., FALCOFF, R., SANCEAU, J. and FALCOFF, E. FEBS Lett. (1977), 82, 71-76.

14. KERR, I.M., BROWN, R.E. and BALL, L.A. Nature (London) (1974), 250, 57-59.

15. HOVANESSIAN, A.G., BROWN, R.E. and KERR, I.M. Nature (London) (1977), 268, 537-542.

16. KERR, I.M. and BROWN, R.E. Proc. Nat. Acad. Sci. U.S.A. (1978), 75, 256-260.

17. LEBLEU, B., SEN, G.C., SHAILA, S., CABRER, B. and LENGYEL, P.

Proc. Natl. Acad. Sci. U.S.A. (1976), 73, 3107-3111.

18. ROBERTS, W.K., HOVANESSIAN, A.G., BROWN, R.E., CLEMENS, M.J. and KERR, I.M. Nature (London) (1976), 264, 477-480.

19. ZILBERSTEIN, A., FEDERMAN, P., SHULMAN, L. and REVEL, M. FEBS Lett. (1976), 68, 119-124.

20. FARRELL, P.J., BALKOW, K., HUNT, T., JACKSON, R.J. and TRACHSEL, H. Cell (1977), 11, 187-200.

21. DARNBROUGH, C., LEGON, S., HUNT, T. and JACKSON, R.J. J. Mol. Biol. (1973), 76, 379-403.

22. KAEMPFER, R. Biochem. Biophys. Res. Commun. (1974), 61, 541-547.

23. SEN, G.C., LEBLEU, B., BROWN, G.E., KAWAKITA, M., SLATTERY, E. and LENGYEL, P. Nature (London) (1976), 264, 370-373.

24. BROWN, G.E., LEBLEU, B., KAWAKITA, S., SHAILA, S., SEN, G.C. and LENGYEL, P. Biochem. Biophys. Res. Commun. (1976), 69, 114-122.

25. PENMAN, D., BECKER, Y. and DARNELL, J.E. J. Mol. Biol. (1964), 8, 541-555.

26. CALIGUIRI, L.A. and MOSSER, A.G. Virology (1971), 46, 375-386.

27. BALTIMORE, D. J. Mol. Biol. (1966), 18, 421-428.

28. LOWRY, O.H., ROSEBROUGH, N.J., FARR, A.L. and RANDALL, R.J. J. Biol. Chem. (1951), 193, 265-275.

29. FALCOFF, E., FALCOFF, R., LEBLEU, B. and REVEL, M. Nature New Biol. (1972), 240, 145-147.

INTERFERON-INDUCED ACTIVATION OF AN ENDONUCLEASE BY 2' 5' OLIGO (A)

C. BAGLIONI, P.A. MARONEY, G.E. CHATTERJEE and M.A. MINKS

Department of Biological Sciences, S.U.N.Y. at Albany
Albany, New York 12222, U.S.A.

INTRODUCTION

Exposure of animal cells to interferon results in the establishment of an antiviral state manifested by the inhibition of virus replication (1). In interferon-treated cells transcription and translation of viral templates are specifically inhibited by the activation of cellular defence mechanisms, the molecular basis of which is still unclear (2).

The biochemical characteristics of extracts from interferon-treated cells have been compared with those of extracts from control cells in order to elucidate the molecular mechanisms underlying the antiviral state. Among the many differences reported between these extracts, the most significant appear to be (see chapter 12): 1) an enhanced inhibition of protein synthesis by double-stranded RNA (dsRNA) (3-4);

2) a decreased methylation of the 5'-terminal guanosine of added viral mRNA (5-6);

3) an increase in three enzymatic activities which are enhanced upon addition of dsRNA to cell extracts.

These are: i) a protein kinase (7-9); ii) an endonuclease (10-12); and, iii) an enzyme which synthesizes oligonucleotides with the structure $pppA(2'p5'A)_n$ - 2'5'oligo(A) - from ATP (13-16).

The relevance of these biochemical differences to the establishment of the antiviral state is presently unclear.

We will discuss here the mechanism of activation by dsRNA of the endonuclease of HeLa cells treated with human fibroblast interferon. Briefly, the synthesis of 2'5'oligo(A) is increased in extracts of interferon-treated HeLa cells upon addition of dsRNA. This oligonucleotide mediates the activation of the endonuclease which cleaves mRNA. The enzyme which synthesizes 2'5'oligo(A), designated 2'5'oligo(A) polymerase, is induced by interferon. A striking correlation between an increase in 2'5'oligo(A) polymerase activity and inhibition of encephalomyocarditis (EMC) virus RNA synthesis has been observed. This correlation suggests an involvement of the polymerase and of the endonuclease activated by 2'5'oligo(A) in the inhibition of viral RNA synthesis.

I. DEGRADATION OF mRNA BY A dsRNA-ACTIVATED ENDONUCLEASE

Incubation of 12-16S vesicular stomatitis virus (VSV) poly(A)-containing mRNA with extracts of control or interferon-treated HeLa cells resulted in the degradation of approximately 20% of the RNA (Figure 1). The products of degradation sedimented at the top of the gradients used for this analysis and were soluble in 5% trichloroacetic acid (data not shown). This suggested that a portion of the added mRNA was degraded by exonucleases to free nucleotides, whereas the bulk of the added mRNA was not significantly cleaved by either endonucleases or exonucleases since it sediments identically to mRNA not incubated with cell extract (Figure 1A). However, when poly(I) poly(C) was added to the incubations together with an ATP generating system, a different pattern of degradation of VSV mRNA was observed (Figure 1C and D). The mRNA sedimented as a broader peak of lower molecular weight and the labeled material at the top of the gradient was also increased. This effect of dsRNA was enhanced in extracts from interferon-treated cells, which showed more extensive degradation of mRNA. The material at the top of gradients consisted of nucleotides (or oligonucleotides) soluble in 5% trichloroacetic acid (Figure 1D).

The conclusion was that dsRNA activates an endonuclease in extracts of interferon-treated cells (10-11). Lengyel and co-workers reported that the endonuclease activity can also be shown in extracts of untreated Ehrlich ascites and HeLa cells, but that treatment with interferon enhances the degradation of mRNA by cell extracts several fold (11). The identification of endonucleolytic activity was based on the observation that the rate of formation of acid-soluble RNA cleavage products is not increased in proportion to the extensive degradation of mRNA shown by gradient analysis (10). However, other less likely explanations for this pattern of mRNA degradation cannot entirely be eliminated (i.e. an exonuclease which partially degrades mRNA producing large fragments.

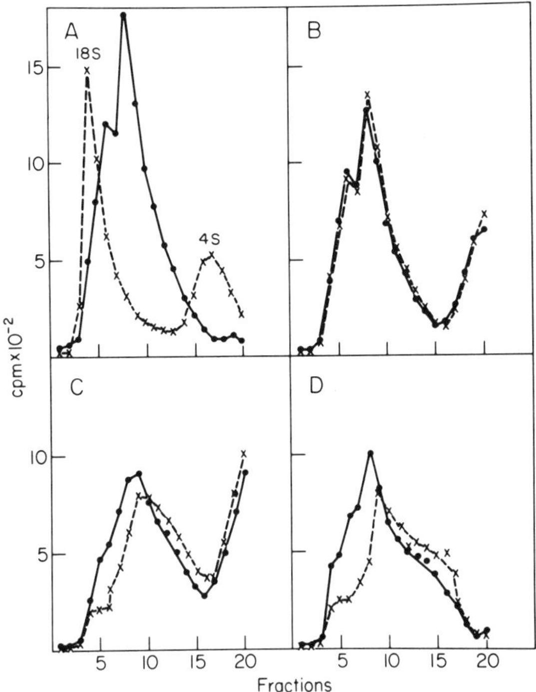

Figure 1. Degradation of vesicular stomatitis virus (VSV) mRNA in extracts of control and interferon-treated cells. A, VSV mRNA (●—●) and markers (x---x); B,C, and D, VSV mRNA incubated in extracts of control cells (●—●) and cells treated 17 hr with 100 units/ml of interferon (x---x) without (B) and with (C and D) added dsRNA. Each 50 μl incubation contained 35 μl of cell extract, 120 mM KOAc, 1.5 mM Mg $(OAc)_2$, 4 mM fructose 1.6 bisphosphate, about 10,000 cpm of (^{3}H) mRNA prepared from VSV-infected cells (12), and 10 μg/ml of poly(I) poly(C) in C and D. The samples were incubated 30 min at 30°C, diluted into 0.5 ml of gradient buffer (12) and centrifuged 17 hr at 31,000 rpm in a SW41 rotor on 5-20% sucrose gradients. Fractions of 0.6 ml were collected and counted directly in A, B and C or counted after precipitation with 5% trichloroacetic acid in D.

To eliminate exonucleolytic attack as a cause of mRNA degradation we followed the fate of mRNA stably attached to ribosomes. If an inhibitor of polypeptide chain elongation is added to the incubations to prevent ribosome movement along mRNA, breakdown of polysomes can only result from endonucleolytic cleavage of the mRNA strand connecting ribosomes. Figure 2 shows that polysome breakdown takes place in extracts of control and interferon-treated HeLa cells incubated with dsRNA in the presence of the inhibitor of elongation sparsomycin. Incubation of extracts for up to one hour in the absence of added dsRNA did not cause appreciable changes in the polysome pattern, suggesting that HeLa cell extracts have a very low level of endonuclease activity in the absence of added dsRNA. The polysome breakdown in extracts of interferon-treated cells was more extensive than in extracts of control cells, suggesting that the dsRNA-activated endonuclease was capable of cleaving cellular mRNA. Clearly, it was important to quantitate the degradation of viral vs. cellular mRNA. When the poly(A)-containing mRNA was measured by chromatography on oligo (dT)-cellulose (18), it was shown (Table 1) that 10 to 20% viral or cellular poly(A)-containing mRNA was lost upon 30 min incubation in the absence of added dsRNA. In the presence of dsRNA, these

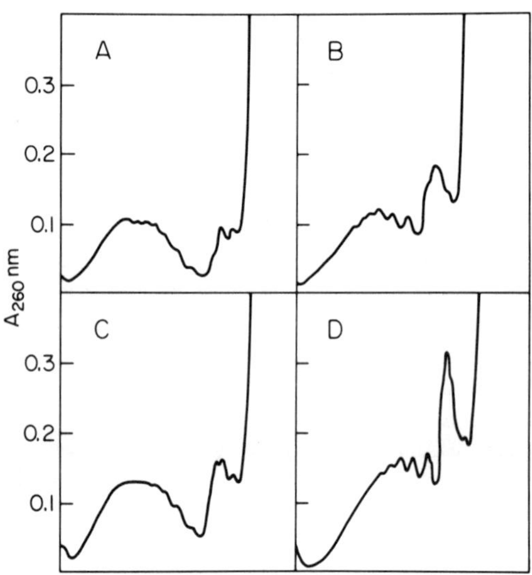

Figure 2. Polysome breakdown in extracts of interferon-treated and control cells incubated with dsRNA. The cells were treated 17 hr with 100 units/ml of interferon. Both control and interferon-treated cells were incubated 1 hr with 1 μg/ml of cycloheximide to increase polysome size according to Fan and Penman (17). Extracts prepared from these cells were incubated 1 hr at 30°C with the components described in Figure 1, 1 mM ATP, 0.1 mM sparsomycin and no added dsRNA in A and C or 10 μg/ml of poly(I) poly(C) in B and D. The samples were centrifuged 90 min at 40,000 rpm on 15-40% sucrose gradients and the A_{260} analyzed with a recording spectrophotometer.

Table 1

EFFECT OF dsRNA ON mRNA CLEAVAGE IN EXTRACTS OF CONTROL AND INTERFERON-TREATED HeLa CELLS

Experiment	Cell extract	Poly(I) poly(c) (μg/ml)	mRNA	$\frac{cpm_t}{cpmt_o}$
Exp. 1	Control	0	HeLa	0.80
	Interferon	0	"	0.84
	Control	1	"	0.67
	Interferon	1	"	0.53
	None	10	"	0.95
Exp. 2	Control	0	VSV	0.90
	Interferon	0	"	0.86
	Control	1	"	0.70
	Interferon	1	"	0.53
Exp. 3	Control	1	HeLa-polysomal	0.64
	Interferon	1	"	0.46

The RNA retained by oligo(dT)-cellulose was measured in 30 min incubations prepared as described in Figure 1. The fraction of RNA cleaved during the incubation is calculated relatively to a sample not incubated. Different incubations contained about 1200 cpm of HeLa poly(A)+ mRNA, 6000 cpm of VSV mRNA, or 3000 cpm of HeLa polysomal poly(A)+ mRNA. The assay of polysomal mRNA cleavage has previously been described (12).

mRNAs were both cleaved in extracts of control cells and more extensively in extracts of interferon-treated cells. Also labeled cellular mRNA bound to polysomes was cleaved, confirming the results of Figure 2. This suggested that the endonuclease activated by dsRNA does not effectively discriminate between viral and cellular mRNA, since both are cleaved by the endonuclease.

Cleavage of mRNA was clearly detected with 1 μg/ml of added poly(I) poly(C) in extracts from interferon-treated cells and with higher concentrations of added dsRNA in extracts of control cells (Figure 3). The degradation of mRNA proceeded faster in incubations with interferon-treated cell extracts (Figure 4). The general conclusion was that dsRNA activate the endonuclease more effectively in extracts of interferon-treated cells.

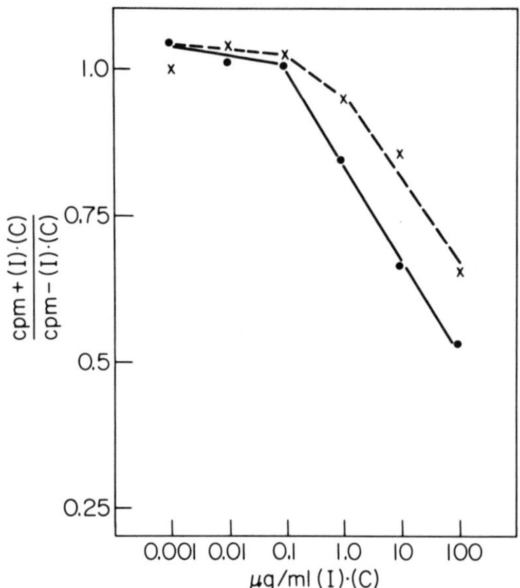

Figure 3. Dependency of mRNA degradation on the concentration of dsRNA in extracts of cells treated 17 hr with 100 units/ml of interferon (●——●) and of control cells (x---x). Assays prepared as described in Figure 1 were incubated for 30 min with the indicated concentration of poly(I) poly(C) or no added dsRNA. The poly(A)-containing RNA was then determined by chromatography on oligo dT)-cellulose as previously described (18). The ordinate shows the ratio of the poly(A)-containing RNA measured in incubations with the concentration of dsRNA indicated in the abscissa to that of an incubation with no added dsRNA.

II. THE ACTIVATION OF ENDONUCLEASE BY 2'5'OLIGO(A)

Kerr and his co-workers have recently reported that dsRNA promotes the formation from ATP of a low molecular weight inhibitor of protein synthesis in interferon-treated L cell extracts (13). This inhibitor can also conveniently be synthesized by an enzymatic activity adsorbed to poly(I) poly(C) conjugated to agarose and can be isolated by chromatography on DEAE-cellulose (13). The inhibitor has been identified as 2'5'oligo(A) (15).

When we measured the synthesis of 2'5'oligo(A) in extracts of interferon-treated and control HeLa cells incubated with dsRNA (Figure 5), a striking parallelism was observed between the

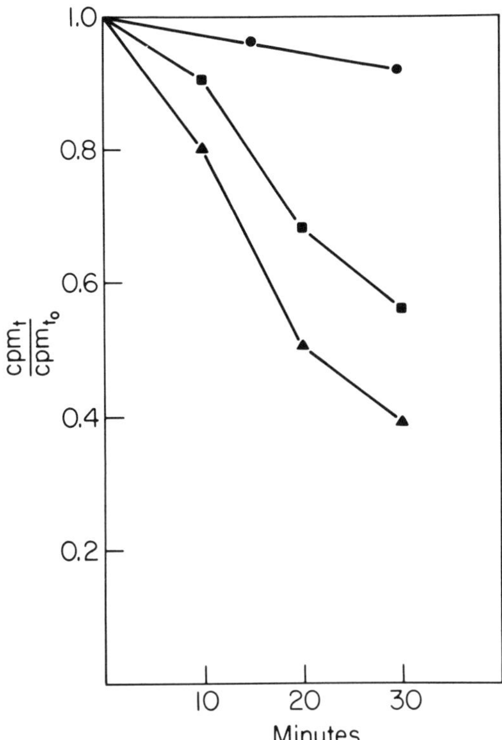

Figure 4. Time course of mRNA cleavage with no added dsRNA (●——●) and with 10 μg/ml of poly(I) poly(C) in extracts of cells treated 17 hr with 100 units/ml of interferon (▲——▲) and of control cells (■——■). The assays were assembled as described in Figure 1 and incubated at 30°C. The poly(A)-containing RNA was then measured as described in Figure 3. The ratio of poly(A)-containing RNA of a sample incubated for the time indicated in the abscissa to that of a sample not incubated is shown in the ordinate. Cleavage of mRNA with no added dsRNA proceeded identically in both cell extracts.

activation of endonuclease by dsRNA and the synthesis of 2'5'oligo(A). Extracts from interferon-treated cells were more active than control cell extracts in both dsRNA-dependent degradation of mRNA and synthesis of 2'5'oligo(A). Moreover, Lengyel and co-workers (11) have shown that the activation of endonuclease requires the presence of dsRNA and ATP, but that the actual cleavage of mRNA can take place in a subsequent incubation

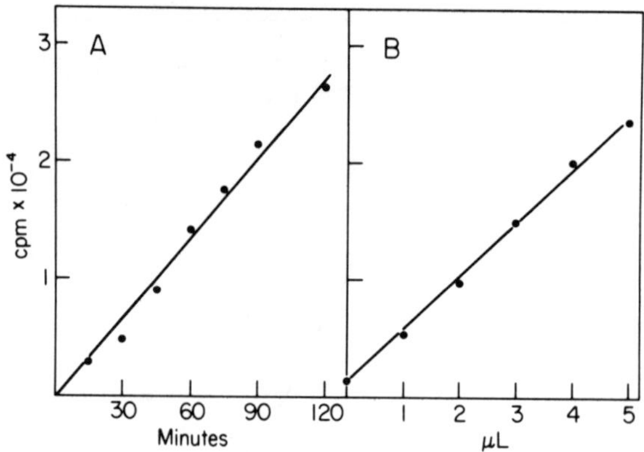

Figure 5. Time course of 2'5'oligo(A) synthesis (A) and dependency on the amount of interferon-treated cell extract (B). Each 25 µl reaction contained 5 µl of cell extract 25 mM $Mg(OAc)_2$, 5 mM ATP, 3mM fructose 1.6 bisphosphate, 120 mM KOAc, 10 µg/ml poly(I) poly(C) and 0.2 µCi of (^3H) ATP. The samples were incubated at 30°C for the indicated time in A and the 2'5'oligo(A) synthesized determined by chromatography on DEAE-cellulose (20). An extract of cells treated 17 hr with 100 units/ml of interferon was used in A, whereas a mixture of interferon-treated and control cell extract was used in B. The incubations in B were for 75 min. The µl of interferon-treated cell extract are indicated in B in the abscissa. The cpm of (^3H) 2'5'oligo(A) synthesized are indicated in the ordinate. About 120,000 cpm of (^3H) ATP were used for each incubation.

after both dsRNA and ATP have been removed. This result strongly implicated a compound formed in the first incubation as an activator of the endonuclease which was detected in the second incubation. The 2'5'oligo(A) synthesized from ATP in the presence of dsRNA fulfills the criteria for such an activator. In fact, addition of 2'5'oligo(A) to cell extracts resulted in the cleavage of mRNA. This explains why 2'5'oligo(A) was such a potent inhibitor of protein synthesis.

The activation of endonuclease by 2'5'oligo(A) has been shown by the cleavage of either viral or cellular mRNA (12). Several observations recently made in different laboratories can be summarized as follows: 1) endonuclease activation requires the continuous presence of 2'5'oligo(A), since removal of this compound causes the endonuclease to revert to an inactive latent state (20); 2) the endonuclease activity is related to the amount of 2'5'oligo(A) present and nM concentrations of this compound are

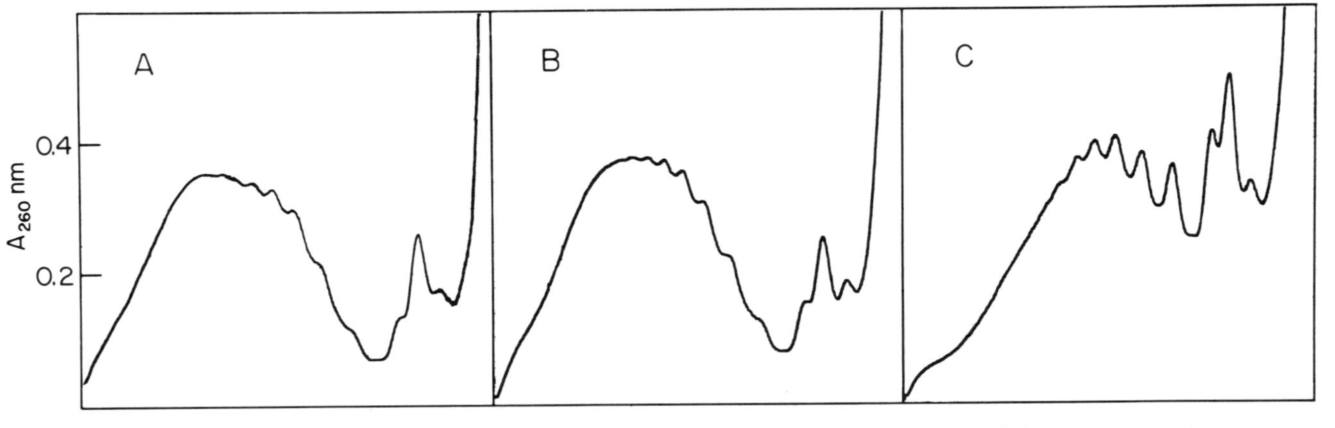

Figure 6. Polysome breakdown in cell extract incubated with 2'5'oligo(A). The treatment of the cells and the composition of the incubations are described in Figure 2, with the only exception that dsRNA was substituted by 2'5'oligo(A) to a final concentration equivalent to 1 μM ATP. A, not incubated; B, 1 hr incubation; C, 1 hr incubation with added 2'5'oligo(A).

effective in the activation (12); 3) the endonuclease degrades both cellular and viral mRNAs, whether free or polysome-bound (12); 4) the endonuclease is present in both interferon-treated and control cells, and even in rabbit reticulocytes (16).

Incubation of an HeLa cell extract with 2'5'oligo(A) resulted in extensive polysome breakdown (Figure 6), whereas no significant changes in polysome pattern could be detected in a control incubation.

III. CORRELATION BETWEEN 2'5'OLIGO(A) POLYMERASE ACTIVITY AND INHIBITION OF VIRAL RNA SYNTHESIS

The relevance of the 2'5'oligo(A) polymerase and of the 2'5'oligo(A)-activated endonuclease to the inhibition of viral RNA synthesis has not yet been established. The major difficulty is that of proving a cause-effect relationship between phenomena taking place in intact cells and biochemical changes measured in cell extracts.

However, when the activity of 2'5'oligo(A) polymerase was measured in cells treated with different interferon concentrations for a constant time (or treated with a constant interferon concentration for a variable time), a dose-response relationship between interferon concentration and 2'5'oligo(A) polymerase activity was observed (Figure 7). The increase in polymerase activity is approximately proportional to the interferon concentration between 10 and 40 units/ml, but a smaller increase is obtained with higher interferon concentrations. The inhibition of EMC virus RNA synthesis follows a similar pattern: it is detected in cells treated with more than 10 units/ml and increases progressively with higher interferon concentrations (Figure 7).

The experiment of Figure 8 shows that 2'5'oligo(A) polymerase activity increased with time of interferon treatment. Little increase was noticeable in the first 4 hours of treatment, but a linear increase in polymerase activity occurred up to 20 hr.

The inhibition of EMC virus RNA synthesis followed a similar kinetic: no inhibition was noticeable after 3 hr of interferon treatment, but a rapid inhibition of viral RNA synthesis occurred afterwards.

This suggested that few hours of interferon treatment were necessary for the increase in 2'5'oligo(A) polymerase activity and that inhibition of viral RNA synthesis was manifested concomitantly with this increase. When actinomycin D was used to block RNA synthesis and cycloheximide to block protein synthesis at different times after addition of interferon, it was clear that active RNA

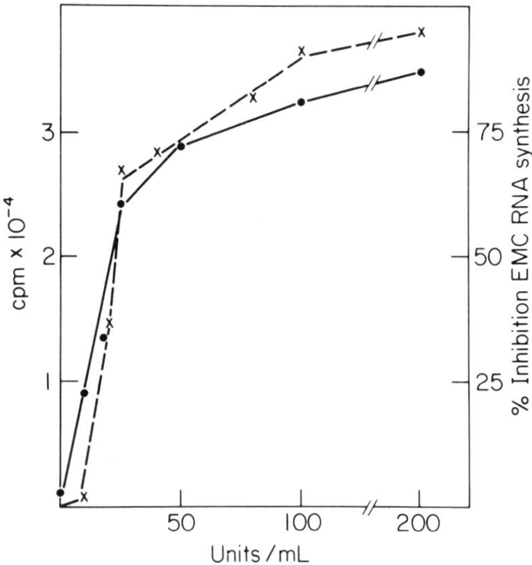

Figure 7. 2'5'oligo(A) polymerase activity and inhibition of EMC RNA synthesis in extracts of cells treated with different interferon concentrations. The cells were treated 17 hr with the interferon concentrations indicated in the abscissa. The 2'5'oligo(A) polymerase activity was assayed as described in Figure 5. ●——●, (H) 2'5'oligo(A) synthesized. Encephalomyocarditis virus RNA synthesis was measured by incubation with (^{3}H) uridine in cells treated with interferon as indicated above; 5 hr after infection actinomycin D was added to 5 μg/ml for 10 min and the cells were incubated with (^{3}H) uridine for 30 min as previously described (21). The inhibition of viral RNA synthesis in interferon-treated cells relative to control infected cells is indicated (x---x).

synthesis was required for at least 4 hours after addition of interferon and active protein synthesis afterwards to obtain an increase in 2'5'oligo(A) polymerase activity and inhibition of EMC virus RNA synthesis (21).

IV. CONCLUSIONS

The degradation of mRNA promoted by dsRNA in extracts of control and interferon-treated cells has been explained by the activation of 2'5'oligo(A) polymerase, synthesis of 2'5'oligo(A) and activation of a latent endonuclease by this oligonucleotide. The increase of 2'5'oligo(A) polymerase activity in interferon-

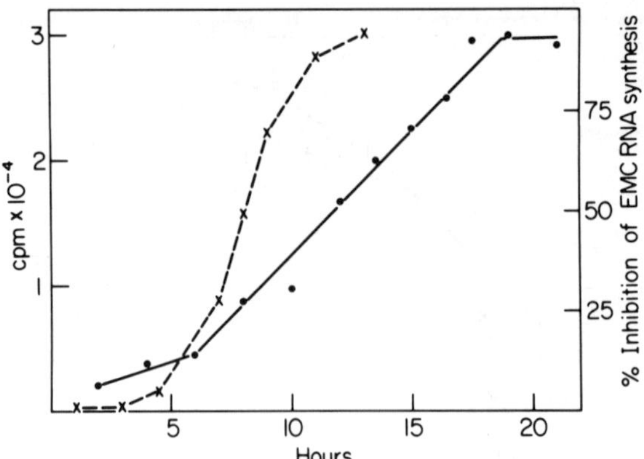

Figure 8. Increase in 2'5'oligo(A) polymerase activity and inhibition of EMC RNA synthesis with time of interferon treatment. Interferon was added at t_o and aliquots of the culture were taken at the indicated times to prepare cell extract and to determine EMC RNA synthesis. The 2'5'oligo(A) polymerase activity was assayed as described in Figure 5. Inhibition of EMC RNA synthesis was measured as described in Figure 7. The incubations measuring EMC RNA synthesis were carried out at the indicated time points 5 hr after infection. •——•, (^{3}H) oligo(A) synthesized; x---x, % inhibition of EMC RNA synthesis.

treated cells is the first clearcut biochemical change correlated with the inhibition of viral RNA synthesis. It seems possible that an increase in the amount of this enzyme prepares the cell to challenge an infecting virus; when the latent polymerase activity is elevated above a critical threshold, the virus fails to replicate. The polymerase may be activated by double-stranded replicative forms of RNA viruses and the 2'5'oligo(A) may in turn activate the endonuclease, causing degradation of viral templates. The weak points in this proposed scheme of events are: 1) activation of 2'5'oligo(A) polymerase by replicative forms of viruses has not yet been demonstrated; 2) synthesis of 2'5'oligo(A) in interferon-treated virus-infected cells has not been shown; 3) the specificity of the endonuclease for viral templates (discrimination) has not been satisfactorily explained. Moreover, the proposed mechanism of interferon action only explains the inhibition of replication of RNA viruses, but cannot explain the effect of interferon on DNA viruses. In conclusion, much remains to be done to demonstrate the involvement of the enzymes discussed in the inhibition of virus replication. It seems likely however, that the mechanism proposed is one of those involved in the establishment of the antiviral state.

ACKNOWLEDGEMENTS

The work reviewed in this chapter has been supported by Grants A1-11887 and HL-17710 of the National Institute of Health.

REFERENCES

1. FINTER, N.B. Interferon and interferon inducers (1973), (2nd edn) North Holland, Amsterdam.

2. FRIEDMAN, R.M. Bact. Rev. (1977), 41, 543-567.

3. KERR, I.M., BROWN, R.E. and BALL, L.A. Nature (London) (1974), 250, 57-59.

4. KERR, I.M., BROWN, R.E., CLEMENS, M.J. and GILBERT, C.S. Eur. J. Biochem. (1976), 69, 551-561.

5. SEN, G.C., LEBLEU, B., BROWN, G.E., REBELLO, M.A., FURUICHI,Y., MORGAN, M., SHATKIN, A.J. and LENGYEL, P. Biochem. Biophys. Res. Commun. (1975), 65, 427-434.

6. SEN, G.C., SHAILA, S., LEBLEU, B., BROWN, G.E., DESROSIERS, R.C. and LENGYEL, P. J. Virol. (1977), 21, 69-83.

7. LEBLEU, B., SEN, G.C., SHAILA, S., CABRER, B. and LENGYEL, P. Proc. Natl. Acad. Sci. U.S.A. (1976), 73, 3107-3111.

8. ZILBERSTEIN, A., FEDERMAN, P., SHULMAN, L. and REVEL, M. FEBS Letters (1976), 68, 119-124.

9. ROBERTS, W.K., HOVANESSIAN, A.K., BROWN, R.E., CLEMENS, M.J. and KERR, I.M. Nature (London) (1976), 264, 477-480.

10. SEN, G.C., LEBLEU, B., BROWN, G.E., KAWAKITA, M., SLATTERY, E. and LENGYEL, P. Nature(London) (1976), 264, 370-373.

11. RATNER, L., SEN, G.C., BROWN, G.E., LEBLEU, B., KAWAKITA, M., CABRER, B., SLATTERY, E. and LENGYEL, P. Eur. J. Biochem. (1977), 79, 565-577.

12. BAGLIONI, C., MINKS, M.A. and MARONEY, P.A. Nature(London) (1978), 273, 684-687.

13. HOVANESSIAN, A.R., BROWN, R.E. and KERR, I.M. Nature(London) (1977), 268, 537-540.

14. KERR, I.M., BROWN, R.E. and HOVANESSIAN, A.G. Nature(London) (1977), 268, 540-542.

15. KERR, I.M. and BROWN, R.E. Proc. Natl. Acad. Sci. U.S.A. (1978), 75, 256-260.

16. CLEMENS, M.J. and WILLIAMS, B.R.G. Cell (1978), 13, 565-572.

17. FAN, H. and PENMAN, S. J. Mol. Biol. (1970), 50, 655-670.

18. HICKEY, E.D., WEBER, L.A. and BAGLIONI, C. Biochem. Biophys. Res. Commun. 80, 377-383.

19. MINKS, M.A., BENVIN, S., MARONEY, P.A. and BAGLIONI, C. Eur. J. Biochem. (1978), submitted for publication.

20. MINKS, M.A., BENVIN, S., MARONEY, P.A. and BAGLIONI, C. Nucleic Acids Res. (1978), submitted for publication.

21. BAGLIONI, C., MARONEY, P.A. and WEST, D.K. Cell (1978), submitted for publication.

SECTION VI:

REPLICATION OF THE VIRAL RNA

THE MECHANISM OF REPLICATION OF PICORNAVIRUS RNA

R. PEREZ-BERCOFF

Laboratory of Virology and Molecular Biology, Institute of Anatomy,
Gloriastrasse 19, 8006 Zürich, Switzerland

INTRODUCTION

The process of replication of the viral RNA is probably the most extensively studied aspect of the molecular biology of picornaviruses. Beginning (just to set a date) in the early sixties with the reports by Sanders (1) and Darnell (2) on the time-course of RNA synthesis, experimental data have accumulated at such a pace that an attempt to provide the reader with a more or less complete list of references would prove not only impossible (because of the unavoidable omission of important contributions), but would also be out of the scope of this chapter. The purpose of this presentation being to review our current understanding of the mechanism of RNA synthesis in picornavirus-infected cells, I shall only refer to those few papers strictly pertinent to the problems under consideration.

I. GENERAL OUTLINE OF THE PROCESS OF PICORNAVIRUS RNA SYNTHESIS

The genome of picornaviruses is a single-stranded RNA molecule about 7000 nucleotides long (see chapter 2). The synthesis of such a macromolecule proceeds by sequential condensation of nucleoside-5'-triphosphates in the 5' to 3' direction (Figure 1).

Replication of the viral RNA is a two-step transcriptional process, involving the synthesis of a complementary ("minus") strand as a first step, followed by the synthesis of new RNA chains of the original polarity, using this time the "minus" strand as template.

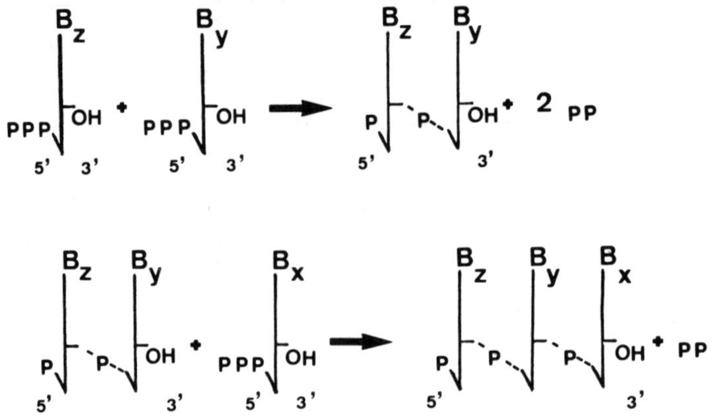

$$PPP\text{-}B_z\text{-}OH + PPP\text{-}B_y\text{-}OH \rightarrow P\text{-}B_z\text{-}P\text{-}B_y\text{-}OH + 2\,PP$$

$$P\text{-}B_z\text{-}P\text{-}B_y\text{-}OH + PPP\text{-}B_x\text{-}OH \rightarrow P\text{-}B_z\text{-}P\text{-}B_y\text{-}P\text{-}B_x\text{-}OH + PP$$

Figure 1. RNA synthesis.

Synthesis of the viral RNA occurs exclusively in the cytoplasm of infected cells (8, 9), namely in a membrane-bound structure known as the replication complex (section V).

Infection with picornaviruses results in a strong (though selective) inhibition of the cellular RNA synthesis (3, 4, 83, 84), and the appearance of a new, virus-induced RNA synthesizing activity. The latter was soon shown to be insensitive to the action of Actinomycin D (5), an antibiotic which prevent DNA-dependent RNA synthesis by intercalating in the dG-dC sequences of the template DNA (6, 7). These observations suggested the possibility of treating infected cultures with the antibiotic in order to suppress host-cell activity and measuring the incorporation of radioactive precursors into viral RNA. Since the precursors added to the medium do not equilibrate instantaneously with the intracellular pool of nucleotides, a correction for this fact must be introduced, at least for the very early times.

II. TIME-COURSE OF RNA SYNTHESIS

RNA synthesis in picornavirus-infected cells proceeds according to a well defined temporal pattern (Figure 2): Beginning about one hour after infection, the synthesis of the viral RNA proceeds at an exponential rate for about two hours. Then, when 20-25% of the total viral RNA has been produced, there is an abrupt change and for 60-90 minutes viral RNA accumulates at a linear rate. At the end of this second phase (4-5 hours after infection, see below), the bulk of the viral RNA has been synthesized and thereafter the rate of synthesis declines.

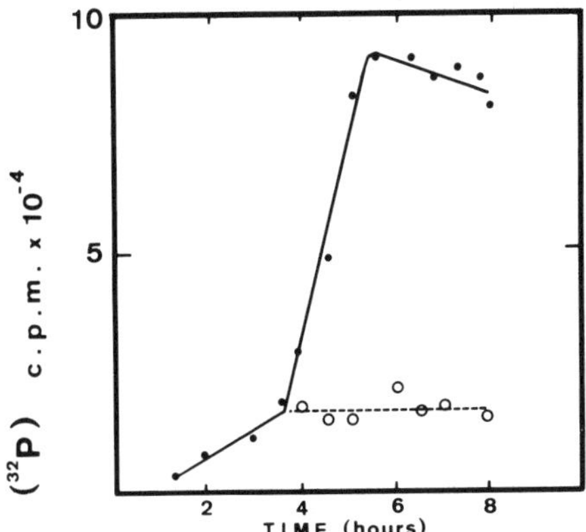

Figure 2. Time-course of RNA synthesis in cells infected with a temperature-sensitive mutant of a picornavirus at permissive (●——●) and non-permissive temperature (o------o).

The shape of this curve, however, depends to some extent on the metabolic state of the cells: in synchronized cultures, viral RNA synthesis was shown to proceed at the highest rate when infection started at the end of the S-phase, that is, at a time when the rate of cellular DNA synthesis was maximal (85).

Infection at very high multiplicities of infection tends to displace the curve towards the earlier times (12).

There is no clear explanation for the mechanism(s) responsible for the two changes in the rate of RNA synthesis.

It is thought that the linear phase ceases due to the damage inflicted to the cellular synthetic machinery unable, at this time, to restore the pool of nucleotides or the energy supply. Alternatively, leakiness due to the altered permeability of the cell membrane has also been implicated.

There is less agreement on the reasons for the switch from the exponential to the linear rate. It has been suggested that this might result from the accumulation of a critical number of template RNA molecules (ca. 4 x 10^4/cell), in coincidence with the time when newly synthesized "plus" strands start to be encapsidated (10). However, the fact that supraoptimal temperatures have been reported to prevent the switch from exponential to linear rate of RNA synthesis suggests the possible involvement of a more complicated mechanism (11).

At the time of maximal activity (that is, when RNA accumulates at linear rate) the synthesis of a full-length RNA molecule takes about one minute (70). At this time, a single infected cell would produce about 3000 RNA molecules per minute (12), a figure which puts the final yield at about 2 x 10^5 viral RNA molecules per cell (10).

The rate of synthesis of each molecular species varies considerably during the infectious cycle: viral s-s RNA synthesis is maximal during the linear phase. There is a constant production of d-s replicative form (RF) which, therefore, accumulates towards the end of the cycle. By contrast, the replicative intermediate (RI) is the major molecular species only during the early (exponential) phase of RNA synthesis (50).

III. VIRUS-INDUCED INTRACELLULAR RNA STRUCTURES

Three virus-induced RNA structures can be isolated from the infected cells. They are:

A) the newly synthesized viral RNA;
B) the double-stranded (d-s) Replicative Form (RF), and
C) the partly single- and partly double-stranded Replicative Intermediate (RI).

Although in the past there was some controversy on this point, there is now little doubt that these RNA species are involved in the viral RNA synthesis, and that they do represent true intermediates in the replication process (see below).

A. The Single-stranded Viral RNA

The virus-induced, single-stranded RNA is by far the major molecular species found in infected cells, and it shares in common with the RNA extracted from virions the same biological and physico-chemical properties: Both are single-stranded with little (if any) stable secondary structure, and can be completely degraded by RNases A and T_1;their sedimentation coefficient (35-37 S in 50 mM NaCl) is markedly affected by the ionic environment (13); they precipitate in high salt buffer (1 M NaCl or 2 M LiCl), adsorb to unmodified cellulose, and can be eluted with buffer containing 15-20% ethanol. These properties have been extensively used to purify the viral s-s RNA out of the mixtures of nucleic acids extracted from infected cells (14). The intracellular s-s RNA has the same base composition (3) and gives the same RNase T_1-fingerprint (15) as the viral RNA.

The intracellular viral RNA is intrinsically <u>infectious</u>.

Actually, the first demonstration that the infectivity of an animal virus resided exclusively in its RNA was provided by Colter et al. in 1957 (16): Experimental evidence was then produced that the RNA extracted from mengovirus-infected Ehrlich ascites cells was infectious. This observation was soon extended to poliovirus RNA (17). Retrospectively considered, it was a very happy circumstance indeed that this pioneer work was carried out in the picornavirus system, for we now know that similar experiments performed with other RNA-containing animal viruses (e.g. influenza) would never have produced such a clearcut result.

The infectivity of the viral s-s RNA (be it from virions or extracted from infected cells) is abolished by nucleases, but is not affected by treatment of the host-cell with Actinomycin D, α-amanitin, or cordycepin (18), (see below).

B. The Replicative Form

The Replicative Form (RF) of picornaviruses is a very stable double-stranded molecule, formed by an intact viral RNA chain hydrogen-bonded to a full-length complementary strand.

The crucial discovery of this structure by Montaigner and Sanders (19), put an end to speculations on the possible involvement of mechanism(s) of RNA replication by-passing the use of complementarity. Their finding was soon extended to RNA-containing phages (20), and other picorna- and plant-viruses as well (21-23).

From the beginning of the infectious cycle a small but defined fraction of the virus-induced RNAs is in a d-s form. At early times the RNase-resistant material accounts for not more than 0.5% of the virus-induced RNAs, but at the end of the cycle (6-8 hours post infection), the d-s molecules may represent up to 10% of the total RNA synthesized at that time.

With a molecular weight of about 5×10^6, the RF of picornaviruses moves as a sharp peak at 18-20 S in sucrose gradients, and the ionic environment has little influence on its sedimentation behaviour.

The buoyant density of RF in Cs salt gradients (about 1.62 g/ml for mengovirus RF, ref.21, 24) is less than that of s-s viral RNA.

RF is soluble in high salt, strongly adsorbs to cellulose, and can be eluted only with buffers lacking ethanol (25).

Owing to the rigid architecture of the molecule (a consequence of the high degree of secondary structure), the RF elutes with the excluded volume when chromatographed in agarose, and migrates slower than expected for its molecular weight during electrophoresis in polyacrylamide/agarose composite gels.

Cations are necessary to stabilize the secondary structure of RF. Accordingly, the degree of RNase-resistance strictly depends on the ionic strength: In high salt (2 x SSC), RF is totally resistant to digestion with RNases A or T_1, whereas in 10 mM NaCl the same enzymes would hydrolyse more than 95-98% of it (24).

As expected for a completely d-s molecule, the melting transition is very sharp ($< 10^{\circ}C$), as revealed by hyperchromic effect (26, 27) or the influence of salt on the susceptibility to nucleases (21). The highly ordered structure of RF can also be inferred from its melting point ($94.5^{\circ}C$ in the case of mengovirus RF, ref. 24), or the typical X-ray diffraction pattern (28).

The RF of picornaviruses is thought to be a linear molecule: there is no proof of "sticky ends", and the experimental evidence behind the claims of circular forms is (in the opinion of this reviewer) extremely tenuous (29-30b).

As already mentioned, the d-s RF tends to accumulate towards the end of the infectious cycle, at a time when viral protein and RNA synthesis clearly decline. This led to speculations that RF might represent a simple by-product, a sort of "dead" molecule without any physiological role of its own in the process of RNA synthesis. For the time being there is no definite proof in one sense or the other (see section VIII).

The RF extracted from picornavirus-infected cells has been shown to be biologically active:

i) like other d-s polyribonucleotides, UV-inactivated RF is an excellent inducer of interferon (23, 31);

ii) the RF of picornaviruses strongly inhibits _in vitro_ protein synthesis in cell-free systems (32-34), and actively promotes the dissociation of the fMet-$tRNA_i^{Met}$/40 S ribosomal subunit complex (35). Although it was originally suspected these findings could provide a reasonable explanation of the mechanism of shut-off of the host-cell protein synthesis, it was soon realized that the added RF indiscriminately blocked the translation of both cellular and viral mRNAs (for an ample discussion see chapters 4 and 10 of this book).

iii) in spite of its well defined secondary structure, the RF of picornaviruses is infectious (18, 19, 27): It is able to initiate a replication cycle which, eventually, leads to the production of progeny virus. But unlike that of the s-s viral RNA, the infectivity of RF is strictly dependent upon the cellular macromolecular synthesis. Treatment of the otherwise susceptible cells with Actinomycin D, α-Amanitin or Cordycepin completely abolishes the infectivity of RF (18). Similarly, enucleated HeLa cells support the replication of s-s, but not d-s poliovirus RNAs (86).

Radioactive RF has been used to infect cells and search for the modifications of the parental molecules in each sub-cellular compartment. Such studies revealed that soon after infection with RF, the input d-s molecules are found in the cytoplasm as part of a Replicative-Intermediate-like structure (18). Treatment of the host-cell with interferon abolishes the infectivity of RF, but does not prevent the intracellular processing of RF into RI. In contrast, exposure of the cells to Actinomycin D inhibits the infectivity of RF and the intracellular transition to RI as well (36). As it was found that a cellular RNA polymerase specifically binds to RF (37), it was postulated that RF served as abnormal template for a cellular RNA polymerase to transcribe the first viral messenger (18, 37).

C. The Replicative Intermediate

A third RNA structure can be isolated from picornavirus-infected cells: the complex Replicative Intermediate (RI), partially single- and partially double-stranded.

First found as an intermediate in the replication of the small RNA-containing phages (38, 39), the RI was soon identified as the actual site of synthesis of the viral RNA.

It consists of a RNase-resistant, d-s "core" plus several s-s nascent chains of different length, attached to the template by a hydrogen-bonded region (40-42).

Intact RI sediments in sucrose gradients as a polydisperse band between 18-70 S. However its sedimentation behaviour depends to a considerable extent on the extraction procedure: phenolization at 60°C causes the release of nascent chains and, accordingly, the RI so obtained sediments between 10-40 S (42).

A substantial portion of RI being s-s, the complex precipitates in high salt (2 M LiCl) together with the viral s-s RNA. The RI adsorbs to cellulose powder and elutes with buffers lacking ethanol, as does the d-s RF (25). Due to its complex structure, RI is excluded by molecular sieving in agarose columns and migrates very slowly (if at all) when subjected to electrophoresis on loose polyacrylamide/agarose gels.

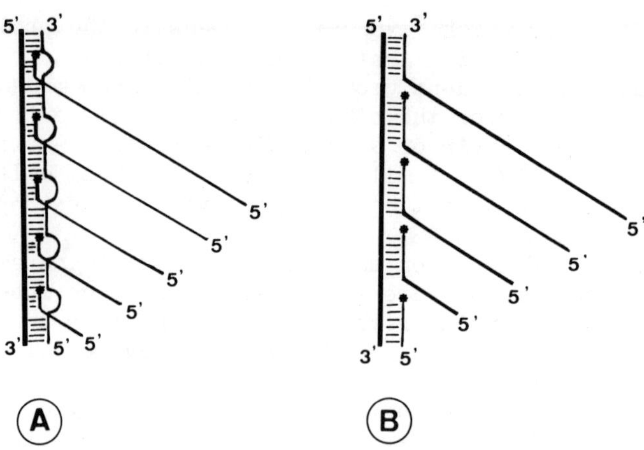

Figure 3. Proposed structures of the RI: A) conservative and B) semi-conservative models. The strand serving as template is represented by a thick line; the RNA-polymerase (*) is attached to the 3' end of the nascent strands (thin lines). For details see the text .

About 38% of RI withstands nuclease digestion. From this figure it was possible to estimate as 6.5 the average number of nascent chains attached to each RI molecule (42). Electronmicroscopic examination of spread RI molecules led to similar conclusions (43).

Further evidence for RI being formed from a d-s core and several s-s chains linked to the core by base-pairing came from analysis of the two-step melting curve (44) of RI and its sedimentation behaviour after nuclease digestion (40, 41).

The structure of RI has not yet been completely elucidated, and two mutually exclusive models of replication have been proposed: a "conservative and a semi-conservative" one. In the former (Figure 3A), a d-s "core" serves as template, one of the resident chains being locally displaced by the progeny RNA strands just at the growing point.

The "semi-conservative" model proposes that the nascent chains are hydrogen-bonded to a s-s template over a short area and, as transcription proceeds, they displace each other (Figure 3B).

Digestion of RI with RNase A originates a d-s "core" (undistinguishable by sedimentation analysis from authentic d-s RF), and a collection of shorter fragments (41, 43). After denaturation of the d-s "cores", full-length molecules and short fragments can be separated by centrifugation through sucrose density gradients (43). About 90% of the full-length molecules present in the d-s "cores" of RI hybridized to viral RNA, but there was no complementarity between the latter and the fragments derived from the d-s "cores".

A first conclusion to be drawn from these results is that most (if not all) of the RI molecules derive from a structure involved in the transcription of virion-like RNA ("plus" strand), a conclusion supported by our fingerprint analysis of the *in vitro* synthesized RNA (see section VII).

They also suggest that the d-s "cores" of the RI and the authentic d-s RF, although similar, might not be the same structure, since denaturation of the latter results in the separation of two full-length complementary strands (43). The d-s RF, therefore, is likely not to originate in the partial degradation of RI as has been repeatedly suggested.

Additional support for this viewpoint comes from the substantially different kinetics of formation of RF and RI in the presence of guanidine, an inhibitor of viral RNA synthesis (see section VI), or cycloheximide. This finding led Noble and Levintow (94) to suggest that "the two extracted forms represent distinct structural and functional entities".

Unfortunately, the above reported experiments do not allow discrimination between the two proposed models of RI: In fact, the nuclease digestion of the RI would result not only in the degradation of the s-s progeny strands, but also in the introduction of nicks in the resident chain, at the points where this has been displaced by the nascent strands.

In any case, the RI represents the backbone of the structure where the synthesis of viral RNA occurs: after very short pulses (2-3 minutes) most of the radioactive label is found in association with RI (40). A direct precursor-product relationship, however could not be proven in the intact cell because of the size of the pools of precursor nucleotides. As "pulse-chase" experiments were performed *in vitro* using isolated replication complexes (see section VII), it was proven beyond any doubt that the RI represents a true intermediate in the synthesis of the viral RNA (45, 46). As far as RF is concerned, the outcome was less clear and its possible role will be discussed later (section VIII).

D. Poly(A) and Poly(U) Tracts in Intracellular RNAs

The genomic RNA of picornaviruses contains a poly(A) tract of variously estimated length at the 3' end of the molecule (chapter 2).

The poly(A) stretch directs transcription of a poly(U) tract at the 5' end of the "minus" strand, which in turn serves as template to reproduce the poly(A) tail in the newly synthesized viral RNA (59, 87-89). Thus, unlike the poly(A) tract of eukaryotic mRNAs, the poly(A) stretch of picornavirus RNA is not post-transcriptionally added but is genetically coded.

Intriguingly enough, however, the length of the poly(A) stretches associated with the complex intracellular structures involved in the replication of the viral RNA,namely RF and RI, were reported to be longer than the corresponding homopolymeric tracts isolated from virion RNA (87, 89).

As an oligo(A)-primed, terminal adenylate transferase of cellular origin was found in tight association with the same membranes the replication complex is bound to, the longer poly(A) stretches were thought to result from either a mechanism of "slippage" or by end addition of adenylate residues (89).

However, since the length of the poly(A) tracts isolated from the different viral RNA structures were deduced essentially from their electrophoretic mobility in polyacrylamide gels under non-denaturing conditions, and there is no report on the ratios of internal AMP to 3' terminal adenosine residues of the "longer" poly(A) tails, it is not yet possible to exclude that the reported difference in length might actually result from hybridization of the poly(A) to other RNA fragments.

E. Do RF and RI Really Exist?

At this point it seems pertinent to discuss the relevance of this description, because the double-strandness of the <u>intracellular</u> structures from which RF and RI were extracted has been questioned.

The general outline of the argument was that once the proteins and membranes associated with the intracellular RNAs were removed by phenol (or detergent) treatment, complementary strands would collapse together and artifactually form a double-stranded structure which did not actually exist in the infected cell (47).

In support of this view, it was usually claimed that extraction of RNA in the presence of diethylpyrocarbonate (DEP) yielded little d-s RNA (48). Afterwards, however, Ehrenfeld showed that DEP lowered considerably the melting temperature of d-s RF (49), and it was evident that the <u>absence</u> (rather than the presence) of d-s RNA was an artifact of the extraction procedure. Furthermore, incubation of complementary strands in the presence of phenol and/

or detergents does not produce any d-s RNA (47).

In the light of the currently available evidence, "it appears that the complementary strands of RNA within the cell, if not literally hydrogen-bonded to each other, must at least be extensively aligned in a specific configuration. In other words, there is little reason to doubt that the extracted structures are accurate reflection of their native intracellular counterpart, hence, valid objects for the study of the replication of viral RNA" (50).

IV. THE ASYMMETRY OF TRANSCRIPTION

Replication of viral RNA involves the synthesis of a complementary strand, in agreement with the Watson-Crick model of replication of DNA.

It was soon realized, however, that some kind of regulatory mechanism must intervene, because the synthesis of virus-induced RNAs was shown to be quite an asymmetrical process: The bulk of the RNA found in the cytoplasm of picornavirus-infected cells is virion-like (i.e. "plus" strand), and only a very minor fraction of the newly synthesized RNA would hybridize to the RNA extracted from virions. Several explanations of this phenomenon have been offered:

i) the two distinct rates of RNA synthesis already described might correlate with the synthesis of RNAs of opposed polarity: during the _exponential_ phase the "minus" strand would be the predominant product and the reverse would be true during the _linear_ phase of synthesis;

ii) the rate of _degradation_ of the "minus" strand might increase during the late stages of the infective cycle;

iii) transcription of the viral RNA and its complementary strand could take place in two different structures, namely: the replication complex "plus" and the replication complex "minus";

iv) two different virus-coded RNA polymerases might be responsible for the synthesis of the two strands.

The asymmetry of the virus-directed RNA synthesis appears, therefore, as a major issue, and for the time being we can only speculate about the possible mechanism(s) involved in such an effective regulation of the production of the "minus" strand (see below).

V. THE REPLICATION COMPLEX

All viral structures involved in picornavirus RNA synthesis are tightly associated with the smooth cytoplasmic membranes (51). Synthesis of viral RNA occurs exclusively in the replication complex (RC), a complex structure including the RNA template and a virus-coded RNA polymerase (52).

The RC was first identified as a membrane-associated structure in the cytoplasm of poliovirus-infected cells. Treatment with deoxycholate liberated a fast-sedimentation component (average 250 S), which contained most of the label after short pulses with radioactive uridine. Ribosomes were not associated with the complex, as EDTA did not modify its sedimentation behaviour (52).

A. The Enzyme(s)

Two enzymic activities have been heretofore demonstrated in the RC:

i) a virus-coded, RNA-dependent, actinomycin-resistant RNA polymerase (53-56). Tightly bound to its template, for more than 12 years attempts to separate the active enzyme from the RC were unsuccessful. Recently, Traub and co-workers were able to purify the replicase of EMC virus in a template-dependent form (57). The properties of the solubilized enzyme are described in full in the following chapter of this book. All the activity is associated with a non-capsid peptide (M.W: 56,000), in agreement with earlier reports (58).

ii) a second enzymic activity has been recently identified in association with the RC: it is an oligo(U)-primed, poly(A)-dependent poly(U) polymerase (59). The enzyme needs both the poly(A) template and the oligo(U) primer, in addition to the UTP substrate, in order to condense UMP residues at the 3' end of the primer. The reaction seems to be rather specific, since GMP is not incorporated when poly(A) is used as template (59).

As synthesis of RNA proceeds in the 5' to 3' direction, transcription of the viral RNA must initiate at the 3' end of the template, that is, on the poly(A) tract. A poly(U) polymerase hence appears quite likely to be the enzyme responsible for initiating the synthesis of the "minus" strand. The need for an oligo(U) primer, however, makes that enzyme dependent on another one (be it of cellular or viral origin) which in turn must provide the primer. Alternatively, the requirement of a pre-formed oligo(U) primer might simply reflect the inability of the _in vitro_ system to initiate transcription in the absence of a UTP residue linked to the small viral protein VPg (chapter 9).

Since the oligo(U) primed poly(U) polymerase seems to be able to transcribe the heteropolymeric portion of the viral RNA, and all the enzymic activity is associated with a peptide of apparent M.W. 56,000, it is possible that these two distinct activities actually correspond to a single enzyme (J.B. Flanegan, personal communication).

B. The Nature of the RNA Template

There have been conflicting reports regarding the nature of the RNA which serves as template in the RC: Based on electron microscopic examination, it was concluded that the template for viral RNA synthesis was mainly s-s RNA (60). On biochemical grounds, Öberg and Phillipson suggested that the nascent chains are hydrogen bonded to the template over a very short area, perhaps held in place just by the polymerase (48).

The question has been recently re-examined, and there is little doubt left that a double stranded portion of the RI is required for the synthesis of the viral RNA (61, 62).

The 5' end of the nascent chains seems to be "capped" (63, 64) by a terminally-linked protein, covalently bound to the 5' end of the genomic RNA of picornaviruses (65-69). This led to the suggestion that the terminally-linked protein might be needed as a primer for the viral replicase (64, and for a more detailed discussion, see chapter 9), but there is not yet experimental evidence to support the notion of a direct involvement of this protein in the process of RNA replication.

C. RNA and Protein Synthesis

The synthesis of picornavirus RNA strictly depends on the concomitant synthesis of viral proteins: so much so that inhibition of the latter results in an almost immediate decline of the production of viral RNA.

It was therefore proposed that either the viral polymerase is extremely unstable and continuous protein synthesis is needed to restore the level of active enzymes (10), or that proteolytic inactivation of the viral replicase leads to the decline of the rate of RNA synthesis observed late in the infectious cycle (81).

A recent analysis of the viral peptide actually associated with poliovirus RC suggested that an inactive precursor to the viral synthetase might be rendered functional either by undergoing some undetectable structural modification or by stably associating with a viral or host factor (82). Obviously, we do not know whether such a factor would also confer strand specificity for transcription (see section VIII).

VI. INHIBITORS OF RNA SYNTHESIS

The search for substances which could distinguish between cell- and virus-directed synthesis led eventually to the identification of a group of compounds able to inhibit specifically the replication of the RNA of some picornaviruses. Guanidine and (to a certain extent) hydroxy-benzyl-benzimidazol (HBB) are the best studied among them (87-89, 101).

Two lines of evidence point to a virus-coded function being the target of these compounds:

i) <u>a defined spectrum of action</u>: guanidine and HBB are effective only on polio-, some entero- and Foot-and-Mouth disease viruses: Mengo, EMC and cardioviruses are practically refractory (90), and

ii) <u>the gentics of sensitivity</u>: guanidine-resistant and even guanidine-dependent mutants of poliovirus were identified which would grow in the presence of the drug in the same host-cell in which the sensitive wild-type would be inhibited (89).

While the mechanism of action of guanidine and HBB is not yet completely understood, the following points are already firmly established:

i) <u>HBB and guanidine have no effect on the host-cell</u>: at the doses employed to block completely virus replication, HBB and guanidine have no noticeable effect on the host-cell metabolism as far as this can be monitored by the rate of cell division, RNA and protein synthesis, oxygen consumption and glucose utilization (88, 90);

ii) <u>Complementation</u>: double infection with a guanidine-sensitive and a guanidine-resistant (or dependent) mutant virus in the presence of the drug, results in the "rescue" of the sensitive one (91, 92);

iii) <u>Guanidine and HBB block viral RNA synthesis</u>: Ten minutes after addition of guanidine, total viral RNA synthesis is strongly depressed. There is still some controversy regarding the intimate mechanism which ultimately leads to the inhibition of viral RNA synthesis: On the one side, it was proposed that the major effect of guanidine may be to prevent the release of the nascent chains, which would accumulate in a RI-like structure called "the guanidon" (93). But analysis of the RNA synthesized in the presence of the drug showed that the production of s-s RNA was almost abolished, whereas d-s RF accumulated, though at a reduced rate (94, 95). The conclusion, in thise case, was that guanidine blocked the initiation of RNA synthesis.

iv) The effect of guanidine and HBB is reversible: and viral RNA synthesis resumes immediately after the removal of the drugs. Since d-s RF is the only species of poliovirus RNA that accumulates in the presence of guanidine and RNA synthesis restarts as soon as the compound is removed, it was suggested that the accumulated RF contained potentially active replication structures (ref. 94 and see section 8).

v) Guanidine and HBB have no effect in vitro: in the presence of the drugs, viral RNA synthesis proceeds unaltered in the replication complex isolated from poliovirus-infected cells (96).

vi) Antagonists: several methylated compounds were identified which antagonize the inhibitory (or dependent) effect of guanidine (97). Methionine, valine, choline, leucine were also reported to inhibit the effects of guanidine, but the effect depends on the host-cell, as the replication of poliovirus in cells from different origins showed a marked difference in its response to the antagonists (98).

vii) Guanidine has no direct effect on viral protein synthesis: In the presence of the drug, poliovirus-directed protein synthesis has been reported to continue unaltered (99) or at a moderately slower rate (100). Under such conditions, poliovirus mRNA was normally translated, and the same set of viral proteins was produced with the one exception of VP_2 (and, possibly VP_4 as well). Since these two polypeptides are cleaved from VP_0 after assembly of the virion (see chapter 1), the observed absence of VP_2 can be explained as an indirect effect of the inhibition of RNA synthesis and the concomitant reduction of virion formation. This can also explain the increased accumulation of empty capsids (top component) (99).

VII. RNA SYNTHESIS IN VITRO

A. Crude Replication Complexes

The replication complex can be isolated with the particulate fraction of the cytoplasm of picornavirus-infected cells (71). Under proper conditions the RCs are able to continue *in vitro* the synthesis of RNA, providing a unique tool to study the mechanism of RNA synthesis with all the advantages (and all the limitations, too) of an *in vitro* cell-free system.

The *in vitro* enzymic activity associated with the RC provides a direct measure of the rate of synthesis of the viral RNA in the infected cells at the time the complexes were isolated (71).

Analysis of the RNAs synthesized *in vitro* showed that the isolated RCs produced the same molecular species found in the infected cells, namely 35 S viral s-s RNA, d-s RF, and RI (45, 46).

As already mentioned (section III. C), "pulse-chase" experiments are only possible in the *in vitro* system. The kinetics of appearance of radioactive precursors into the different RNA products proved that synthesis of the viral RNA occurred in the RI which seems to be the direct precursor to 35 S s-s viral RNA. The role RF might play in the process of RNA replication could not be clearly defined (38, 44, 45).

To summarize, the isolated RCs reproduced *in vitro* the pattern of viral RNA synthesis observed in the infected cell.

Apparently there is no re-initiation of RNA synthesis *in vitro*: the isolated RCs seem to be able only to complete the RNA chains already in the course of synthesis (45).

B. Detergent-treated Replication Complexes

The very early attempts to purify the viral replicase succeeded at least in freeing the RC from the smooth cytoplasmic membranes to which it was bound (72-76). The basic approach consisted of incubating crude preparations with detergents, followed by velocity sedimentation through sucrose density gradients. The general outcome of these experiments was that after detergent treatment the active RCs separated into two components, a slow and a fast-sedimenting one, associated with d-s RF and multi-stranded RI, respectively.

Further analysis of the RNAs synthesized *in vitro* by these two components showed that the slow-sedimenting one incorporated labelled precursors into a d-s structure, whereas the RNAs produced in the fast-sedimenting component were partially single- and partially double-stranded (72, 77). The conclusions drawn about the polarity of the RNAs synthesized are (in our personal view) less justified (see below).

C. Analysis of the RNAs Synthesized in Vitro

The isolated RCs seemed to provide an ideal tool to investigate the process of transcription and the mechanism(s) responsible for its asymmetry. Obviously, however, the suitability of an *in vitro* system to investigate regulatory mechanism(s) can be a matter of endless discussion. In the case under consideration it can be questioned whether the procedure used for the isolation of the RC or the conditions employed for the *in vitro* RNA synthesis would provide a distorted picture of the process occurring in the intact cell, either by specifically selecting for or by stimulating the expression of one type of structure to the detriment of another. Every *in vitro* study can be subjected to such criticisms and we should bear in mind these reservations.

Several attempts have been made to characterize the products of the *in vitro* synthesis:

i) on the basis of their electrophoretic mobility (46), sedimentation behaviour, and sensitivity to nucleases (45), it was concluded that the isolated RC incorporated radioactive precursors into the same RNA species found in the infected cells;

ii) the nature of the *in vitro* products was further investigated, the basic approach being to hybridize radioactively-labelled *in vitro* synthesized RNA with RNA extracted from virions. The degree of complementarity was estimated from the percentage of radioactivity rendered resistant to nuclease digestion (77, 78). The conclusion was that 20% of all the *in vitro* product was complementary RNA, that is: "minus" strand. However, since the hybridization experiments were carried out in the presence of the RNA which served as template for the transcription of the labelled one, self re-annealing reached very high values, preventing, therefore, drawing too firm a conclusion.

iii) as the recently developed techniques of fingerprint analysis allow unmistakeable identification of even minute amounts of RNA, a different approach became possible. Accordingly, the labelled RNAs synthesized *in vitro* by the RCs, isolated from L cells at different times after infection with mengovirus, were analyzed by two-dimensional gel electrophoresis after complete digestion with RNase T_1. The RC isolated as early as 120 minutes after infection (at a time when the first virus-directed activity could be detected), or as late as 7 hours post-infection, synthesized only virion-like RNA. Not even traces of other RNAs could be identified (79). The obvious conclusion was that under the conditions used the isolated RC can only synthesize "plus" strand.

At this point we might note that the replicase of the small RNA-containing phage Qβ synthesizes *in vitro* a mixture of "plus" and "minus" strands when synthesis is carried out in the presence of a "host factor". In its absence, however, the same enzyme seems to be unable to use the "plus" strand as a proper template (80). Although the above reported experiments were purposely performed with crude RC, the authors cannot exclude that a similar (though hypothetical) factor conferring strand specificity, was lost.

The question as to where and how the "minus" strand is synthesized, thus remains open.

VIII. THE "SINGLE-RUN" MODEL OF RNA REPLICATION

Although for the time being there is only indirect experimental evidence to support it, it is tempting to postulate a hypothetical model of picornavirus RNA replication based on the existence of a preferential initiation site for transcription at the 3' end of the "minus" strand. The model is illustrated in Figure 4.

Basically, in the absence of a better template, the viral replicase would bind to, and initiate the transcription of, the viral RNA. Conceivably, the very first step of this process would be the synthesis of the poly(U) tract (Figure 4, A), followed by the transcription of the heteropolymeric portion of the "plus" strand (Figure 4, B). But as soon as the first complementary strand is completed, its 3' end would offer a better initiation site and the viral replicase, hence, would switch to this competitive template (Figure 4, C).

If RNA synthesis stops at this point (or the rate of synthesis slows down) the RNA structure resulting after a single run of synthesis would be authentic d-s RF. The observed accumulation of d-s RF towards the end of the infectious cycle (in concomitance with the total decline of RNA and protein synthesis), or in the presence of guanidine or cycloheximide, might therefore reflect the inability of the RNA synthesizing machinery to proceed with the next step of transcription rather than an increased rate of degradation of RI. Thus, the tight association of RNA and protein synthesis would arise from the stringent need of a constant supply of a virus-coded peptide indispensable for re-initiation to occur. The limiting factor, in this case may be used either to re-form an active molecule of viral replicase, or to "prime" the initiation, a role (at the time this review is written) tentatively assigned to VPg (see chapter 9).

The preferential initiation site on the "minus" strand might result from a particular sequence at the 3' end of the complementary RNA, or the conformation of the double-stranded complex so formed may be such that the 3' end of the "minus" strand becomes the only possible initiation site (Figure 4, C an D).

In any case, other molecules of replicase can bind at the same site and start transcription (Figure 4, E): The RC so formed would

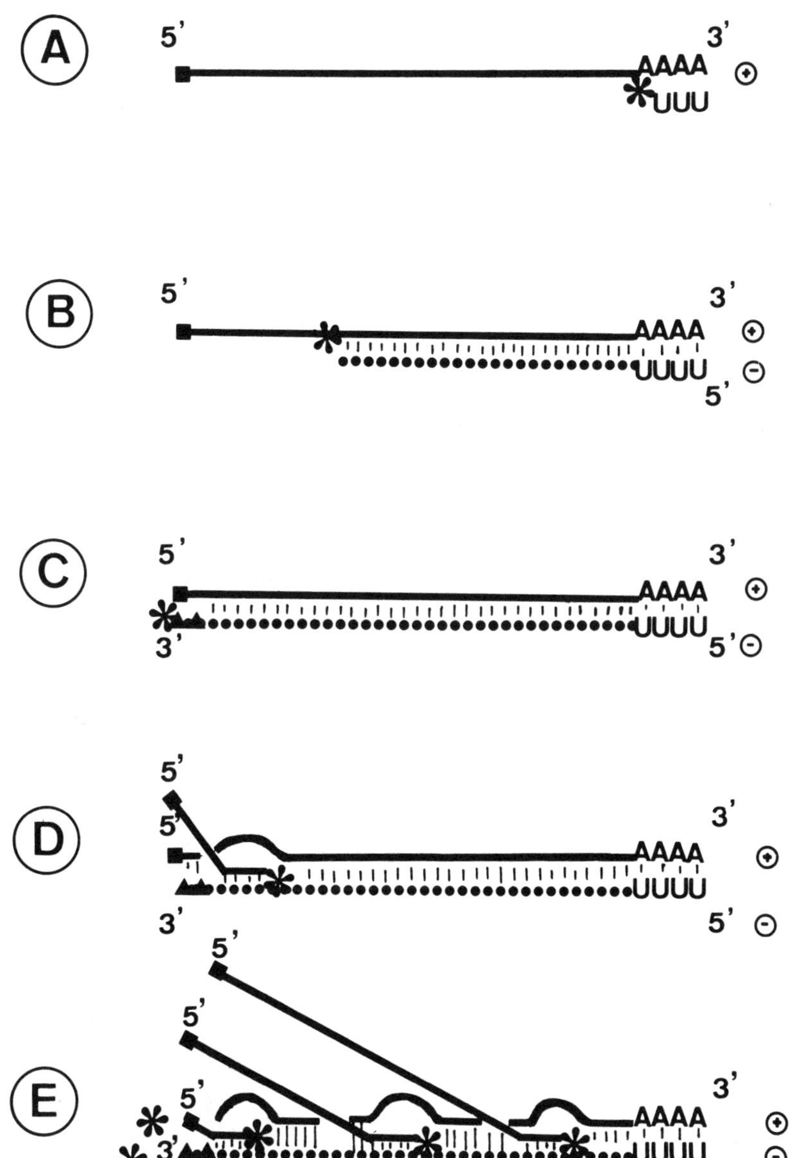

Figure 4. <u>The "Single-Run" model of replication of viral RNA</u> (details are discussed in the text). Symbols (*)= the viral replicase. VPg (■) "caps" the 5' end of all "plus" strands. The preferential initiation site (▲▲▲) at the 3' end of the "minus" strand.

only produce "plus" strands.

Transcription through this "single run" mechanism would result in an auto-regulated process, since at the end of a single run of synthesis using the "plus" strand as template the complementary RNA synthesized would successfully compete for the available enzyme, thus preventing over-production of "minus" strand. Synthesis of "minus" strand at a very low rate would occur all through the replication cycle, depending on the amount of the available enzyme and the relative efficiency of initiation on the two competing templates.

The "single-run" model of RNA synthesis re-evaluates the central role of RF in the replication process: The d-s RF instead of a by-product is conceived as a sort of very short-lived precursor, almost immediately converted into RI. At the normal rate of RNA synthesis (120-150 nucleotides per second), this relationship would be practically impossible to put in evidence. But indirect support for the proposed model comes from the observed immediate conversion of RF into RI during the reversal of guanidine blockade (see section VI), the kinetics of accumulation of d-s RF, and the relationship between RNA and protein synthesis.

Since the viral replicase has been recently solubilized and purified (see next chapter of this book), it will be soon possible to check the proposed model by readily designable experiments.

REFERENCES

1. SANDERS, F.K. Nature (London), (1960), 185, 802-805.

2. DARNELL, J.E. and LEVINTOW, L. Virology (1961), 13, 271-279.

3. BALTIMORE, D. and FRANKLIN, R.M. Proc. Nat. Acad. Sci. U.S.A. (1962), 48, 1383-1390.

4. ZIMMERMAN, E.F., HEETER, M. and DARNELL, J.E. Virology (1963), 19, 400-408.

5. REICH, E., FRANKLIN, R.M., SHATKIN, A.S. and TATUM, E.L. Science (1961), 134, 556-557.

6. SOBELL, H.M., JAIN, S.C., SAKORE, T.D. and NORDMAN, C.E. Nature New Biology (1971), 231, 200-205.

7. HASELKORN, R. Science (1964), 143, 683-684.

8. FRANKLIN, R.M. and ROSNER, J. Biochem. Biophys. Acta. (1962),

55, 240-241.

9. HAUSER, H. Z. Naturfoschung (1962), 17b, 158-160.

10. BALTIMORE, D. The replication of picornaviruses. In The Biochemistry of Viruses. Levy, H.B. ed. (1969), pp. 101-175, Marcel Dekker Inc., New York.

11. LWOFF, A. Cold Spring Harbor Symp. Quant. Biol. (1962), 27, 159-174.

12. BALTIMORE, D., GIRARD, M. and DARNELL, J.E. Virology (1966), 29, 179-189.

13. MONTAIGNER, L. and SANDERS, F.K. Nature (London) (1963), 197, 1178-1181.

14. BISHOP, J.M. and KOCH, G. Virology (1969), 37, 521-534.

15. PEREZ-BERCOFF, F. and GANDER, M. Unpublished results.

16. COLTER, J.S., BIRD, H.H. and BROWN, R.A. Nature (London) (1957), 179, 859-860.

17. ALEXANDER, H.E., KOCH, G., MOUNTAIN, I.M., SPRUNT, K. and VAN DAMME, O. J. Expt. Med. (1958), 108, 493-506.

18. PEREZ-BERCOFF, R., CIOE, L., MEO, P., CARRARA, G., MECHALI, M., FALCOFF, E. and RITA, G. J. Gen. Virol. (1974), 25, 53-62.

19. MONTAIGNER, L. and SANDERS, F.K. Nature (London) (1963), 199, 664-667.

20. WEISSMANN, C., BORST, P., BURDON, R.H., BILLETER, M.A. and OCHOA, S. Proc. Nat. Acad. Sci. U.S.A. (1964), 51, 682-689.

21. BISHOP, J.M., SUMMERS, D.F. and LEVINTOW, L. Virology (1965), 54, 1273-1281.

22. BALTIMORE, D., BECKER, Y. and DARNELL, J.E. Science (1964), 143, 1034-1036.

23. MANDEL, H.G., MATTHEWS, R.E.F., MATUS, A. and RALPH, R.K. Biochem. Biophys. Res. Comm. (1964), 16, 604-609.

24. MECHALI, M., PEREZ-BERCOFF, R., CARRARA, G. and FALCOFF, E. Biochimie (1973), 55, 361-363.

25. FRANKLIN, R.M. Proc. Nat. Acad. Sci. U.S.A. (1966), 55, 1504-1511.

26. AMMAN, J., DELIUS, H. and HOFSCHNEIDER, P.H. J. Mol. Biol. (1964), 10, 557-561.

27. BISHOP, J.M. and KOCH, G. J. Biol. Chem. (1967), 242, 1736-1743.

28. LANGRIDGE, R., BILLETER, M.A., BORST, P., BURDON, R.H. and WEISSMANN, C. Proc. Nat. Acad. Sci. U.S.A. (1964), 52, 114-119.

29. AGOL, V.I., ROMANOVA, L.I., CUMAKOV, I.M., DUNAYEVSKAYA, L.D. and BOGDANOV, A.A. J. Mol. Biol. (1972), 72, 77-89.

30. CLEMENTS, J.B. and MARTIN, S.J. J. Gen. Virol. (1971), 12, 221-232.

30b. YOGO, Y. and WIMMER, E. Nature New Biology (1973), 242, 171-174.

31. FALCOFF, R. and FALCOFF, E. Biochem. Biophys. Acta. (1970), 199, 147-158.

32. EHRENFELD, E. and HUNT, T. Proc. Nat. Acad. Sci. U.S.A. (1971), 68, 1075-1078.

33. HUNT, T. and EHRENFELD, E. Nature New Biol. (1971), 230, 91-94.

34. ROBERTSON, H.D. and MATHEWS, M.B. Proc. Nat. Acad. Sci. U.S.A. (1973), 70, 225-229.

35. DARNBROUGHT, G., HUNT, T. and JACKSON, R.J. Biochem. Biophys. Res. Comm. (1972), 48, 1556-1564.

36. PEREZ-BERCOFF, R. et al. Virology (1979), in press.

37. PEREZ-BERCOFF, R., CARRARA, G., DOLEI, A., CONCIATORE, G. and RITA, G. Biochem. Biophys. Res. Comm. (1974), 56, 876-883.

38. FENWICK, M.L., ERIKSON, R.L. and FRANKLIN, R.M. Science (1964), 146, 527-530.

39. ERIKSON, R.L., FENWICK, M.L. and FRANKLIN, R.M. J. Mol. Biol. (1964), 10, 519-529.

40. BALTIMORE, D. and GIRARD, M. Proc. Nat. Acad. Sci. U.S.A. (1966), 56, 741-748.

41. BISHOP, J.M., KOCH, G., EVANS, B. and MERRIMAN, M. J. Mol.

Biol. (1969), 46, 235-249.

42. BALTIMORE, D. J. Mol. Biol. (1968), 32, 359-368.

43. SAVAGE, T., GRANBOULAN, N. and GIRARD, M. Biochimie (1971), 53, 533-543.

44. PLAGELMAN, P.G.W. and SWIM, H.H. J. Mol. Biol. (1968), 35, 13-35.

45. GIRARD, M. J. Virol. (1969), 3, 376-384.

46. McDONNELL, J.P. and LEVINTOW, L. Virology (1970), 999-1006.

47. WEISSMANN, C., FEIX, G. and SLOR, H. Cold Spring Harbor Symp. Quant. Biol. (1968), 33, 83-100.

48. OEBERG, B. and PHILIPSON, L. J. Mol. Biol. (1971), 58, 725-737.

49. EHRENFELD, E. Biochem. Biophys. Res. Comm. (1974), 56, 214-219.

50. LEVINTOW, L. The reproduction of picornaviruses. In Comprehensive Virology, Fraenkel-Conrat, H. and Wagner, R.R. eds., (1974), Vol.3. pp. 109-169, Plenum Press, New York.

51. PENMAN, S., BECKER, Y. and DARNELL, J.E. J. Mol. Biol. (1964), 8, 545-555.

52. GIRARD, M., BALTIMORE, D. and DARNELL, J.E. J. Mol. Biol. (1967), 24, 59-74.

53. BALTIMORE, D. and FRANKLIN, R.M. Biochem. Biophys. Res. Comm. (1962), 9, 388-392.

54. BALTIMORE, D. and FRANKLIN, R.M. J. Biol. Chem. (1963), 238, 3395-3400.

55. BALTIMORE, D., EGGERS, H.J., FRANKLIN, R.M. and TAMM, I. Proc. Nat. Acad. Sci. U.S.A. (1963), 49, 843-849.

56. BALTIMORE, D. Proc. Nat. Acad. Sci. U.S.A. (1964), 51, 450-456.

57. TRAUB, A., DISKIN, B., ROSENBERG, H. and KALMAN, E. J. Virol. (1976), 18, 375-382.

58. LUNDQUIST, R.E., EHRENFELD, E. and MAIZEL, J.V. Jr., Proc. Nat. Acad. Sci. U.S.A. (1974), 71, 4773-4777.

59. FLANEGAN, J.B. and BALTIMORE, D. Proc. Nat. Acad. Sci. U.S.A. (1977), 3677-3680.

60. THACH, S.S., DOBERTIN, D., LAWRENCE, C., GOLINI, F. and THACH, R.E. Proc. Nat. Acad. Sci. U.S.A. (1974), 71, 2549-2553.

61. LUNDQUIST, R.E. and MAIZEL, J.V. Jr., Virology (1978), 85, 434-444.

62. MEYER, J., LUNDQUIST, R.E. and MAIZEL, J.V. Jr., Virology (1978), 85, 445-455.

63. FLANEGAN, J.B., PETTERSON, R.F., AMBROS, V., HEWLETT, M.J. and BALTIMORE, D. Proc. Nat. Acad. Sci. U.S.A. (1977), 74, 961-965.

64. NOMOTO, A., DETJEN, B., POZZATTI, R. and WIMMER, E. Nature (London)(1977), 268, 208-213.

65. LEE, Y.F., NOMOTO, A., DETJEN, B. and WIMMER, E. Proc. Nat. Acad. Sci. U.S.A. (1977), 74, 89-96.

66. PETTERSON, R.F., FLANEGAN, J.B., ROSE, J.K. and BALTIMORE, D. Nature (London) (1977), 268, 270-272.

67. SANGAR, DV., ROWLANDS, D.J., HARRIS, T.J. and BROWN, F. Nature (London) (1977), 278, 648-650.

68. HRUBY, D.E. and ROBERTS, R.K. J. Virol. (1978), 25, 413-415.

69. PEREZ-BERCOFF, R. and GANDER, M. FEBS Lett. (1978), 96, 306.

70. DARNELL, J.E., GIRARD, M., BALTIMORE, D., SUMMERS, D.F. and MAIZEL, J.V. Jr., The synthesis and translation of poliovirus RNA. In The Molecular Biology of Viruses, Colter, J.S. and Parachych, W. eds., (1967), pp. 375-401. Academic Press, New York and London.

71. EHRENFELD, E., MAIZEL, J.V. and SUMMERS, D.F. Virology (1970), 40, 840-846.

72. ARLINGHAUS, R.B. and POLLATNICK, J. Proc. Nat. Acad. Sci. U.S.A. (1969), 62, 821-828.

73. ARLINGHAUS, R.B., SYREWICZ, J.J. and LOESCH, W.T. Arch. Gesamte Virusforsch. (1972), 38, 17-28.

74. CALIGUIRI, L.A. and TAMM, I. Science (1969), 166, 885-886.

75. CALIGUIRI, L.A. and TAMM, I. Virology (1970), 42, 100-111.

76. CALIGUIRI, L.A. and MOSER, A.G. Virology (1971), 46, 375-386.

77. CALIGUIRI, L.A. Virology (1974), 58, 526-535.

78. MITCHELL, W.R. and TERSHAK, D.R. Virology (1973), 56, 386-389.

79. PEREZ-BERCOFF, R. and GANDER, M. In preparation.

80. KAMEN, R., KONDO, M., ROEMMER, W. and WEISSMANN, C. Eur. J. Biochem. (1972), 31, 44-51.

81. KORANT, B.D. In Proteases in Biological Control, Reich, E., Rifkin, D. and Shaw, E. eds., (1975), pp.621 644. Cold Spring Harbor Lab.

82. LUNDQUIST, R.E. and MAIZEL, J.V. Jr., Virology (1978), 89, 484-493.

83. APRILETTI, J.W. and PENHOET, E.E. J. Biol. Chem. (1978), 253, 603-611.

84. MILLER, H.I. and PENHOET, E.E. Proc. Soc. Exp. Biol. Med. (1972), 140, 435-438.

85. EREMENKO, T., BENEDETTO, A. and VOLPE, P. J. Gen. Virol. (1972), 16, 61-68.

86. DETJEN, B.M., LUCAS, J. and WIMMER, E. J. Virol. (1978), 27, 582-586.

87. RIGHTSEL, W.A., DICE, J.R., McALPINE, R.J., TIMM, E.A., McCLEAN, I.W., DIXON, G.J. and SCHABEL, F.M. Science (1961), 134, 558-559.

88. LODDO, B., FERRARI, W., BROTZU, G. and SPANEDDA, A. Nature (London) (1962), 193, 97-98.

89. LODDO, B., FERRARI, W,, SPANEDDA, A. and BROTZU, G. Experientia (1962), 18, 518-519.

90. TAMM, I. and EGGERS, H.J. Science (1963), 142, 24-33.

91. CORDS, C.E. and HOLLAND, J.J. Proc. Nat. Acad. Sci. U.S.A. (1964), 51, 1080-1083.

92. IKEGAMI, N., EGGERS, H.J. and TAMM, I. Proc. Nat. Acad. Sci. U.S.A. (1964), 52, 1419-1426.

93. BALTIMORE, D. In Medical and Applied Virology. Proceedings of the Second International Symposium, Sanders, M. and

Lenette, E.H. eds., (1968), pp. 340-346, Green St. Louis.

94. NOBLE, J. and LEVINTOW, L. Virology (1970), 40, 634-642.

95. CALIGUIRI, LA. and TAMM, I. Virology (1968), 35, 408-417.

96. BALTIMORE, D., EGGERS, H.J., FRANKLIN, R.M. and TAMM, I. Proc. Nat. Acad. Sci. U.S.A. (1964), 49, 843-849.

97. LODDO, B., GESSA, G.L., SCHIVO, M.L., SPANEDDA, A., BROTZU, G. and FERRARI, W. Virology (1966), 28, 707-712.

98. PHILIPSON, L., BENGTSSON, S. and DIMTER, Z. Virology (1966), 29, 317-329.

99. JACOBSON, M.F. and BALTIMORE, D. J. Mol. Biol. (1968), 33, 369-378.

100. CALIGUIRI, L.A. and TAMM, I. Virology (1968), 36, 223-231.

101. CROWTHER, D. and MELNICK, J.L. Virology (1961), 15, 65-74.

THE RNA-DEPENDENT RNA POLYMERASE (REPLICASE) OF ENCEPHALOMYOCARDITIS VIRUS

H. ROSENBERG, B. DISKIN, E. KALMAR and A. TRAUB

Department of Biochemistry, Israel Institute for Biological Research,
Ness-Ziona, Israel

INTRODUCTION

Infection of animal cells with a picornavirus results in the formation of an RNA dependent RNA polymerase (replicase) which is apparently responsible for the biosynthesis of the viral RNA genome. The discovery of the picornavirus replicase (1-3) and similar enzymatic activities in E. coli infected with RNA bacteriophages (4-6) were made about fifteen years ago. Today, there is detailed knowledge of the structure and properties of the replicases of bacteriophage Qβ (7, 8) and f2 (9), but only partial information on the picornavirus replicase (10-14). This rather slow progress is due mainly to the difficulties encountered in the isolation of a stable RNA dependent replicase from a eukaryotic cell-virus system.

A comparison of the bacterial and animal systems with respect to development of an enzyme isolation method indicates the disadvantage of the latter. There is a much smaller amount of replicase in animal cells infected with a picornavirus than in E. coli infected with bacteriophage Qβ. Thus, about 180-300 units of Qβ replicase are found in 1 g of infected E. coli (15), but only 1-2 units of EMC virus replicase in 1 g of infected BHK cells (13). The RNA replication complex of a picornavirus is bound to smooth cytoplasmic membranes (10, 16). In order to obtain a soluble RNA dependent activity it is necessary to dissociate the enzyme from the membranes by means which adversely affect the activity of the enzyme. It is also more laborious and costly to grow and infect large quantities of animal cells than to carry out a large scale infection of E. coli with an RNA bacteriophage. So far, because of these limitations, the isolation of a picornavirus replicase was carried out with very dilute solutions of enzyme-protein, a condition responsible for rapid loss of activity of the purified enzyme. Flanegan and Baltimore (14) have recently shown that poliovirus-infected HeLa cells contain a poliovirus-specific poly(U) polymerase that copies a poly(A) template complexed to an oligo(U) primer, and have suggested that the picornavirus replicase may in fact be a primer-dependent enzyme. If this proves to be the case, it may explain the failure of previous attempts at enzyme purification as being due also to the lack of a replicase assay system containing an RNA template with a pre-formed primer.

Although a stable preparation of a picornavirus replicase has not yet been obtained, it may be instructive for future work to summarize some of the experience gained so far in attempts to purify and study the properties of the replicase of encephalomyocarditis (EMC) virus. We will therefore describe in some detail our experimental approach to the isolation of an RNA-dependent polymerase from baby hamster kidney cells (BHK 21) infected with EMC virus, and also results of experiments aimed to identify EMC virus-specific proteins in the purified enzyme (13).

I. THE EXPERIMENTAL APPROACH

The cells are grown and infected in suspension culture. To synchronize the infection a dense cell suspension is mixed with the virus inoculum, kept for one hour in an ice bath and then diluted ten-fold with warm medium and incubated for 7.5 h at 37°C. A cell extract is prepared by homogenizing the cells with cold distilled water in a Dounce homogenizer followed by centrifugation at 20,000g. Most of the EMC replicase activity is found in the supernatant which is only now buffered with 0.02M Tris-HCl, pH 8.0, 0.01M KCl, 0.001M $MgCl_2$ and 0.1mM dithoithreitol (DTT). The presence of a trace amount of Mg^{++} ions (0.1mM) during the homogenization step results in extensive association of the replicase with the nuclear and mitochondrial fractions.

Caliguiri and Tamm (16) have shown that in HeLa cells infected with poliovirus the replicase is associated with the smooth cytoplasmic membranes. Since a similar situation is found in the case of the EMC replicase in infected BHK cells, in the next step the smooth cytoplasmic membranes are isolated from the 20,000g supernatant by fractionation in a 20 to 60% discontinuous sucrose gradient. The distribution of the replicase activity in the sucrose gradient layers is shown in Table 1. The greater part of the activity is found in the second layer from the top (fraction 2) at a sucrose concentration of 18%. (When uninfected cells are homogenized and fractionated in the same way, fraction 2 is found to contain the smooth membranes of the endoplasmic reticulum). Fraction 2 contains only a small amount of the total protein, 30% of the phospholipids and about 50-80% of the replicase activity.

Table 1. Distribution of EMC replicase, protein and lipid P in the isopycnic sucrose gradient fractions[a]

Fraction	Sucrose in fraction (%)	Replicase (units)[b]	Protein (mg)	SP ACT (units/mg protein)	Lipid P (μmol)
Cytoplasmic extract		5.200	81.60	0.063	13.86
Fraction 1	10.2	0.120	0.75	0.160	0.69
Fraction 2	18.0	2.640	1.70	1.376	5.10
Fraction 3	26.0	0.450	6.48	0.069	4.14
Fraction 4	28.0	0.165	20.00	0.008	0.83
Fraction 5	32.0	0.172	23.10	0.007	0.96
Fraction 6	40.0	0.185	13.50	0.013	0.82
Fraction 7	46.5	0.160	7.80	0.020	0.86
Fraction 8	51.0	0.158	6.46	0.026	0.045
Pellet	--	0.086	0.65	0.014	--

[a]Prepared from 3×10^9 infected cells.
[b]One unit of activity is the amount of enzyme required for the incorporation of 1 nmol of GMP in 20 min at 37°C.

This step achieves about a 30-fold purification. An electron micrograph of a smooth membrane pellet collected from fraction 2 is shown in Figure 1. It manifests a unique structure made of multi-layers of smooth membranes arranged in closed concentric circles in a "fingerprint"-like pattern.

Infection of HeLa cells with poliovirus elicits extensive proliferation of smooth cytoplasmic membranes (16). This occurs also in BHK cells infected with EMC virus. The results of an experiment in which smooth cytoplasmic membranes were isolated from aliquots of infected cells at various times after infection are shown in Figure 2. The membrane pellets were assayed for activity of EMC replicase and the protein and phospholipid content determined. For the first 4 hours of infection there is no change in the amount of cellular smooth membranes. However, between 4 and 8 hours, at the time when the viral replicase is being formed, there occurs also a steep rise in the formation of smooth membranes. The protein content of the smooth membranes rises by about five-fold and the phospholipids by ten-fold. The diagram in Figure 3 depicts the increase of membrane material during infection and indicates also the changing ratios of protein to phospholipid which takes place in the smooth membranes synthesized between 4 and 8 hours after infection: from 5.0 to 1.7. The low protein content of the new

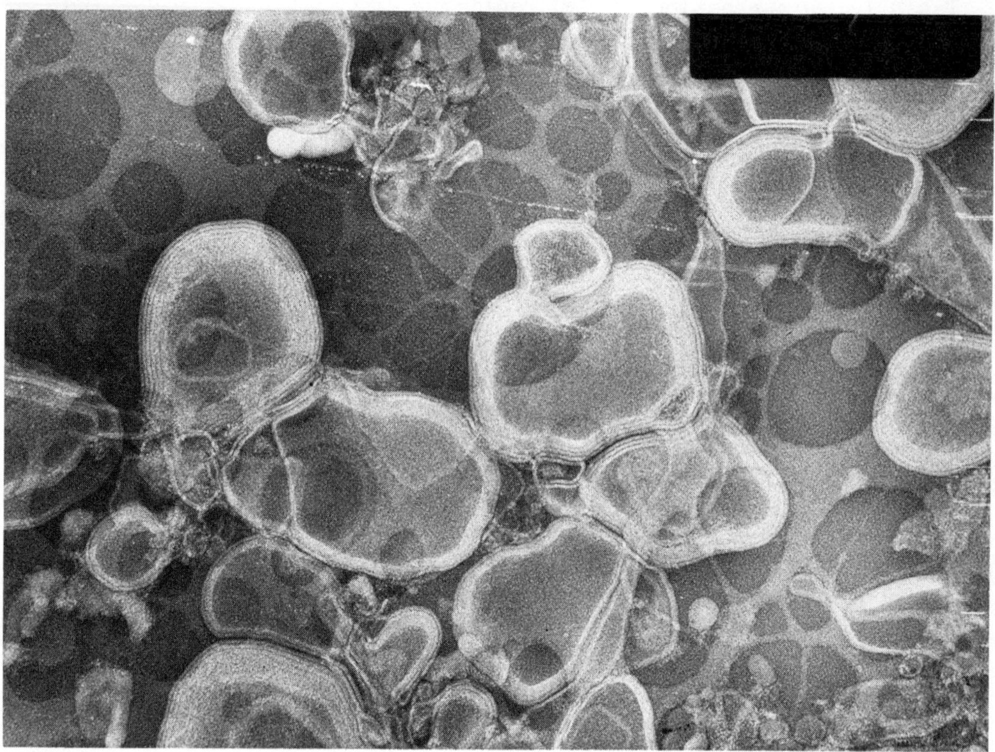

Figure 1. Electron micorgraph of smooth cytoplasmic membranes isolated from BHK cells infected for 7.5 h with EMC virus. A negative stained preparation made with 2% phosphotungstic acid (x 165,000).

membranes suggests that their assembly is carried out by a system which is unbalanced or deficient with respect to protein. This is conceivable in view of the fact that infection with EMC virus causes rapid inhibition of synthesis of cellular proteins. Thus, when the cell is triggered, as a result of virus infection, into rapid proliferation of smooth membranes, the new membranes are assembled from proteins which have existed in the cell prior to infection. The fact that under these unusual circumstances the viral replicase is inserted into the newly formed smooth membranes raises the question of whether other proteins which are also "foreign" to smooth membrane structure are incorporated into the new membranes during EMC virus infection. In this way they may be removed from the cellular pool. One may, therefore, suggest that a better understanding of the pathology of picornavirus infection may benefit from an analysis and identification of the proteins of

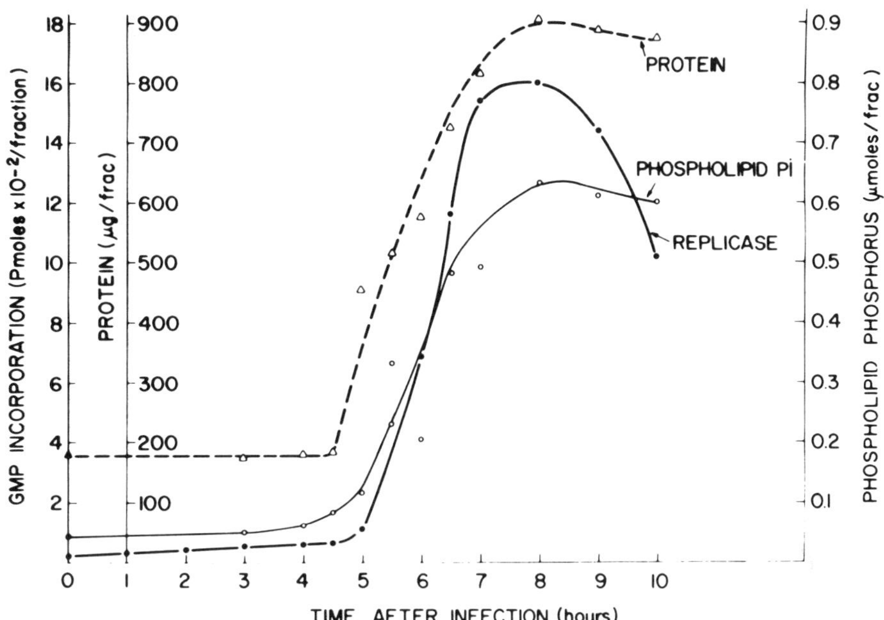

Figure 2. Time course of formation of smooth cytoplasmic membranes and EMC replicase in BHK cells infected with EMC virus. Smooth membranes were isolated from aliquots of 1.5 x 10^9 cells and analyzed for protein, phospholipid and replicase activity.

the smooth membranes that are assembled during virus infection.

The EMC RNA replication complex is bound tightly to the smooth membranes. The association is such that it protects it from the action of micrococcal nuclease and trypsin, suggesting that the replication complex is enclosed inside the membrane. Inactivation of the enzyme by m. nuclease takes place only after removal of most of the membrane lipids from the replication complex. This is done by pre-treating the membranes with 0.05% Triton X-100 and 0.02M DTT, followed by sedimentation through a 5-20% sucrose gradient containing 0.05% Triton X-100 and 0.02M DTT (Figure 4). The pellet obtained in this way contains most of the replicase activity and only a small amount of the phospholipids. The activity in the pellet can now be abolished with m. nuclease, indicating that the template RNA is exposed.

At this stage it was expected that pellets obtained after removal of most of the phospholipids would serve as starting material for the isolation of an RNA-dependent replicase by methods

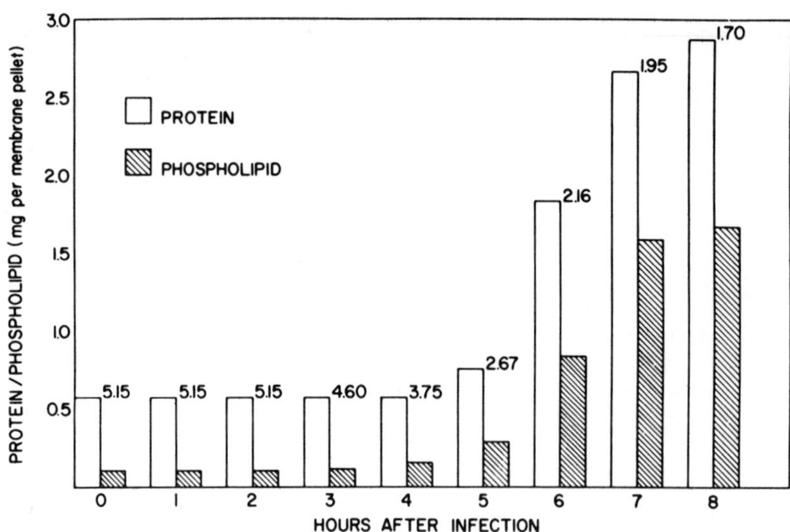

Figure 3. Protein and phospholipid contents of smooth cytoplasmic membranes isolated from aliquots of 1.5×10^9 infected cells during the time course of infection with EMC virus.

used for the isolation of Qβ replicase such as high salt-liquid polymer fractionation, column chromatography, etc. However, experiments with this material resulted with elution of a smear of endogenous activity, indicating the presence of heterogeneous forms of RNA-enzyme complexes. It was later found that in order to obtain an RNA or a poly(C) dependent activity, it is necessary first to remove completely all the lipids from the EMC RNA replication complex. This is done in the following way: Smooth membrane pellets suspended in 0.05% Triton X-100, 0.02M DTT, 0.01M KCl, 0.0015M $MgCl_2$ and 0.01M Tris-HCl, pH 8.0 are brought to 0.15% sodium dodecyl sulfate (SDS) and immediately extracted with 2 volumes of the fluorocarbon Genetron 113 (1.1.2-trichloro-trifluoroethane). After phase separation the aqueous phase is found to contain a lipid-free RNA replication complex which serves as a good source for the preparation of an RNA-dependent replicase. This is done by one of two alternative chromatographic procedures: The first, by passing the aqueous phase through a small column of a strong anion exchanger like Dowex-1 or QAE Sephadex at pH 7.0 (0.01M Tris 7.0-0.01M $MgCl_2$-5% glycerol-TMG buffer). Under these conditions the RNA replication complex is held by the column, probably through the exposed RNA template, while most of the other proteins are removed with the starting buffer. Elution with 0.6M NaCl results in removal of a very small amount of protein which manifests a poly(C) or an RNA-dependent activity (Figure 5). Each fraction must be assayed

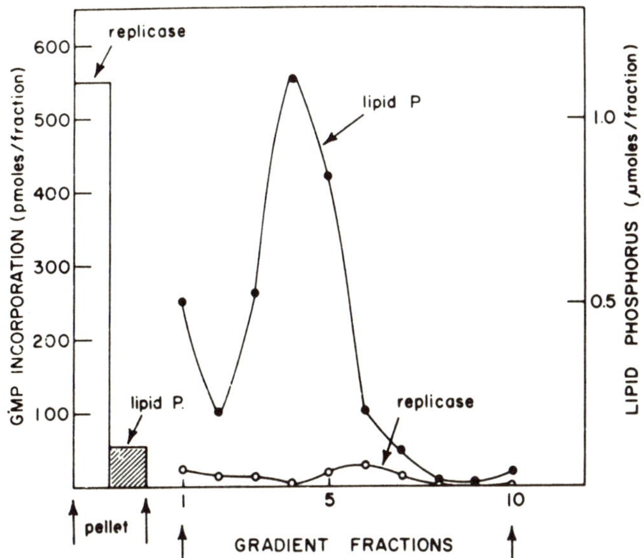

Figure 4. Separation of EMC RNA replication complex from smooth cytoplasmic membranes. Membrane pellets were treated with 0.05% Triton X-100 and 0.02M dithiothreitol and sedimented for 16 h at 50,000 rpm through a 4.5 ml 5-20% sucrose gradient containing 0.05% Triton X-100 and 0.02M dithiothreitol (Spinco rotor type SW 50).

immediately upon collection for there occurs a very rapid loss of activity.

The second approach is to dissociate the enzyme from the RNA by high salt-dextran-polyethylene glycol phase partition of the Genetron extract. However, in this case the high salt polyethylene glycol phase which contains the enzyme cannot be dialyzed for it loses activity. The problem here is how to carry out an ion exchange column chromatography of proteins which are in solution with 3M NaCl. This is done by gradient sievorptive chromatography (17). A column is packed, end to end, with DEAE-Sephadex A-25, and a gradient of 0.4 to 0.01M NaCl is introduced from the lower part of the column, extending to about one third of the column length. The high salt-polyethylene glycol phase which contains the replicase is introduced from the lower part of the column. The sodium and chloride ions diffuse into the pores of the A-25 Sephadex beads while the proteins move over the beads along the 0.4-0.01M NaCl gradient, and are adsorbed at different points along the decreasing slope of the gradient. Elution is followed with a buffer containing 1M NaCl. This method which combines rapid removal of salt together with column chromatography, proved to be effective

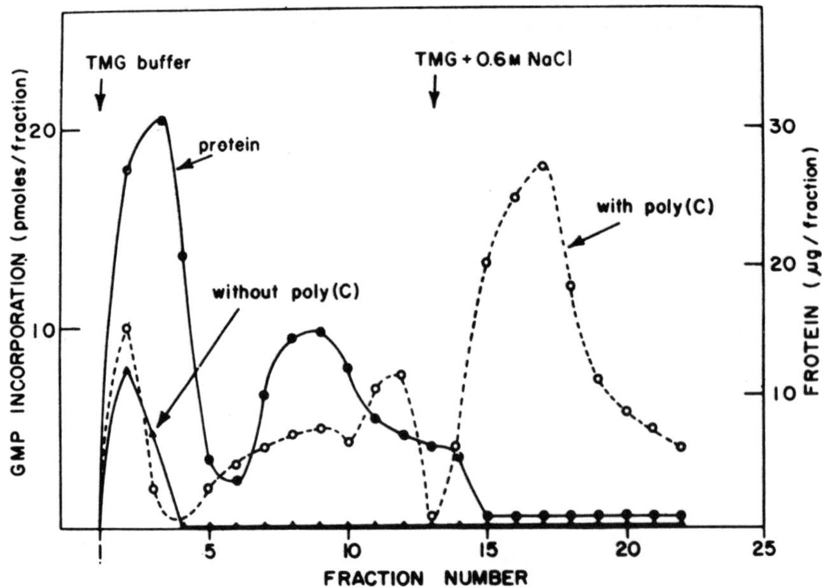

Figure 5. Dowex-1 chromatography of poly(C)-dependent EMC replicase (11). 5 μl samples from each 1 ml fraction were assayed in the presence of 20 μg poly(C).

for the separation of EMC replicase (Figure 6). There is good separation between the bulk of the proteins and the enzyme which elutes between 0.27 and 0.3M NaCl. An advantage of this method is that the mode of enzyme separation is very reproducible, namely, if one uses the same column and same gradient conditions the activity elutes at the same fractions. This is important for work with a very labile enzyme which must be assayed upon collection.

The replicase prepared by either of the two procedures can be further purified by sedimentation through a 10-30% glycerol gradient. As shown in Figure 7 the peak of activity has a sedimentation coefficient of about 6S. A summary of the purification procedure is presented in Table 2.

So far, because of the minute amount of protein and the extreme lability of the enzyme we have not been able to carry the purification any further. Although there is not any evidence that the 6S peak represents a single homogeneous protein, we have analyzed the 6S peak (after precipitation with TCA) by SDS polyacrylamide gel electrophoresis. There are protein bands of approximate molecular weight of 75,000, 65,000, 56,000, 45,000 and

Table 2. Purification of EMC replicase.

Step	Protein (mg)	Activity (units)[b]	Sp act (units/mg protein)	Yield (%)	Purification (fold)
Cytoplasmic extract	384	18	0.046	100	1
Fraction 2	10.2	14.60	1.43	81	31
Membrane pellet	8.6	12.40	1.67	69	36
Lipid-free membrane extract	4.2	6.80	1.62	38	35
Sievorptive column eluate	0.14	2.36	17.0	13	370
Glycerol gradient fraction	0.016	0.59	36.8	3	800

[a] From 12 g of infected BHK cells

[b] One unit of activity is the amount of enzyme required for the incorporation of 1 nmol of GMP in 20 min at 37°C.

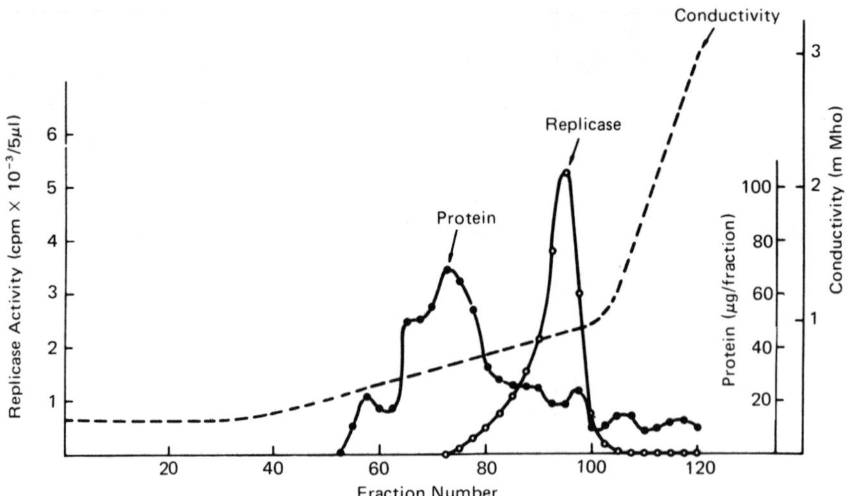

Figure 6. Purification of EMC replicase by gradient sievorptive chromatography (13). 5 µl samples from each 1 ml fraction were assayed for activity in the presence of 2 µg EMC RNA.

35,000 (Figure 8). Since different preparations manifested different proportions of the protein bands, it is impossible at present to determine which of the bands belong to the replicase and which is a contaminant. However, there is suggestive evidence that the 56,000 and 75,000 dalton bands represent EMC virus-specific proteins. The evidence comes from the work of Lundquist, Ehrenfeld and Maizel on the polymerase of poliovirus (12) and from our own work (13) with a purified EMC replicase labeled with (^{35}S) methionine. This will be discussed later.

II. PROPERTIES OF THE ENZYME

We would now like to describe some of the properties of the enzyme and the requirements of the reaction. We want first to emphasize that the data are of a preliminary nature. Because of the extreme lability of the enzyme and the rapid loss of activity immediately after collection of the fractions, it is impossible at present to carry out quantitative experiments or study the kinetics of the reaction. The data given is therefore of a qualitative nature and not conclusive.

Table 3 summarizes the reaction conditions necessary for the incorporation of GMP into an acid-insoluble product by 1 µg of

Table 3. Properties of EMC replicase

Reaction system[a]	GMP (pmol/1 μg of protein)
Complete	12.8
Omit Mg^{2+}	1.8
Omit Mg^{2+} and Mn^{2+}	3.2
Omit EMC RNA	0.32
Omit 3 NTP[b]	0.36
Add α-amanitin (5 μg)	13.2
Add rifaminin (10 μg)	12.6
Add RNAse (5 μg)	0.28
Add DNAse (5 μg)	13.0

[a] The complete reaction mixture (70 μl) contained: Tris-hydrochloride (pH 8), 50mM; $MgCl_2$, 4mM; KCl, 100mM; ATP, CTP, UTP, 2mM each; (^{3}H)-GTP, 0.06nmol (15 Ci/mmol); actinomycin D, 5 μg; EMC RNA, 2 μg; enzyme, 0.1 μg of protein.

[b] NTP, Nucleoside triphosphates

Table 4. Activity of EMC replicase in presence of different polyribonucleotide templates[a]

Template	Radioactive XTP	XMP (pmol/1 μg of protein)
EMC RNA (2 μg)	(^{3}H)GTP	11.5
Qβ RNA (2 μg)	(^{3}H)GTP	10.8
BHK rRNA (2 μg)	(^{3}H)GTP	12.3
Poly(C) (20 μg)	(^{3}H)GTP	9.5
Poly(A) (20 μg)	(^{3}H)UTP	0.32
Poly(U) (20 μg)	(^{3}H)GTP	0.28

[a]Reaction conditions are as described in the legend to Table 3 except that the ribohomopolymer templates were assayed without the three other ribonucleoside triphosphates.

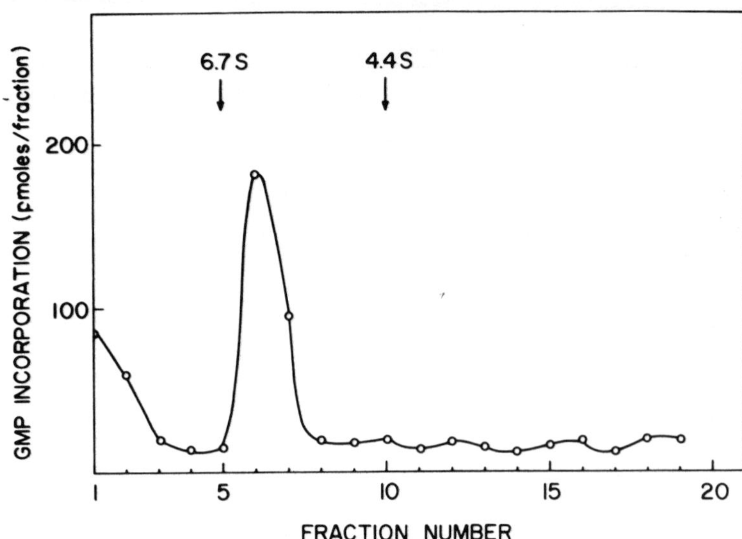

Figure 7. Glycerol gradient centrifugation of EMC replicase. 3 ml of a pooled active fraction from the sievorptive chromatography step were loaded onto a 10-30% glycerol gradient and centrifuged for 18 h at 27,000 rpm in Spinco rotor type SW27. 10 μl of each 1 ml fraction were assayed in the presence of 2 μg EMC RNA. (Markers: 6.7S alcohol dehydrogenase; 4.4S bovine serum albumin).

enzyme-protein. There is strict dependency on Mg^{++}, the four nucleoside triphosphates and template RNA. Mn^{++} cannot effectively replace Mg^{++}. The incorporation is not affected by actinomycin D, α-amanitin, rifampicin or DNAse. It is sensitive to RNAse. The pH optimum is around 8.0, The response of the replicase to different RNA templates is shown in Table 4. Qβ RNA and BHK ribosomal RNA were found to be as effective as EMC RNA. There is synthesis of poly(G)in the presence of poly(C), but poly(A) and poly(U) do not act as templates. The results suggest that the isolated enzyme is probably a core-replicase which has lost a protein factor responsible for the recognition and specific initiation of EMC RNA.

III. ARE PEPTIDES M.W: 56,000 or 75,000 THE VIRAL REPLICASE?

We would like now to consider the evidence for the suggestion made before that the 56,000 and 75,000 dalton proteins found in the purified preparation of the replicase may in fact be an EMC

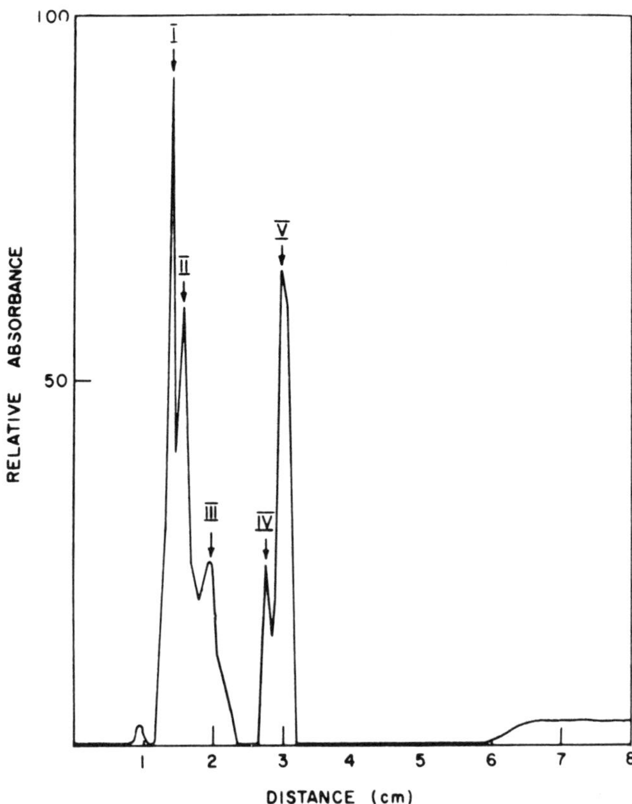

Figure 8. SDS polyacrylamide gel electrophoresis of EMC replicase in 10% gels run according to Weber and Osborn (21). I, 75,000; II, 65,000; III, 56,000; IV, 45,000; V, 35,000 daltons.

virus-specific protein. The studies of Butterworth et al. (18) on the biosynthesis of EMC virus proteins in infected HeLa cells have indicated that three primary gene products, A,F,C, with cumulative molecular weights of about 220,000 are generated during translation of the EMC RNA genome (Figure 9. For a detailed discussion see chapter 7 of this volume). A and C then undergo post-translational cleavages while F remains uncleaved. The proteins produced by the cleavage of A include all the four capsid proteins, α, β, γ, δ, and also a stable 12,000 dalton polypeptide, H. C is cleaved so as to give rise to stable proteins I, E and G. Thus, a total of 15 EMC proteins are found in the infected cell. Six of these are unstable precursor proteins, and nine are stable end-products.

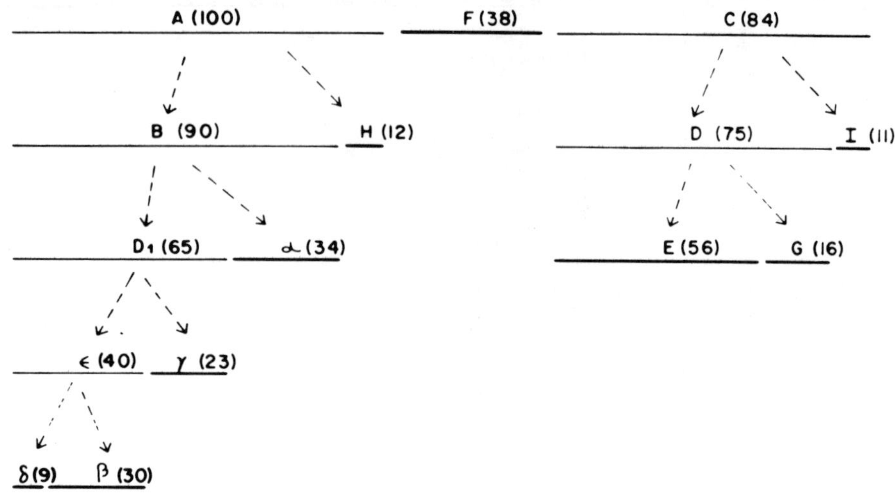

Figure 9. Model for biosynthesis of EMC virus proteins according to Butterworth *et al.* (18).

For the identification of the EMC proteins in HeLa cells, Butterworth *et al.* used pulse-chase experiments done at a time of infection when host protein synthesis is blocked, and only the viral proteins are being formed. The stable EMC proteins were identified after 1 h chase with cold amino acids. In an attempt to obtain a purified EMC replicase containing an assumed EMC stable protein we carried out a similar experiment. A suspension of 1.5×10^9 infected BHK cells was maintained for 4 h in a methionine-deficient medium, containing actinomycin D (10 μg/ml). At 4 h, 2 mCi of (^{35}S) methionine were added, and the incubation continued to 6.5 h. The cells were collected and resuspended in a medium containing cold methionine. After a 1 h chase the cells were harvested and from them we prepared a whole cell lysate, isolated the smooth cytoplasmic membranes and from these purified the replicase by QAE Sephadex chromatography and glycerol gradient sedimentation. Samples of radioactive cell lysate, smooth membrane and purified enzyme were analyzed by SDS polyacrylamide gel electrophoresis. A profile of the radioactive proteins in a cell lysate is shown in Figure 10A. It contains three minor peaks of molecular weight 95,000, 75,000 and 65,000 which correspond to EMC virus unstable proteins B, D and D1; three of the four coat proteins: α (34,000), β (30,000) and γ (26,000) and relatively large amounts of stable EMC virus proteins E (56,000) and F (38,000). A profile of the EMC proteins found in the isolated smooth membranes is shown

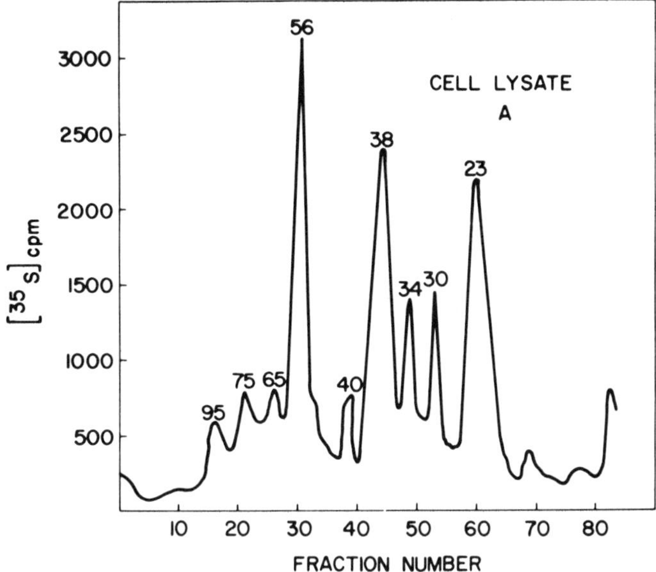

A

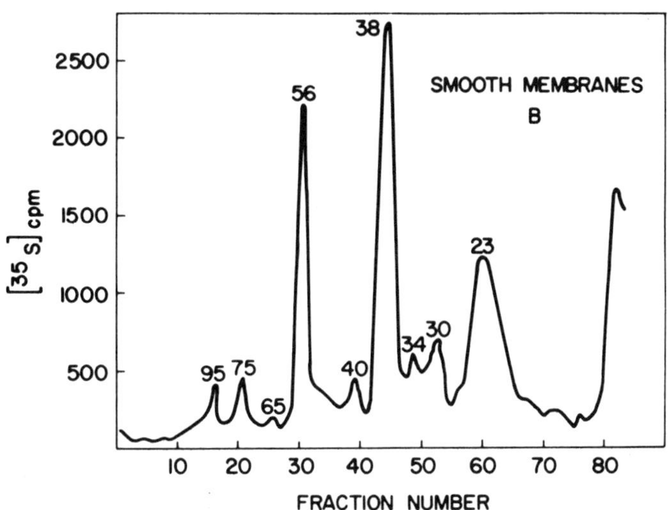

B

Figure 10. SDS polyacrylamide gel electrophoresis of (^{35}S) methionine labeled EMC virus-specific proteins. (A) Cell lysate; (B) Smooth cytoplasmic membranes.

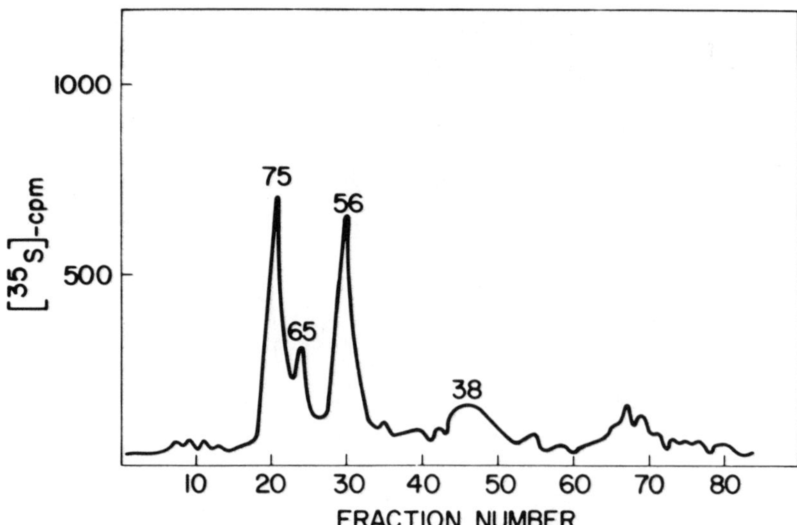

Figure 11. SDS polyacrylamide gel electrophoresis of (^{35}S) methionine labeled preparation of purified EMC replicase (A 6S glycerol gradient enzyme peak).

in Figure 10B. Qualitatively it is similar to the one found in whole cell lysate, but it is relatively more enriched with stable proteins E (56,000) and F (38,000) than with capsid proteins α, β, γ. The pattern of EMC virus proteins found in the 6S glycerol gradient peak of the purified replicase is shown in Figure 11. There are two major bands of molecular weight 75,000 (D) and 56,000 (E) and two minor ones of 65,000 and 38,000. In comparison to the viral proteins found in the smooth membranes, the purified enzyme is enriched with unstable polypeptide D, suggesting that one of the subunits of EMC replicase may be an unstable viral protein. This would be in accord with other findings which have indicated the unstable nature of the picornavirus replicase (19, 20). The case for polypeptide E is supported by previous work of Lundquist, Ehrenfeld and Maizel (12) on the polymerase of poliovirus. They have isolated from infected HeLa cells in which only the poliovirus proteins were labeled with (^{35}S) methionine a 25S ribonucleoprotein particle that carried the activity of the viral polymerase. SDS gel electrophoresis revealed that of all the poliovirus proteins the particle contained predominantly protein 4 which is analogous to EMC protein E. However, an unequivocal proof for a subunit role for polypeptides D and E will depend on further progress in the purification of the enzyme to a stage of a homogeneous protein.

IV. CONCLUSIONS

The work carried out so far on the EMC virus replicase provides a method for the isolation of minute quantities of an unstable RNA dependent replicase which allowed one to conduct a preliminary study of some of the properties of the enzyme. However, information on the ezyme's subunit composition, its mode of action, and on possible additional factors that may determine its specificity toward an EMC virus RNA template will require a stable enzyme preparation at a higher level of purity. There is also a growing feeling that the elucidation of the mechanism of biosynthesis of the picornaviral RNA will probably depend on an understanding of the possible functional relationships between the viral RNA replication complex and the smooth membranes in which it is enclosed.

ACKNOWLEDGEMENT

The research described in this chapter was supported by a grant from the United States-Israel Binational Science Foundation (BSF), Jerusalem, Israel.

REFERENCES

1. BALTIMORE, D. and FRANKLIN, R.M. Biochem. Biophys. Res. Commun. (1962), 9, 388-392.

2. CLINE, M.J., EASON, R. and SMELLIE, R.M. J. Biol. Chem. (1963), 238, 1788-1792.

3. DALGARNO, L. and MARTIN, E.M. Virology (1965), 26, 450-456.

4. WEISSMAN, C., SIMON, L. and OCHOA, S. Proc. Natl. Acad. Sci. U.S.A. (1963), 49, 407-414.

5. AUGUST, J.T., COOPER, S., SHAPIRO, L. and ZINDER, N.D. Cold Spring Harbor Symp. Quant. Biol. (1963), 28, 95-97.

6. HARUNA, I., NOZUK, K., OHTAKA, Y. and SPIEGELMAN, S. Proc. Natl. Acad. Sci. U.S.A. (1963), 50, 905-911.

7. KAMEN, R. Nature (London) (1970), 228, 527-533.

8. KONDON, M., GALLERANI, R. and WEISSMAN, C. Nature (London) (1970), 228, 525-527.

9. FEODOROFF, N.V. and ZINDER, N.D. Proc. Natl. Acad. Sci. U.S.A.

(1971), 68, 1838-1843.

10. EHRENFELD, E., MAIZEL, J.V. and SUMMERS, D.F. Virology (1970), 40, 840-846.

11. ROSENBERG, H., DISKIN, B., ORON, L. and TRAUB, A. Proc. Natl. Acad. Sci. U.S.A. (1972), 69, 3815-3819.

12. LUNDQUIST, R.E., EHRENFELD, E. and MAIZEL, J.V. Proc. Natl. Acad. Sci. U.S.A. (1974), 71, 4773-4777.

13. TRAUB, A., DISKIN, B., ROSENBERG, H. and KALMAR, E. J. Virol. (1976), 18, 375-382.

14. FLANEGAN, J.B. and BALTIMORE, D. Proc. Natl. Acad. Sci. U.S.A. (1977), 74, 3677-3680.

15. KAMEN, R. Biochem. Biophys. Acta. (1972), 262, 88-100.

16. CALIGUIRI, L.A. and TAMM, I. Virology (1970), 42, 100-111.

17. KIRKEGAARD, L.H. Biochemistry (1973), 12, 3627-3632.

18. BUTTERWORTH, B.D., HALL, L., STOLTZFUS, C.M. and RUECKERT, R.R. Proc. Natl. Acad. Sci. U.S.A. (1971), 68, 3083-3087.

19. SCHARFF, M.D., THOREN, M.M., McELVAIN, N.F. and LEVINTOW, L. Biochem. Biophys. Res. Commun. (1963), 10, 127-132.

20. BALTIMORE, D. The replication of picornaviruses. In The Biochemistry of Viruses. Levy, H.B. ed., (1969), pp. 101-176. Marcel Dekker,New York.

21. WEBER, K. and OSBORN, M. J. Biol. Chem. (1969), 244, 4406-4412.

SECTION VII:

VIRUS-DIRECTED SYNTHESIS UNDER NON-PERMISSIVE CONDITIONS

HOST-RESTRICTION OF PICORNAVIRUS INFECTION

MILTON W. TAYLOR and V. GREGORY CHINCHAR

Department of Biology, Indiana University

Bloomington, Indiana 47401, U.S.A.

INTRODUCTION

There is no doubt that a host organism has many lines of defense against virus infection. These include the immune system, both cellular and humoral, and interferon production. Perhaps not so easily recognized are mechanisms for preventing virus infection that occur at the cellular level, both at the surface of the cell, in which specific receptors might be lacking, (viral resistance) or internal barriers to virus replication, due to the genetic characteristic of the particular cell (viral restriction). The purpose of this presentation is to review virus restriction at the cellular level and in particular, to look at restriction of picornaviruses, with specific emphasis on mengovirus restriction.

Host mediated virus restriction has been studied in bacteriophage (1), in oncornaviruses (murine leukemia viruses) (2), adenoviruses (3), avian reovirus (4), and Sendai (5). The mechanisms involved appear to be quite different for each system. None of these studies involves inhibition of virus replication by interferon, a topic discussed elsewhere in this volume.

I. RESTRICTION AT SITE OF VIRUS ATTACHMENT (RESISTANCE)

A major obstacle to a productive virus infection is the cell surface. Only certain cell types have the appropriate lipoproteins on their periphery which can act as receptor material for a particular picornavirus (6). This receptor material serves as an adsorption site for the virus, the initial virus interaction with the host, and the first step in the infective process. The cell types that can be infected by a picornavirus, the host range of that virus, reflect the viral tissue specificities observed in an animal. Poliovirus, for example, is known to replicate in the human brain, spinal cord, and intestine. Examination of homogenates of these susceptible tissues reveals the presence of poliovirus receptor material; while insusceptible tissue homogenates, heart, lung, and kidney, completely lack the ability to adsorb this virus (6).

McLaren and co-workers (7) demonstrated that the resistance of non-primate cells to poliovirus infection resulted from their failure to irreversibly adsorb input poliovirus. In susceptible primate cells (e.g. HeLa cells) more than 90% of the input polio inoculum was adsorbed in 2 hours at 37°C. By contrast, non-primate cells such as mouse L-cells, or primaries of rabbit, dog, and guinea pig, adsorbed less than 10% of the poliovirus inoculum. Furthermore, only about 1% of this adsorbed virus was irreversibly bound to these host cells.

Examination of the host range of coxsackievirus A9 (8) indicated that those cells susceptible to poliovirus type I infection did not adsorb this strain of coxsackievirus to any appreciable extent. Coxsackievirus resistant cell lines including HeLa cells and continuous lines of human amnion or kidney, failed to release coxsackievirus inactivating material (viral receptors) when repeatedly frozen and thawed. The differences in the host range of poliovirus and coxsackievirus A9 indicate that the cellular receptors for different enteroviruses are distinct and separate entities.

Penetration and true eclipse, as reflected experimentally by the progressive loss of virus susceptibility to specific antiserum, was not observed in isolated clones of HeLa cells resistant to poliovirus (9). Poliovirus adsorbed to resistant HeLa cells failed to penetrate, remaining bound to the cell surface. Thus unlike poliovirus-resistant non-primate cells which failed to appreciably adsorb input virus, these resistant HeLa cells retained the capacity for poliovirus adsorption but were unable to irreversibly eclipse adsorbed virus.

The most elegant demonstration of the role of the specific receptor in limiting enterovirus infection is the report (10) that phenotypically-mixed virions resulting from the dual infection of poliovirus type I and coxsackievirus B1, when subsequently added to cells resistant to poliovirus (by virtue of the absence of specific receptors) under conditions limiting specifically the replication of coxsackievirus RNA, yielded only poliovirus type I virions. Since only poliovirus RNA coated in coxsackievirus capsid material would be capable of initiating infection and productively multiplying, these results dramatically illustrate that the specific interaction of virus capsid proteins and cellular receptor sites constitutes a primary barrier to enterovirus infection.

Perhaps one of the more fascinating studies of virus resistance, and changes in cell surface that affects viral adsorption, is the observation (11) that viral transformed cells are much more sensitive to bovine enterovirus-1 infection than the parental non-transformed cell (Table 1). Thus tumor-virus (or spontaneous) transformation results in modifications of the cell surface that allows for virus adsorption. Further analysis demonstrates that the resistance to bovine enterovirus-1 of 3T3 and primary kidney is due to a lack of receptor sites.

Table 1. Response of various mouse cells to BEV-1 infection

Cell Type	Characteristic	C.P.E.
3T3	Continuous cell line	-
3T3 Py3	Transformed cell line	+
3T3 Py6	Transformed cell line	+
Kidney	Primary	-
L-cell	Continuous cell line	+
L1210	Leukemic	+
Sarcoma 1	Tumor cell	+

II. INTRACELLULAR RESTRICTION

A. EMC, ECHO6 and GDVII Restriction

EMC virus replication is restricted in monkey cells (12) and a subline of HeLa cells (13). EMC restriction in monkey cells is characterized by a reduction in both viral yield and viral RNA synthesis. Dubois and Chany (12) studied viral growth in hybrids formed between permissive mouse and restrictive monkey cells and correlated a reduction in permissiveness with the retention of monkey chromosomes. Their data suggests that restriction may be a dominant trait. The inhibition of EMC virus growth in a subline of HeLa cells can be overcome by either an increase in the multiplicity of infection or by coinfection with poliovirus (13). The helper function of poliovirus occurs even if poliovirus is unable to replicate. This result suggests that an increase in the amount of a (homologous or heterologous) viral component can relieve restriction by removing an inhibitory HeLa cell factor.

The enterovirus, echovirus 6, shows a unique type of restriction. Echovirus 6 exists as two plaque size variants - a small variant, m, and a large variant m^+ (14, 15). While the small plaque variant forms plaques on both human and monkey cell lines with equal efficiency, the large variant is restricted in its ability to plaque on monkey cells (16). Restriction of m^+ on monkey cells is not associated with a reduction in viral yield since m^+ produces a normal viral yield in restrictive cells. Recently an "enhancer" has been prepared from permissive cells which allows m^+ to plaque on restrictive cell lines (17, 18). Enhancer is effective whether it is used to pre-treat the virus or the cells. Neither the mechanism of restriction, nor the mode of action of enhancer are known.

GDVII, a mouse encephalomyocarditis virus, is restricted in HeLa cells. Both viral yield and viral RNA synthesis are reduced in infected restricted cells. However, the restriction in viral yield is more than ten-fold greater than the restriction of viral RNA synthesis. This observation that the viral yield is more severely depressed than the yield of viral RNA was also seen in the mengovirus-MDBK system (20). A non-restrictive variant of GDVII was obtained following alternate viral passage on restrictive and permissive hosts. The authors were unable to identify a mechanism of restriction (19).

B. Mengovirus Restriction in MDBK Cells

MDBK (Madin-Darby Bovine Kidney) cells are an established line of male bovine cells. Although restrictive for mengovirus replication, they have been used as a productive host for the WSN strain of influenza virus (21) and for vesicular stomatitis virus (22).

Mengovirus replicates efficiently in HeLa (human), L (mouse) and BHK (syrian hamster) cells but replicates poorly in rabbit embryo fibroblasts, PtK-1 cells (kangaroo rat), and MDBK cells (23). Restriction of viral growth is the most severe in MDBK cells with viral yields of only 10-20 PFU/cell seen. Yields in permissive L cells range from 500-1000 PFU/cell.

That viral penetration and uncoating are normal in the restricted system is indicated by: 1) The irreversible eclipse of the virus, 2) the cessation of host protein synthesis, and 3) the simultaneous initiation of viral RNA synthesis in both permissive and restrictive hosts (24, 25). Indeed, no major differences occur (are detectable?) in the infected restrictive cell compared to the permissive host until four hours post-infection, at which time there appears to be a cessation of viral RNA accumulation. Initial data (25) supported the hypothesis that this was due to RNA degradation superimposed upon ongoing RNA synthesis. However, others in this laboratory have been unable to confirm this.

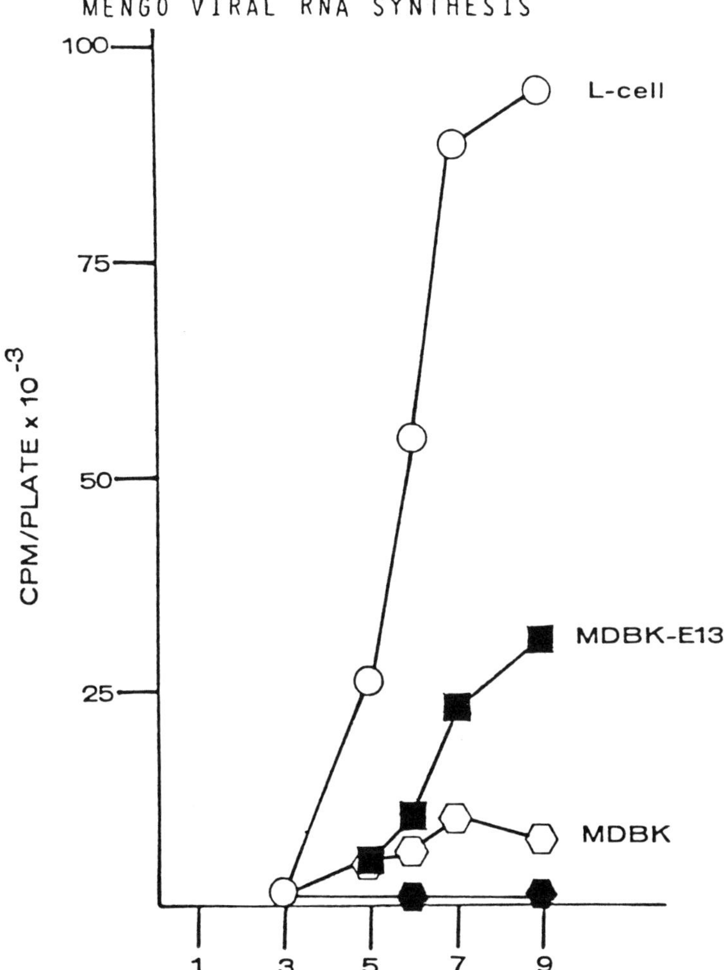

Figure 1. Hours post-infection

Attempts to demonstrate a premature release of lysosomal enzymes that could account for degradation of viral single-stranded RNA were negative. Lysosomal enzyme release, as measured by release of β- glucuronidase activity, accompanied the onset of cytopathic effects of 7-8 hours after infection. The restriction process was not affected by treating cells with hydrocortisone, which has been reported to stabilize lysosomes.

We have analyzed in more detail those events occurring in the restricted MDBK cell using a plaque purified (medium) virus.

In the course of selecting mutants of MDBK for hybridization purposes, we have isolated an "intermediate" strain of MDBK (-E13), that produces about 100 infective virus particles per cell. The time of onset of viral RNA accumulation in MDBK and L-cells is essentially the same, however, it is clearly seen that a greater than 10-fold difference exists in the capacity to accumulate viral RNA between L-cells and MDBK (Figure 1). It should be noted that in E13, a proportional amount of RNA accumulates. In the restrictive cell lines the levels of RNA synthesis is still three times the background level. It should be noted that there are differences between the initial rates of RNA synthesis, depending upon whether plaque purified stocks are used or not.

Examination of the rate of viral RNA synthesis also shows a marked difference between permissive and restrictive cell line. L-cells initially show an exponential rate of viral RNA synthesis up to six hours post-infection. However, in MDBK,in agreement with accumulation data, there is very little viral specific synthesis. There was no increase in viral RNA synthesis on increasing the multiplicity of infection up to 100 PFU/cell.

An analysis of the classes of RNA made early on infection of MDBK show that all three expected classes, RF, RI, and SS RNA are synthesized (24, 25).

Having shown that restriction in virus yield is accompanied *in vivo* by a restriction in viral RNA synthesis and accumulation, we ascertained whether this condition exists *in vitro*.

Initial experiments were concerned with demonstrating the existence of an RNA dependent, RNA polymerase activity in virus infected cells. Accordingly mengovirus infected cells were fractionated and fractions assayed for replicase activity (Figure 2). Optimum conditions were worked out for replicase activity, including enzyme concentration, nucleotide concentrations, pH optimum, Mg^{++} concentration, etc. Utilizing these optimum conditions, the levels of replicase activity in mengo-infected L and MDBK was measured at various times. The results indicate that the peak of replicase activity in mengo-infected L-cells occurs at 6 hours post-infection, and is about eight-fold that seen in mengo infected MDBK cell (Figure 2).

Finally, mixing experiments were performed to examine whether anything inhibitory exists in the MDBK extract or stimulatory in the L-cell extracts. These experiments were done with both membrane-bound and soluble replicase. Results from these experiments indicated that there was no inhibitory factor in the MDBK extracts, nor stimulatory factors in the L-cell extracts.

All of this data points to an inability of mengo RNA to

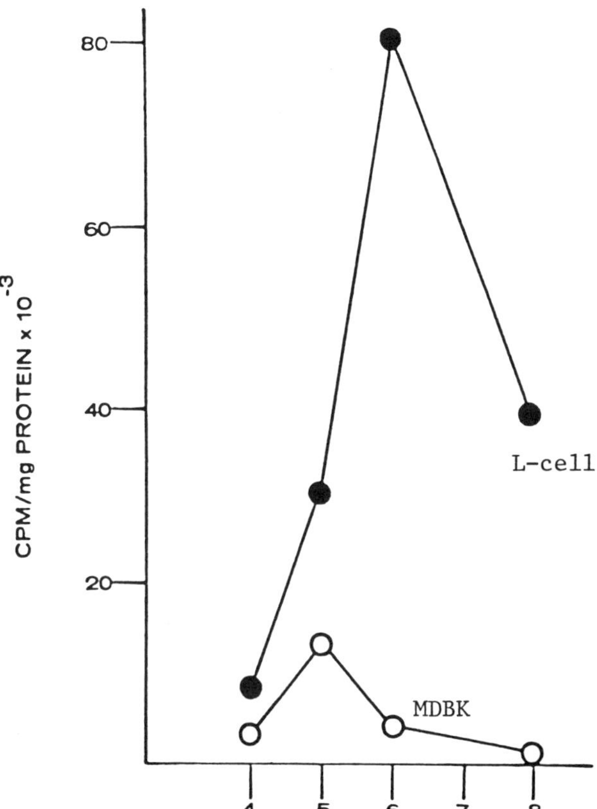

Figure 2. Hours post-infection

replicate in MDBK cells. Either there is not enough template available (RNA degradation?), or insufficient replicase is present to replicate the template. Our experiments do not differentiate between these alternatives. Another possibility is that there is a factor missing in the restrictive host that is needed for viral replication. The mixing experiments only prove that this stimulatory or inhibitory factor is not in the cell sap.

Some viral protein synthesis (translation) must be occurring in the restrictive cells early in infection, since the viral replicase is required to initiate viral RNA replication and we can detect some replicase activity. Since some viral RNA synthesis does occur, cleavage of the viral polypeptide must have occurred to have viral replicase synthesized. To date we have been unable to differentiate between early viral protein synthesis and host protein synthesis, since complete inhibition of host protein synthesis does not occur until 2-3 hours post-infection. All attempts to demonstrate mengoviral specific protein synthesis late in MDBK cell infection by measuring the rate of amino acid incorporation into viral protein and identifying viral proteins on gels, has been unsuccessful. Similar negative results have been reported by others for mengo-MDBK cells (26). In mengo-infected L-cells there is a detectable burst of viral protein synthesis between 4-6 hours post-infection. No such burst of viral protein synthesis is observed in MDBK cells.

In summary, our biochemical evidence would indicate that the early steps of infection are efficient (adsorption, penetration etc.), in the restricted cell. However, viral RNA is poorly used as a template for replication and/or is not capable of being translated efficiently into viral proteins. Since replication and translation are obviously linked in this system it is impossible to tell from our data primary and secondary causes.

As indicated above, these data do not tell us whether "a function" is lacking in MDBK which is essential for viral protein or RNA synthesis, or whether a macromolecule is synthesized which prevents the translation of viral proteins or viral RNA synthesis. Since MDBK can be infected productively by other viruses (Bovine enterovirus, influenza), it must be a molecule specific for the mengo-MDBK cell interaction. The greatest enigma in our data is the origin of the early RNA replicase, since very little (or no) protein synthesis occurs.

In order to determine whether restriction is due to a lack of host factor as opposed to inhibition by a host factor of viral functions, it was decided to hybridize MDBK with L-cells. We have introduced an $HGPRT^-$ mutation into the MDBK cells, and utilized a TK^- strain of L-cells. Hybrids were selected in HAT medium (hypoxanthine, aminopterin, thymidine), clones were selected, grown up, and tested for virus yield. Table 2 summarizes the data obtained.

There are obviously two classes of hybrids, those that show normal virus yields, and those that show the restriction phenomenon. These data can obviously be interpreted in two ways, either permissiveness is dominant, and the presence of mouse (L-cell) chromosomes allows for the permissive expression, or restriction is dominant, and the presence of a particular MDBK chromosomes leads to the restrictive phenomenon.

It is well documented that in human x mouse hybrid cells that there is a preferential loss of human chromosomes (27). We, therefore, examined the isozyme pattern and chromosome complement

Table 2. Mengovirus yield in hybrid cell lines (PFU/cell)

	PFU/cell	Pm^+/Pm^-
Permissive parent ($LMTK^-$)	590	
Restrictive parent (MDBK)	27	
Hybrid clones		
44-1	412	Pm^+
43-3	1066	Pm^+
44-8	1040	Pm^+
44-10	100	?
418-2	2430	Pm^+
418-7	925	Pm^+
615-1	33	Pm^-
620-1	23	Pm^-
620-26	369	Pm^+
620-3	429	Pm^+

of each hybrid cell (28). It was clear from these studies that there was a rapid loss of MDBK chromosomes. If MDBK chromosomes are preferentially lost, and the majority of hybrids are permissive, one could argue for dominance of permissiveness. However, the isolation of some restricted hybrids under such circumstances argues for the alternative suggestion, that restriction is dominant. This is in agreement with similar studies done with EMC virus (12).

Rather than attempting to find a host component missing in MDBK that can be provided by L-cells for mengovirus growth, we must assume that MDBK alters the viral genome in such a fashion that it is non-functional (or almost non-functional) in viral protein synthesis or that MDBK produces a"repressor" that prevents both replication and expression of some parts of the viral genome. This is also supported by the isolation of E13, a less restrictive mutant of MDBK.

Similar results to the above have been found in other systems. Hybrids between SV-40 permissive and non-permissive cells (29, 30) were resistant to challenge by SV-40. However, the reason for this is probably totally different from the one involving mengovirus in MDBK, since the presence of an integration region on the chromosome (31) allowed for expression of late viral function.

Hybrids between human and mouse cell have also been used in mapping the polio receptor site (32). Hamster-mouse hybrids, and human-mouse hybrids are susceptible to polyoma virus (33) as long as the parental mouse chromosomes (permissive cell line) were present as majority. A decrease in viral susceptibility occurred with an imbalance of mouse-hamster chromosomes. Cells become completely resistant to polyoma when there is a considerable reduction in mouse chromosomes.

III. POSSIBLE SITES OF VIRUS RESTRICTION

Crucial to the understanding of viral restriction is the realization that mengovirus protein and RNA synthesis are interconnected processes. Viral protein synthesis provides the enzymes (the polymerase molecules) needed for viral transcription, while viral RNA synthesis supplies the templates on which viral protein synthesis proceeds. A third activity, viral protein processing, cleaves the product of viral translation to generate the structural (capsid) and functional (polymerase) proteins needed for viral replication. The intimate relationship between viral protein and viral RNA synthesis demands that an early inhibition of any one of these essential functions - translation, transcription, or viral protein processing - may affect both viral RNA and protein synthesis.

Because restriction begins soon after infection, both viral RNA and protein synthesis are markedly reduced. The resulting low level of viral macromolecular synthesis delays the inhibition of host functions, impairs virion formation, and prevents physical particle accumulation. Although it is not known which of the three major picornavirus synthetic activities - translation, transcription, or protein processing - is the primary site of restriction, the hypothesis is that at least one of these is impaired.

Despite the restrictive (abortive) infection described above, macromolecular events that lead to inhibition of host protein synthesis, and ultimately cell death still occur. This would suggest that one might need these specific macromolecules only in catalytic amounts in order to disrupt the cells. Since viral double-stranded RNA is synthesized in such cells, it becomes an ideal candidate as an effector of inhibition of host protein synthesis (see Lucas-Lenard, this volume).

The sites at which restriction may function are many. In searching for a genetically dominant characteristic one could speculate that aberrant methylation of viral RNA may occur, or that aberrant processing might occur on the long polypeptide, or an inhibitor of viral translation might be present in MDBK (or other restrictive cells) extracts. These possibilities are currently under investigation.

REFERENCES

1. LEWIN, B. Gene expression 3, Plasmids and Phages, (1977) John Wiley and Sons, New York.

2. TENNANT, R.W., SCHLUTER, B., MYER, F.E., OTTER, J.A., YANG, W.K. and BROWN, A. J. Virol. (1976), 20, 589-596,

3. ERON, L., WESTPHAL, H. and KHOURY, G. J. Virology (1975), 15, 1256-1261.

4. SPANDIDOS, D.A. and GRAHAM, A.F. J. Virol. (1976), 19, 968-976.

5. DARLINGTON, R.W., PORTNER, A. and KINGSBURY, D.W. J. Gen. Virol. (1970), 9, 169-177.

6. HOLLAND, J.U. and HOYER, B.H. Cold Spring Harbor Symp. Quant. Biol. (1962), 27, 101-112.

7. HOLLAND, J.J., McLAREN, L.C. and SYVERTON, J.T. J. Exp. Med. (1959), 110, 65-80.

8. McLAREN, L.C., HOLLAND, J.J. and SYVERTON, J.T. J. Expt. Med. (1960), 112, 581-594.

9. HOLLAND, J.J. Virology (1962), 16, 163-176.

10. CORDS, C.E. and HOLLAND, J.J. Virology (1964), 492-495.

11. SEDMAK, J.V., TAYLOR, M.W., MEALEY, J.Jr., CHEN, T.T. Nature New Biology (1972), 238, 7-8.

12. DUBOIS, M.F., CHANY, C. J. Gen. Virol. (1976), 31, 173-181.

13. SHIRMAN, G.A., MASLOVA, S.V., GAVRILOVSKAYA, I.V. and AGOL, V.I. Virology (1973), 51, 1-10.

14. BARRON, A. and KARZON, D. Am. J. Epidemiology (1965), 81, 323-332.

15. SUTO, T., KARZON, D., BUSSELL, R. and BARRON, A. Am. J. Epidemiology (1965), 81, 333-340.

16. SUTO, T., KARZON, D., BUSSELL, R. Am. J. Epidemiology (1965), 81, 341-349.

17. RIGHTHAND, F.V. and KARZON, T. Proc. Soc. Exp. Biol. Med. (1967), 1248-1254.

18. RIGHTHAND, F. and HUGHES, P.F. Inf. Imm. (1974), 9, 134-141.

19. STURMAN, L.S. and TAMM, J. J. Virol. (1969), 3, 8-16.

20. PRATHER, S.O., TAYLOR, M.W. J. Virol. (1975), 15, 872-881.

21. CHOPPIN, P.W. Virology (1969), 39, 130-134.

22. HUANG, A. and BALTIMORE, D. Nature (London) (1970), 226, 325-327.

23. BUCK, C.A., GRANGER, G.A., TAYLOR, M.W. and HOLLAND, J.J. Virology (1967), 33, 36-46.

24. WALL, R.T. and TAYLOR, M.W. J. Virol. (1969), 4, 611-687.

25. WALL, R.T. and TAYLOR, M.W. Virology (1970), 42, 78-86.

26. KIEHN, E.D. and HOLLAND, J.J. J. Virol. (1970), 5, 358-367.

27. WEISS, M.C. and GREEN, H. Proc. Natl. Acad. Sci. U.S.A. (1967), 58, 1104-1111.

28. CHINCHAR, V.G., FLOYD, A.D., CHINCHAR, G.D. and TAYLOR, M.W. Bioch. Genetics, in press.

29. SWETLY, P., BARBANTI-BRODANO, G., KNOWLES, B.B. and KOPROWSKI, H. J. Virol. (1969), 4, 348-355.

30. KNOWLES, B.B., BARBANTI-BRODANO, G. and KOPROWSKI, H. J. Cell Physiol. (1971), 78, 1-8.

31. CROCE, C.M., GRISARDI, A.J. and KOPROWSKI, H. Proc. Natl. Acad. Sci. U.S.A. (1973), 70, 3617-3620.

32. MEDRANO, L., GREEN, H. Virology (1973), 54, 515-524.

33. POLLACK, R.E., SALAS, J., WANG, R., KUSANO, T. and GREEN, H. J. Cell Physiol. (1971), 77, 117-119.

PARTICIPANTS

AUJEAN, Odile. Institut du Radium, Section de Biologie, F75231 Paris Cedex 05, France.

BABICH, Alexander. The Rockefeller University, 1230 York Avenue, New York, N.Y. 10021, U.S.A.

BACHRACH, Howard, L. Plum Island Animal Disease Center, U.S. Department of Agriculture, Greenport, New York 11944, U.S.A.

BAGLIONI, Corrado. Department of Biological Sciences, State University of New York at Albany, Albany, N.Y. 12222, U.S.A.

BARTELING, Simon. Centr. Vet. Inst.-Virologie, Lelystrad, The Netherlands.

BLACK, Donald, N. Animal Virus Research Institute, Pirbright, Woking, Surrey,GU24 ONF, U.K.

BOZZONI, Irene. Centro degli Acidi Nucleici C.N.R., Roma, Italy.

BROWN, Fred. Animal Virus Research Institute, Pirbright, Surrey, GU24 ONF, U.K.

CARRASCO, Luis. Centro de Biologia Molecular, Universidad Autonoma, Canto Blanco, Madrid 34, Spain.

COVA, Lucyna. Faculte de Medicine, Lab. de Virologie, Lyon 8[e] France.

DAHL, Helen. Kaptein W. Wilhelmsen of Frues Bakteriologiske Institut, Rikshospitalet, Oslo 1, Norway.

DOEL, T.R. Animal Virus Research Institute, Pirbright, Woking, GU24 ONF, U.K.

DOMINGO, Esteban. Centro de Biologia Molecular,Universidad Autonoma de Madrid, Canto Blanco, Madrid 34, Spain.

DRZENIEK, Rudolf. Heinrich-Pette-Institut, 2 Hamburg 20, BRD.

EHRENFELD, Elvera. Department of Microbiology, Medical Centre, University of Utah, Salt Lake City, Utah 84132, U.S.A.

FALCOFF, Rebeca. Laboratoire Pasteur, Institut du Radium, F75005 Paris, France.

FALCOFF, Ernesto. Laboratoire Pasteur, Institut du Radium, F75005 Paris, France.

FELLNER, Peter. Searle Research Laboratoires, Lane End Road, High Wycombe, Bucks, HP12 4HL, U.K.

FLANEGAN, James, B. Department of Immun. & Med. Microbiology, College of Medicine, University of FLA, Gainesville, FLA 32610, U.S.A.

GIORGI, Colomba, Lab. Malattie Batteriche e Vir., Ist. Superiore Sanita, I-00161 Roma, Italy.

HEWLETT, Martinez, J. Department of Cell & Devel. Biol., Bioscience West, University of Arizona, Tucson, Arizona 85721, U.S.A.

HILLER, Eckhard, Physiologisches Chem. Institut der Universität Hamburg, 2000 Hamburg 13, BRD.

HOEY, Elizabeth, M. Department of Biochemistry, The Queen's University of Belfast, Belfast, BT9 7BL, U.K.

JACKSON, Richard, J. Department of Biochemistry, University of Cambridge, Tennis Court Road, Cambridge, CB2 1QW, U.K.

JEN, George, Department of Biology, Washington University, Saint Louis, Missouri 63130, U.S.A.

JOHNSON, John, E. Department of Biological Sciences, Purdue University, Lafayette, Indiana 47907, U.S.A.

JONES, Charlotte, L. Biology Department C-016, University of California, San Diego, La Jolla, California 92093, U.S.A.

KELLER, Françoise. Laboratoire de Virologie, 3, rue Koeberle, 67000 Strasbourg, France.

KORANT, Bruce, D. Central Research Department, Du Pont de Nemours and Co., Wilmington, Delaware 19898, U.S.A.

LEONARD, Joan. University of Alberta, Department of Biochemistry, Edmonton, Alberta T6G 2H7, Canada.

LEVANON, A. c/o Institut für Molekularbiol. I der Universität Zürich, 8049 Zürich-Hönggerberg, CH.

LEWIS, John, A. Department of Anatomy and Cell Biology, Downstate Medical Centre, S.U. New York, Brooklyn, N.Y.11203, U.S.A.

LORIA, Roger, M. Medical College of Virginia, Department of Microbiology, Richmond, Virginia 23298, U.S.A.

LUCAS-LENARD, Jean. Biochemistry and Biophysics U-125, University of Connecticut, Storrs, Connecticut 06268, U.S.A.

LUND, Garry. Department of Biochemistry, University of Alberta, Edmonton, Alberta, T6G 2H7, Canada.

MAPOLES, John, E. Biophysics Laboratory, University of Wisconsin, Madison, Wisconsin 53706, U.S.A.

McCLURE, Marcella Ann. Washington University, St. Louis, Mississippi 63110, U.S.A.

MELOEN, R.H. Central. Vet. Institute, Dept. of Virologie, Lelystad, The Netherlands.

MIDULLA, Mario. via Pereira 21, I-00136, Roma, Italy.

MILSTIEN, Julie, B. Building 29A, Room 3C-22, 8800 Rockville Pike Bethesda, Maryland 20014, U.S.A.

NAGY, Eva. Vet. Med. Res. Institut, Hungarian Acad. Sciences H-1581 Budapest, P.O.Box 18, Hungary.

NODO, Makoto. School of Medicine, Department of Microbiology, Keio University, Shinjuku, Tokyo 160, Japan.

OREN, Rachel. Department of Virology, Israel Inst. Biol. Res. P.O.Box 19, Ness-Ziona, Israel.

OTTO, Michael, J. Biological Sciences Group, University of Connecticut, Storrs, Connecticut 06268, U.S.A.

PEREZ-BERCOFF, Raul. Laboratory of Virology and Molecular Biology, Institute of Anatomy, 8006 Zürich, CH.

PERSON, Anne. Institut de Recherche en Biol. Moleculaire, Tour 43, F75221 Paris Cedex 05, France.

PHILLIPS, Bruce, A. Department of Microbiology, Univ. Pittsburgh Medical School, Pittsburgh, PA 15261, U.S.A.

REVEL, Michel, Department of Virology, The Weizmann Institute of Science, Rehovot, Israel.

RICHARDS, Oliver, C. Department of Microbiology, University of Utah, Salt Lake City, Utah 84132, U.S.A.

ROTTIER, Peter, J.M. Department of Molecular Biology, Agricultural University, Wageningen, The Netherlands.

RUCKERT, Ronald, R. Biophysics Laboratory, University of Wisconsin, Madison, Wisconsin 53706, U.S.A.

RUGGERI, Franco. Laborat. Malatie Batt. Virali, Ist. Superiore die Sanita, I-00161 Roma, Italy.

SALZBERG, Samuel. Department of Life Sciences, Bar-Ilan University, Ramat-Gan, Israel.

SCRABA, Douglas, Department of Biochemistry, University of Alberta, Edmonton, Alberta, T6G 2H7, Canada.

SHAPIRA, Adam. Israel Institute for Biological Research, P.O.Box 19, Ness Ziona, Israel.

SHTRAM, Yehuda. Lab. for Molec. Virology, Hebrew University, Hadassah Medical School, Jerusalem, Israel.

SMIRNOV, Yuri, A. The D.I. Ivanovsky Institute of Virology, Academy Med. Sci. USSR, Gamaleya St. 16, Moscow 123098, USSR.

SUMMERS, Donald, D. Department of Microbiology, University of Utah, Salt Lake City, Utah 84132, U.S.A.

TAYLOR, Milton. Department of Biology, Indiana University, Bloomington, Indiana 47401, U.S.A.

TERSHAK, Daniel, R. Department of Microbiology and Cell Biology, Pennsylvania State University, University Park, PA 16807, U.S.A.

THOENE, Ingo, Heinrich Pette Institut, 2000 Hamburg 20, BRD.

TRAUB, Abram. Israel Institute for Biological Research, P.O.Box 19, Ness Ziona, Israel.

VRIJSEN, Raf. Vrije Universiteit, Farmaceutisch Institut, 1640 St. Genesius-Rode, Belgique.

WETZ, Klaus. Institut für Virologie, Freie Universität, Berlin, 1000 Berlin 45, BRD.

WIEGERS, Klaus, J. Heinrich-Pette Institut, 2 Hamburg 20, BRD.

WIMMER, Eckard. Department of Microbiology, School of Basic Health Sciences, State University of New York, Stony Brook, N.Y. 11794, U.S.A.

ZEICHHARDT, Heinz. Inst. für Klinische and Exp. Virologie, Freie Universität Berlin, 1000 Berlin 45, BRD.

ZOLER, Mitchel. Department of Med. Viral Oncology, Roswell Park Memorial Inst., Buffalo, N.Y.14263, U.S.A.

MARATEA
The small fishing harbor

MARATEA

D.N. BLACK; J. PERRAULT & F. BROWN

E. FALCOFF

D. SCRABA; R. PEREZ BERCOFF; I. BOZZONI; J. PERRAULT & M.A. McCLURE

E. NAGY; S. SALZBERG & J.E. MAPOLES

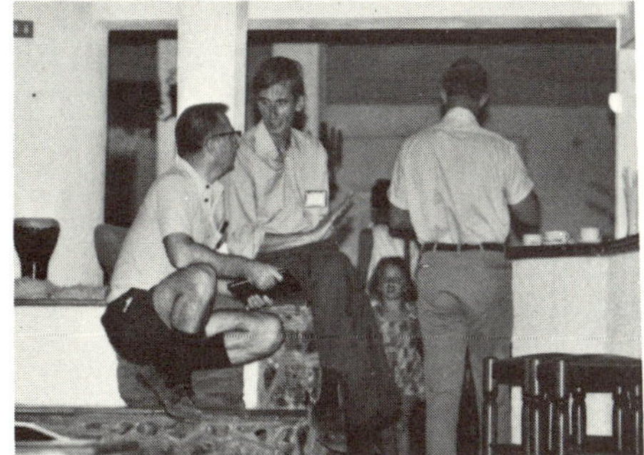

Coffee-break:
R.R. RUECKERT & R.J. JACKSON

J.B. FLANEGAN and E. EHRENFELD

A. PERSON & J.A. LEWIS

F. BROWN & J. PERRAULT (front row); E. WIMMER & R.J. JACKSON (2nd. row); S. SALZBERG (standing)

The youngest participants . . .

and the busiest ones!

INDEX